ADVANCES IN CRYOGENIC ENGINEERING

Related Titles from AIP Conference Proceedings

To learn more about these titles, or the AIP Conference Proceedings Series, please visit
the webpage **http://proceedings.aip.org**

ADVANCES IN CRYOGENIC ENGINEERING

Proceedings of the International
Cryogenic Materials Conference—ICMC

Madison, Wisconsin   16–20 July 2001

VOLUME 48A

EDITORS
Balu Balachandran
Donald Gubser
K. Ted Hartwig

◎ **CD-ROM INCLUDED**

AMERICAN INSTITUTE OF PHYSICS

Melville, New York, 2002
AIP CONFERENCE PROCEEDINGS ■ VOLUME 614

Editors:

Balu Balachandran
Argonne National Laboratory
9700 S. Cass Avenue
Argonne, IL 60439
USA

E-mail: balu@anl.gov

Donald Gubser
Naval Research Laboratory
Code 6300
4555 Overlook Avenue, SW
Washington, D.C. 20375-5343
USA

E-mail: gubser@anvil.nrl.navy.mil

K. Ted Hartwig
Texas A&M University
Department of Mechanical Engineering
College Station, TX 77843-3123
USA

E-mail: thartwig@mengr.tamu.edu

L.C. Catalog Card No. 2002103656
ISBN 0-7354-0060-1 Set
ISSN 0094-243X
Printed in the United States of America

CONTENTS

PART A

CRYOGENIC MAGNETIC MATERIALS

CRYOGENIC MATERIALS TESTING

PHYSICAL AND MECHANICAL PROPERTIES AT CRYOGENIC TEMPERATURES

STRUCTURAL MATERIALS TESTING

CRYOGENIC DEFORMATION AND FRACTURE OF METALS AND ALLOYS

NON-METALLIC MATERIALS

NON-METALLIC MATERIALS—TESTING AND EVALUATION

NON-METALLIC MATERIALS—INSULATION

PART B

ISSUES FOR HIGH-FIELD HTS INSERT COILS

HTS BULK

STRAIN PROPERTIES OF HTS

HTS COATED CONDUCTORS: SUBSTRATE AND BUFFER LAYERS

HTS COATED CONDUCTORS: GRAIN BOUNDARY ISSUES

HTS COATED CONDUCTORS: SUPERCONDUCTOR DEPOSITION

ADVANCES IN HTS COATED CONDUCTORS FOR
ELECTRIC POWER APPLICATIONS

DEVELOPMENT OF LARGE-GRAIN RARE-EARTH DOPED YBCO BULK AT ISTEC

BSCCO CONDUCTORS

HTS MECHANISMS

MgB_2

Nb₃Sn CONDUCTORS

STABILITY AND AC LOSSES

FLUX PINNING AND VORTEX DYNAMICS

PREFACE

The 2001 Joint Cryogenic Engineering Conference and International Cryogenic Materials Conference were held in Madison, Wisconsin, USA from July 16th to 20th. The conference title was "The Wright Place" a play on the name of Frank Lloyd Wright, the famous Wisconsin architect who designed the Monona Terrace Convention Center, in which the conference was held. The conference was attended by an international group of 776 participants from 32 countries.

The conference was a mixture of plenary sessions, oral and poster presentations, a strong cryogenic exhibit, and an evening session devoted to new results on the recently discovered MgB_2 superconductor. There were several joint CEC-ICMC sessions that highlighted the synergy between materials development and cryogenic applications. There were two technical tours, one to the National Magnetic Resonance Facility and the other to the Applied Superconductivity Center, both at the University of Wisconsin-Madison. Attendees were treated to an American football tailgate party, complete with cheerleaders, pep band, and Bucky Badger (the UW-Madison school mascot) prior to the Thursday evening banquet.

On the CEC side, Bob Witt (UW-Madison) served as Conference Chair, while Susan Breon (NASA Goddard Space Flight Center) served as CEC Program Chair and John Pfotenhauer (UW-Madison) served as Assistant Program Chair. Eric Hellstrom (UW-Madison) served as ICMC Chair, while Wilfried Goldacker (Forschungzentrum, Karlsruhe, Germany), Takayo Hasegawa (Showa Electric Wire and Cable Co. Ltd., Japan), and Peter Lee (UW-Madison) served as Assistant ICMC Program Chairs. Michael Capers (Scientific Instruments, Inc.) served as Cryo Expo Chair and Bill Burt (TRW) served as the Publicity Chair.

Centennial Conferences provided superb management to the conference as overseen by Paula Pair. Kim Bass and Annett D'Antonio handled the technical part of the conferences and Monica Park spearheaded the Cryo Expo.

The next joint CEC-ICMC conference will be held in Anchorage, Alaska, USA, September 22-26, 2003. The ICMC Chair is Herb Freyhardt (University of Göttingen, Germany) and the CEC Chair is Tom Peterson (FermiLab). ICMC will hold a regional conference entitled Superconductors for Practical Applications (SPA'2002), June 16-20, 2002 in Xi'an, China that will be chaired by Lian Zhou (Northwest Institute for Nonferrous Metal Research, China).

We thank all those concerned with the publication procedure, particularly Melinda Adams, the managing editor of the entire publication process, as well as the technical editors, reviewers, and authors for help in preparing this volume in a timely fashion.

Bob Witt
Eric Hellstrom
Chairs – CEC and ICMC 2001

ICMC AWARDS

Best Paper Awards ('99 Conference)

K. M. Amm, R.A. Ackermann, P. S. Thompson, A. Mogro-Campero,
and J. M. van Oort
for their paper
"Thermal Conductivity of 34-700 Carbon Fiber Composites
at Cryogenic Temperatures"

H. Kamakura, H. Kitaguchi, H. Miao, K. Togano, K. Tanaka, M. Okada, K. Ohata,
J. Sato, T. Koizumi, and T. Hasegawa
for their paper
"Development of Large Current Carrying Bi-2212 Tapes and Wires"

Student Meritorious Papers ('01 Conference)

The ICMC Board of Directors recognizes students who write high quality
papers. The papers are submitted for evaluation prior to the conference and ranked on
the basis of research merit and quality of writing. The ICMC board grants awards to
the following contributions:

H. Huang: A. O. Pecharsky, V. K. Pecharsky, and K. A. Gschneidner, Jr.
for their paper
"Preparation, Crystal Structure and Magnetocaloric Properties of $Tb_5(Si_xGe_{4-x})$"

Raquel Gonzalez-Arrabal, M. Eisterer, H. W. Weber, D. Litzkendorf,
T. Habisreuther, W. Gawalek, S. Nariki, M. Muralidhar, and M. Murakami
for their paper
"Trapped Field Characterization of Melt-textured Monoliths
at Liquid Nitrogen Temperature"

Travel Assistance Awards

The ICMC Awards Committee also provides financial assistance to selected
students to attend the conference. For the 2001 conference, a special travel assistance
grant was provided to:

Mr. Gideon Wiegerinck
University of Twente
Twente, The Netherlands

BOARD OF DIRECTORS

SPONSORS

Alliant Energy Corporation
Applied Superconductivity Center – University of Wisconsin-Madison
Argonne National Laboratory
Fermi National Accelerator Laboratory
Florida State University – National High Magnetic Field Laboratory
Iowa State University
Los Alamos National Laboratory
Oak Ridge National Laboratory
University of Wisconsin College of Engineering
U.S. Department of Energy

CRYOGENIC MAGNETIC MATERIALS

EFFECTS OF PRASEODYMIUM ADDITIONS ON THE MAGNETOTHERMAL PROPERTIES OF ERBIUM

Y.L. Wu,[1,2] A.O. Pecharsky,[1] V.K. Pecharsky,[1,2] and K.A. Gschneidner, Jr.[1,2]

Ames Laboratory[1] and Department of Materials Science and Engineering,[2] Iowa State University, Ames, IA 50011-3020, USA

ABSTRACT

The effects of the Pr additions (10, 20, 25, and 30 at.%) on the magnetothermal properties of polycrystalline metallic Er have been studied using low temperature, high magnetic field heat capacity (4~120 K, 0~75 kOe), dc magnetization and ac magnetic susceptibility. These properties include the volumetric heat capacity, the magnetocaloric effect (MCE), the magnetic transition temperatures, the paramagnetic Curie temperatures and the effective magnetic moments. It has been found that Pr additions increase the Curie temperature and decrease the Néel temperature of Er, as well as wipe-out the 26 K and 52 K transitions apparently by merging with the former two transitions. This gives rise to only one MCE peak in each of the Er-Pr alloys, from the broad table-like shape for the $Er_{90}Pr_{10}$ alloy to the typical narrow caret-like shape for the $Er_{70}Pr_{30}$ alloy. Pr additions also increase the heat capacity of Er significantly below 18 K. The implications of this higher heat capacity as potential cryocooler regenerator materials are discussed.

INTRODUCTION

Lanthanide-based intermetallic compounds were first used in cryocooler regenerators in 1990, when Sahashi et al. [1] and Kuriyama et al. [2] used Er_3Ni as the low temperature stage regenerator material in a two-stage Gifford-McMahon (GM) cryocooler. These scientists proposed that low temperature regenerator material in common use, Pb, be partially replaced by Er_3Ni. This replacement allowed them to reduce the low temperature limit of GM cryocoolers from 10 K to 4 K. The achieved improvement is due to the higher volumetric heat capacity of Er_3Ni relative to Pb below 25 K.

The efficiency of cryocoolers is proportional to the volumetric heat capacity of the regenerator material especially bellow 20 K, the higher the heat capacity the greater the amount of heat that can be transferred from the cold heat exchanger to the hot heat exchanger per cycle of fluid flow through the regenerator bed. Thus, the utilization of the

CP614, *Advances in Cryogenic Engineering:*
Proceedings of the International Cryogenic Materials Conference - ICMC, Vol. 48,
edited by B. Balachandran et al.
© 2002 American Institute of Physics 0-7354-0060-1/02/$19.00

magnetic contribution to the total heat capacity offers a promising avenue to follow to improve the efficiency of a cryocooler.

Recently, new Er-Pr regenerator materials, which are ductile, oxidation resistant, environmentally friendly and non-toxic were developed and tested [3,4] as a replacement for Pb in low temperature cryocoolers. Due to their higher heat capacities than that of Pb, improvement in performance was observed. In the present study, further research on this kind of materials was carried out to investigate the dependence of the basic magnetothermal properties of Er on the Pr content. High purity Er and Pr (~99.8 at.%) were used instead of the commercial grade lanthanide metals (~96.8 at.%) which were used in the earlier studies [3,4].

EXPERIMENTAL

A total of four $Er_{100-x}Pr_x$ alloys where x = 10, 20, 25 and 30 were prepared by arc melting of stoichiometric mixtures of pure Er and Pr. The Er and Pr were prepared by the Materials Preparation Center of the Ames Laboratory. The Er was 99.86 at.% pure with the major impurities as follows (in ppm atomic): H-994, C-14, N-36, O-137, F-220, and Fe-42. The Pr was about 99.8 at.% pure with the major impurities of: H-(not determined, but typically the H content varies between 500 and 1000 ppm atomic), C-12, N-121, O-238, F-230, and Fe-13. Weight losses during arc melting were negligible and, therefore, the alloy compositions were assumed unchanged. The arc-melted alloys were then heat treated for 7 days at 1210 K in evacuated and back-filled with helium quartz tubes.

The heat capacities at constant pressure as a function of temperature were measured using an adiabatic heat-pulse-type calorimeter [5] from ~3.5 K to ~350 K in magnetic fields of 0, 20, 50, and 75 kOe. The total entropy in different magnetic fields as a function of temperature was obtained by numerically integrating the following equation

$$S(T)_H = \int_0^T \frac{C(T)_H}{T} dT \tag{1}$$

neglecting the zero temperature entropy. The MCE in terms of both the isothermal magnetic entropy change and the adiabatic temperature change, was calculated from the total entropy functions [6] as

$$\Delta S_M (T)_{\Delta H,T} = [S (T)_H - S (T)_{H=0}]_T \tag{2}$$

and

$$\Delta T_{ad} (T)_{\Delta H,S} = [T (S)_H - T (S)_{H=0}]_S \tag{3}$$

The magnetic measurements were carried out using a Lake Shore magnetometer, model 7225. The ac magnetic susceptibility was measured in an ac field of 25 Oe at a frequency of 125 Hz with no bias dc field. The dc magnetization and dc susceptibility were measured over the temperature range from ~5 K to ~120 K in dc fields from 0 kOe to 50 kOe. The MCE in terms of the isothermal magnetic entropy change, was also calculated based on the magnetization data by integrating the Maxwell equation [6]

$$\Delta S_M (T)_{\Delta H,T} = \int_0^H \left(\frac{\partial M(H,T)}{\partial T} \right)_H dH \tag{4}$$

4

RESULTS AND DISCUSSION

Heat Capacity Measurements

The measured heat capacities vs temperature at zero magnetic field of the Er and Er-Pr alloys are shown in Fig. 1 from 4 to 100 K. First of all, the temperature of the heat capacity maximum of pure Er is 19.0 K, which is close to $T_C = 18.7$ K, which was obtained from a solid state electrolysis (SSE) purified Er sample (99.97 at.%)[7,8]. Compared to 20 K (refs. 9, 10, and 11) or 20.4 K (ref. 12), it is demonstrated again that the higher the purity of the Er sample the lower the magnetic transformation temperature, which is consistent with similar observations on other rare-earth metals. Second, Er shows four magnetic transitions in the absence of a magnetic field (Fig. 1a). Two major second order transformations occur at 86.4 K [paramagnetic (P) to a c-axis modulated ferromagnetic (CAM)] and at 52.7 K [(CAM to a complex magnetic structure-antiphase domain (APD) + cone + helix (APD)]; a first order transition at 19.0 K [(APD to a ferromagnetic cone + helix (FC)], and one spin-slip transition at 26.2 K. These data are in good agreement with many previously reported results [7,8,11,13-18]. Third, the Curie temperature increases and the peak height decreases with increasing Pr addition, the Néel temperature decreases with increasing Pr addition, and the magnetic transitions at 52.7 K and 26.2 K in Er seem to disappear. They might be merging with the Néel temperature and Curie temperature transitions, but lower Pr content (<10 at.%) alloys need to be studied to verify this. Table 1 lists the magnetic transition temperatures in Er and Er-Pr alloys, and Fig. 2 shows the relationship between the magnetic transition temperatures and the Pr content. Finally, it is noted that for the four Er-Pr alloys the heat capacities are larger than that of Er below18 K. This is shown in more detail in Fig.1b, where it is seen that alloys containing more than 10 at.% Pr show similar behavior, the heat capacity of $Er_{70}Pr_{30}$ is ~7 times to 18% higher than

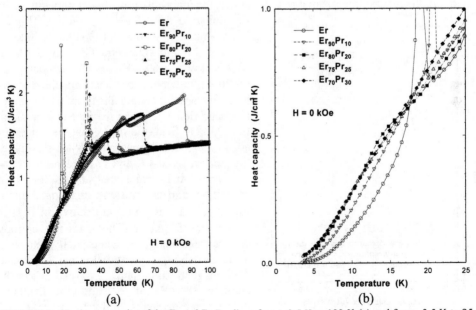

FIGURE 1. The heat capacity of the Er and Er-Pr alloys from ~3.5 K to 100 K (a) and from ~3.5 K to 25 K (b) measured in zero magnetic field.

5

TABLE 1. Magnetic Transition Temperatures (T_C-Curie and T_N-Néel), Paramagnetic Curie Temperatures (Θ_p) and Effective Magnetic Moments (p_{eff}) in Er and Er-Pr Alloys.

	T_C (K)	T_N (K)	Θ_p (K)	p_{eff} Experimental (μ_B)	p_{eff} Theory (μ_B)
Er (MPC)	19.0[a]	86.4[a]	36.0[*]	9.91[*]	9.58
	15.1[b]	85.6[b]	36.2[**]	9.82[**]	
Er$_{90}$Pr$_{10}$	20.7[a]	63.2[a]	40.1[*]	9.27[*]	9.16
	22.2[b]	65.4[b]	39.8[**]	9.25[**]	
Er$_{80}$Pr$_{20}$	32.9[a]	48.9[a]	38.8[*]	8.71[*]	8.72
	33.5[b]	49.2[b]	37.8[**]	8.68[**]	
Er$_{75}$Pr$_{25}$	34.4[a]	43.4[a]	24.9[*]	8.54[*]	8.49
	37.5[b]	42.8[b]	23.3[**]	8.51[**]	
Er$_{70}$Pr$_{30}$	34.5[a]	38.2[a]	29.2[*]	8.20[*]	8.25
	33.6[b]	38.7[b]	29.6[**]	8.08[**]	

a --- from heat capacity data;   b --- from ac susceptibility data.
* --- calculated from dc magnetization data (cooling);   ** --- calculated from ac susceptibility data (cooling).

that of Er at 4 K and 17 K respectively, while the Er$_{90}$Pr$_{10}$ alloy has only a slight advantages over Er (~2.5 times at 4 K and 14% at 17 K). Considering our previous work [3], alloying with Pr is really an excellent way to enhance the heat capacity of Er at lower temperatures (below 20 K), and thus it is a better choice to replace Pb in the cryocooler regenerator.

The measured heat capacities vs temperature at non-zero magnetic fields of the Er and Er-Pr alloys are shown in Fig. 3 from 4 to 100 K, where it is seen that the magnetic field affects the heat capacities of the Er-Pr alloys. The magnetic field shifts the ferromagnetic phase transition towards higher temperatures, and decreases the peak magnitude drastically. At higher fields (e.g., 50 kOe or 75 kOe) the T_C peaks are hardly evident (Figs. 3a and 3b). In general, the magnetic field affects the magnetic transition temperatures in Er similar to that of the Pr additions (Fig.2). For example, T_C increases from 20.7 K (0 kOe) up to 37.7 K (75 kOe), and T_N decreases from 63.2 K down to 60.1 K. At low temperatures (4~20 K), the volumetric heat capacities of Er-Pr alloys are still larger than that of Er under magnetic fields, just like they are at zero-field. In a 20 kOe field, the heat capacities of Er-Pr alloys are 3 to 7 times and 18%-24% higher than that of Er, at 4 K and 17 K, respectively.

The MCE calculated from the heat capacity and magnetization using Eqs. 2 and 4 for ΔS_M, and 3 for ΔT_{ad} are shown in Figs. 4 (ΔS_M) and 5 (ΔT_{ad}). The magnetic entropy change calculated from heat capacity data (Fig. 4) is in fair agreement with that calculated from magnetization data. As noted above the addition 30 at.% of Pr shifts the magnetic ordering temperatures of Er from 19.0 K and 86.4 K to 34.5 K and 38.2 K respectively, and thus the two peaks seem to merge. This gives

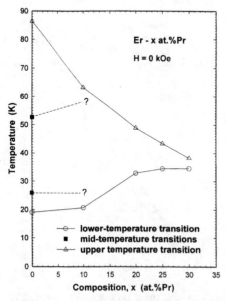

FIGURE 2. The variation of the magnetic ordering temperatures as a function of the Pr concentration.

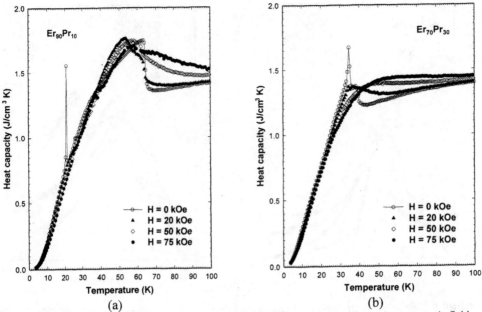

FIGURE 3. The heat capacity of the $Er_{90}Pr_{10}$ alloy (a) and $Er_{70}Pr_{30}$ alloy (b) under various magnetic fields.

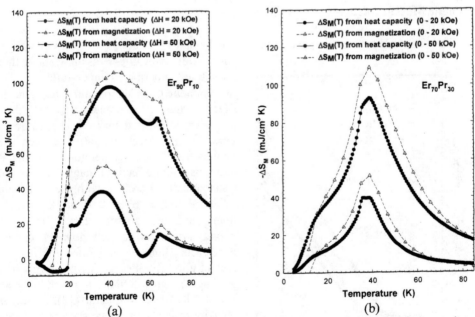

FIGURE 4. The magnetic entropy change of $Er_{90}Pr_{10}$ (a) and $Er_{70}Pr_{30}$ (b) calculated from heat capacity and magnetization data for various magnetic field changes.

rise to only one MCE peak in each of the Er-Pr alloys: from the broad table-like shape for the $Er_{90}Pr_{10}$ alloy (Fig. 4a) to the typical narrow caret-like shape for the $Er_{70}Pr_{30}$ alloy (Fig.4b). In general, the addition of Pr slightly lowers the MCE of Er.

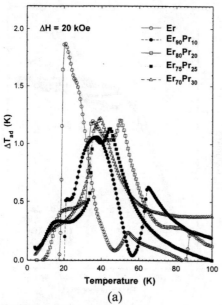 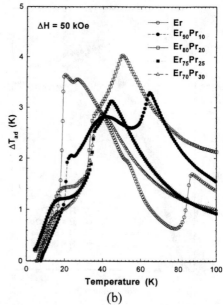

(a)   (b)

FIGURE 5. The adiabatic temperature change of the Er and Er-Pr alloys calculated from heat capacity data for a field change of 0 to 20 kOe (a) and 0 to 50 kOe (b).

Magnetic Measurements

The inverse dc magnetic susceptibilities of Er and Er-Pr alloys are shown in Fig.6.

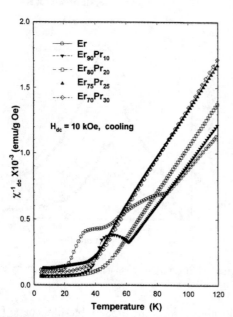

FIGURE 6. The inverse dc magnetic susceptibility of the Er and Er-Pr alloys.

They follow the Curie-Weiss law above their respective ordering temperatures. The paramagnetic Curie temperatures (Θ_p) and the effective magnetic moment (p_{eff}) of these alloys are also listed in Table 1. Although there is no direct relation between the paramagnetic Curie temperatures and the Pr additions, the effective magnetic moment systematically decreases with the increasing Pr concentration from 9.91 to 8.20 μ_B/Er atom, and are slightly larger than the theoretical values for the low Pr content alloys, but slightly smaller for the high Pr content alloy ($Er_{70}Pr_{30}$). The p_{eff} for Er is in good agreement with those reported by Green et al [12] and Astrom et al [13]. Both the behavior of the dc magnetic susceptibility and the positive paramagnetic Curie temperatures are indicative of a ferromagnetic ground state in Er and all four Er-Pr alloys.

8

The ac magnetic susceptibility of the Er-Pr alloys and Er is shown in Fig. 7. Except for the $Er_{70}Pr_{30}$ alloy, Pr additions lower the ac magnetic susceptibility of the other three Er-Pr alloys. Also it is clear that after the Pr alloying, the spin-slip transitions disappear. Table 1 lists the magnetic transition temperatures in Er and Er-Pr alloys obtained from the ac susceptibility measurements, and these data are in good agreement with those obtained from heat capacity measurements. Similar to the dc results, we calculated the paramagnetic Curie temperature and the effective magnetic moments of Er and Er-Pr alloys from the ac susceptibility data, the results are also listed in Table 1. It was found that the differences between the effective magnetic moments calculated from dc data and ac data are small, about 0.2% to 1.5%.

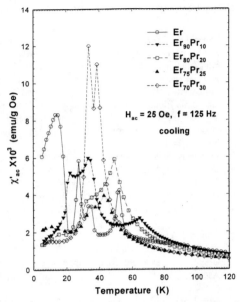

FIGURE 7. The real part of the ac susceptibility of the Er and Er-Pr alloys.

CONCLUSIONS

The study of the effects of Pr additions on the magnetothermal properties of polycrystalline metallic Er shows that, the addition of Pr enhances the heat capacity of Er both at lower temperatures (below ~18 K) and at intermediate temperatures (20 – 60 K).

The Pr additions increase the Curie temperature and lower the Néel temperature of Er, as well as wiping-out the 26 K and 52 K transitions. This makes the two peaks, T_C and T_N, approach each other and thus gives rise to only one MCE peak in the Er-Pr alloys, from the broad table-like shape for the $Er_{90}Pr_{10}$ alloy to the typical narrow caret-like shape for the $Er_{70}Pr_{30}$ alloy. Furthermore, the Pr additions hardly show any effect on the paramagnetic Curie temperatures of the Er-Pr alloys, but the effective magnetic moments systematically decrease because of the lower effective magnetic moment of Pr.

Pr additions increase the heat capacity of Er significantly below 18 K, by as much as 7 times at 4 K and 24% at 17 K for the $Er_{70}Pr_{30}$ alloy, and thus we would expect a significant increase in cooling power over Pb in a cryocooler using an Er-Pr alloy as a regenerator.

ACKNOWLEDGEMENTS

The Ames Laboratory is operated by the U.S. Department of Energy by Iowa State University under Contract No. W-7405-ENG-82. This work was supported by the Office of Basic Energy Sciences, Materials Sciences Division.

REFERENCES

1. Sahashi, M., Tokai, Y., Kuriyama, T., Nakagome, H., Li, R., Ogawa, M., and Hashimoto, T., *Adv. Cryogen. Eng.*, **35**, pp. 1175-1182 (1990).
2. Kuriyama, T., Hakamada, R., Nakagome, H., Tokai, Y., Sahashi, M., Li, R., Yoshida, O., Matsumoto, K., and Hashimoto T., *Adv. Cryogen. Eng.*, **35**, pp. 1261-1269 (1990).
3. Gschneidner, K.A., Jr., Pecharsky, A.O., and Pecharsky, V.K., *Cryocoolers 11*, Kluwer Academic/Plenum Publishers, New York, 2001, pp. 433-442.
4. Wysokinski, T.W., Barclay, J.A., Gschneidner, K.A., Jr., Pecharsky, V.K., and Pecharsky, A.O., "Comparative evaluation of erbium and lead regenerator materials for low temperature cryocoolers," to be published.
5. Pecharsky, V.K., Moorman, J.O., and Gschneidner, K.A., Jr., *Rev. Sci. Instr.*, **68**, pp. 4196-4207 (1997).
6. Pecharsky, V.K. and Gschneidner, K.A., Jr., *J. Appl. Phys.* **86**, pp. 565-575 (1999).
7. Pecharsky, V.K. and Gschneidner, K.A., Jr., *Phys. Rev.* **B47**, pp. 5063-5071 (1993).
8. Gschneidner, K.A., Jr., Pecharsky, V.K. and Fort, D., *Phys. Rev. Letters* **78**, pp. 4281-4284 (1997).
9. Cable, J.W., Wollan, E.O., Koehler, W.C., and Wilkinson, M.K., *Phys. Rev.* **140**, pp. A1896-1902 (1965).
10. Cowen, J.A., Stolzman, Barry, Averback, R.S., and Hahn, H., *J. Appl. Phys.*, **61**, pp. 3317-3319 (1987).
11. Ito, T., Legvold, S., and Beaudry, B.J., *Phys. Rev.* **B30**, pp. 240-243 (1984).
12. Green, R.W., Legvold, S., and Spedding, F.H., *Phys. Rev.* **122**, pp. 827-830 (1961).
13. Astrom, H.U., Chen, D-X, Benediktsson, G., and Rao, K.V., *J. Phys.: Condens. Matter*, **2**, pp. 3349-3357 (1990).
14. Ali, N. and Willis, F., *Phys. Rev.* **B42**, pp. 6820-6822 (1990).
15. Vajda, P. and Daou, J.N., *J. Less-Common Met.*, **101**, pp. 269-284 (1984).
16. Blinov, A.G., Boyarskii, L.A., Savitskii, E.M., Tarasenko, A.P., and Chistyakov, O.D., *Sov. Phys. Solid State*, **25**, pp. 564-567 (1983).
17. Helgesen, G., Hill, J.P., Thurston, T.T., and Gibbs, D., *Phys. Rev.* **B52**, pp. 9446-9454 (1995).
18. Gibbs, D., Bohr, J., Axe, J.D., Moncton, D.E., and D'Amico, K.L., *Phys. Rev.* **B34**, pp. 8182-8185 (1986).

PREPARATION, CRYSTAL STRUCTURE AND MAGNETOCALORIC PROPERTIES OF Tb$_5$(Si$_x$Ge$_{4-x}$)

H. Huang,[1,2] A. O. Pecharsky,[1] V. K. Pecharsky,[1,2] K. A. Gschneidner, Jr.[1,2]
Ames Laboratory[1] and Department of Materials Science & Engineering[2]
Iowa State University, Ames, Iowa, 50011-3020, U.S.A.

ABSTRACT

The crystal structure and magnetothermal properties of the potential magnetic refrigerant materials, Tb$_5$(Si$_x$Ge$_{4-x}$) alloys where $x \geq 2$, have been studied. The crystal structure of the Tb$_5$(Si$_x$Ge$_{4-x}$) alloys varies from the orthorhombic Gd$_5$Si$_4$-type to the monoclinic Gd$_5$(Si$_2$Ge$_2$)-type as silicon concentration decreases from $x = 4$ to $x = 2$. The dc magnetic susceptibility and magnetization data indicate that a paramagnetic to ferromagnetic phase transition occurs in these alloys. The heat capacity data measured in magnetic fields varying from 0 to 100 kOe show that a second order ($x = 4$, 3) or a first order phase transition ($x = 2$) occurs in the alloys. The magnetic entropy change and the adiabatic temperature change calculated from the heat capacity data indicate that the Tb$_5$(Si$_x$Ge$_{4-x}$) alloys are promising materials for magnetic refrigeration from about 100 to 250 K when their chemical composition varies from $x = 2$ to $x = 4$.

INTRODUCTION

Magnetic refrigeration is an emerging technology that offers a potential for high energy efficiency. The high efficiency occurs because the compression-expansion part of the vapor cycle refrigeration is replaced by the magnetizing and demagnetizing of a magnetic material. The latter processes approach 100% Carnot efficiency. In addition magnetic refrigeration provides two important environmental benefits. First, because of the reduced energy consumption, the amount of greenhouse gases produced to generate electricity by burning coal or other fossil fuels is reduced. Second, since magnetic refrigerator uses a solid magnetic media as the refrigerant and water or another non-toxic fluid as the heat transfer medium, CFCs, HCFCs, and HFCs are no longer needed and thus these well-known sources of ozone depletion and global warming gases are eliminated. The recent discovery of a new family of giant magnetocaloric effect materials [1-4] and successful design and testing of proof-of-principle reciprocating magnetic refrigerator [5-7] demonstrate great potential for widespread commercialization of energy efficient and environmentally friendly magnetic refrigeration technology.

CP614, *Advances in Cryogenic Engineering:*
Proceedings of the International Cryogenic Materials Conference - ICMC, Vol. 48,
edited by B. Balachandran et al.
© 2002 American Institute of Physics 0-7354-0060-1/02/$19.00

Magnetic refrigeration utilizes the magnetocaloric effect (MCE), a distinct property of magnetic materials which is manifested as the change of materials' magnetic entropy during the isothermal application/removal of a magnetic field, or as the change of their temperature when a magnetic field is altered adiabatically. Similarly to a conventional vapor compression-based refrigerator, the performance of a magnetic refrigerator system is largely dependent on the performance and properties of the working body – magnetic refrigerant. It is, therefore, quite important that the best magnetic refrigerant solids are found and, eventually, exploited in future devices. This requires experimental characterization of magnetocaloric properties in a variety of magnetic materials. The magnetocaloric properties of many different materials including intermetallic compounds, amorphous alloys, lanthanide-based manganese pervoskite-type oxides, and nanocomposite materials have been studied to date (for a complete set of references see recent review articles, Refs. 8-10). It appears, however, that lanthanide-based intermetallic compounds are the best choice for near and below room temperature magnetic refrigeration due to their larger than average magnetocaloric effect. Among intermetallic compounds, the $Gd_5(Si_xGe_{4-x})$ system stands out since it displays reversible giant magnetocaloric effect when $x < \sim 2$ [1-4]. Therefore, experimental studies of magnetocaloric properties in related systems containing other magnetic lanthanides present both basic and applied interest.

In 1967, Holtzberg et al. [11] studied the structure and magnetic properties of Tb_5Si_4. It has the orthorhombic Sm_5Ge_4-type crystal structure and orders ferromagnetically at ~225 K. In 1981, Serdyuk et al. [12] reported the zero magnetic field specific heat, magnetization, magnetic susceptibility, and electrical conductivity of Tb_5Si_4. They confirmed the existence of a ferromagnetic-paramagnetic phase transition and reported Curie temperature, which is close to that found in Ref. 11. Recently, Spichkin et al. [13] investigated the series of $(Gd_xTb_{5-x})Si_4$ alloys with respect to their magnetic and thermal properties. Morellon et al. [14] reported the crystal structures and magnetic entropy changes calculated from isothermal magnetization data for Tb_5Si_4, $Tb_5(Si_2Ge_2)$ and Tb_5Ge_4. A large magnetic entropy change of ~21.8 (J/kg K) ($\Delta H = 50$ kOe) was achieved in $Tb_5(Si_2Ge_2)$ at about 105 K [14]. In this paper we report on the results of our study of the crystallography, magnetic and thermal properties of $Tb_5(Si_xGe_{4-x})$ alloys where $x \geq 2$.

EXPERIMENTAL DETAILS

Three samples with the Tb_5Si_4, $Tb_5(Si_3Ge)$ and $Tb_5(Si_2Ge_2)$ compositions were prepared by arc-melting stoichiometric mixtures of the components on a water-cooled copper hearth in an argon atmosphere. The high purity Tb was prepared by the Materials Preparation Center of the Ames Laboratory and was ~99.67 at % pure with the major impurities (in at. ppm): O – 199, C – 105, N – 12. The silicon and germanium were purchased from CERAC Inc. and were ~99.99 wt % pure. Each ingot had the total weight of about 20 g and was re-melted six times with the button being turned over each time to ensure sample homogeneity. Weight losses after the arc-melting were less than ~0.5 wt %, and the alloy compositions were assumed unchanged. The prepared alloys were quite brittle and could be easily ground into powder. The x-ray powder diffraction data were collected on an automated Scintag powder diffractometer using Cu-K_α radiation. All samples were found to be essentially single-phase materials within the limitations of the x-ray technique (typically ~5 vol % of the impurity phase) and, therefore, none of the alloys were heat treated. Magnetic measurements were performed using a Lake Shore ac/dc susceptometer/magnetometer. The dc magnetization and magnetic susceptibility were measured from ~5 K to ~320 K in magnetic fields ranging from 0 to 50 kOe. The isothermal magnetic entropy change as a function of temperature, $\Delta S_M(T)_{\Delta H,T}$, was calculated from the magnetization data using the Maxwell relation:

$$\Delta S_M(T)_{\Delta H,T} = \int_0^H \left(\frac{\partial M}{\partial T}\right)_H dH \tag{1}$$

The heat capacities were measured using an adiabatic heat pulse calorimeter [15] from ~3.5 K to ~ 350 K in magnetic fields of 0, 20, 50, 75, and 100 kOe. The accuracy of the heat capacity data was better than ~1%. The total entropy (the zero temperature entropy was neglected) in different magnetic fields, H, as a function of temperature was calculated from the heat capacity as

$$S(T)_H = \int_0^T \frac{C(T)_H}{T} dT \tag{2}$$

The MCE in terms of both magnetic entropy change and adiabatic temperature change was calculated from the total entropy functions as

$$\Delta S_M(T)_{\Delta H,T} = \left[S(T)_H - S(T)_{H=0}\right]_T \tag{3}$$

$$\Delta T_{ad}(T)_{\Delta H,S} = \left[T(S)_H - T(S)_{H=0}\right]_S \tag{4}$$

More details on the calculations and error analysis of the magnetocaloric effect from heat capacity data can be found in Ref. 16.

TABLE 1. Crystallographic Data of the $Tb_5(Si_xGe_{4-x})$ Alloys.

Composition	Space group	a (Å)	b (Å)	c (Å)	γ (degrees)	Reference
Tb_5Si_4	Pnma	7.4267(6)	14.628(1)	7.6944(6)	90	This work
	Pnma	7.413	14.625	7.699	90	[11]
	Pnma	7.4234(5)	14.621(1)	7.6938(6)	90	[13]
$Tb_5(Si_3Ge)$	Pnma	7.4409(4)	14.6498(9)	7.7235(4)	90	This work
$Tb_5(Si_2Ge_2)$	$P112_1/a$	7.5258(7)	14.681(1)	7.7138(7)	93.111(4)	This work
	$P112_1/a$	7.5088(6)	14.653(1)	7.7147(7)	93.001(1)	[14]

TABLE 2. Magnetic Properties of the $Tb_5(Si_xGe_{4-x})$ Alloys.

Composition	Temperature of low temperature transition(K)	T_c (K)	Θ_P (K)	p_{eff} (μ_B)	Reference
Tb_5Si_4	85	225.1	216.9	10.27	This work
		225	216	9.53	[11]
		222	211	9.51	[12]
		223.2	227	9.76	[13]
$Tb_5(Si_3Ge)$	80	209.6	205.4	10.23	This work
$Tb_5(Si_2Ge_2)$	74	100.1	101.1	10.21	This work

EXPERIMENTAL RESULTS AND DISCUSSIONS

A. Crystal Structure

The crystal structures of the three samples were determined from powder diffraction data and refined using the full profile least squares (Rietveld) method. Based on the Rietveld least squares refinement (Table 1), Tb_5Si_4 has the orthorhombic Gd_5Si_4-type crystal structure. Its lattice parameters agree well with previous results. $Tb_5(Si_3Ge)$ has the same orthorhombic crystal structure. The increase of its lattice parameters is consistent with the formation of a substitutional solid solution based on Tb_5Si_4. The third alloy, $Tb_5(Si_2Ge_2)$, has a monoclinic $Gd_5(Si_2Ge_2)$-type [17] crystal structure. This series of the samples is similar to other series of $R_5(Si_xGe_{4-x})$ alloys where R = Gd, Nd, and Dy [18] in that when x decreases from 4 to 2, the crystal structure changes from orthorhombic to monoclinic.

B. Magnetic and Thermal Measurements

The inverse dc magnetic susceptibility of the alloys follows the Curie-Weiss law above their respective ordering temperatures. The paramagnetic Curie temperatures, Θ_P, and effective magnetic moments, p_{eff}, are listed in Table 2. The effective magnetic moments of Tb ions slightly decrease with Si concentration from $10.27\mu_B$ to $10.21\mu_B$, and are slightly

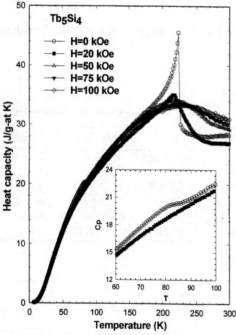

FIGURE 1. The heat capacity of the Tb_5Si_4 alloy in various magnetic fields. The inset shows the heat capacity of the Tb_5Si_4 alloy in 0 and 20 kOe magnetic field from 60-100K.

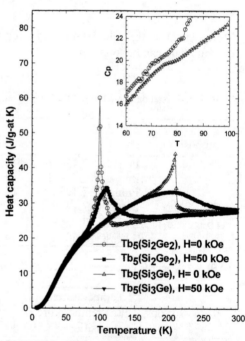

FIGURE 2. The heat capacities of $Tb_5(Si_3Ge)$ and $Tb_5(Si_2Ge_2)$ in 0 and 50 kOe magnetic fields. The inset shows the heat capacities of $Tb_5(Si_3Ge)$ and $Tb_5(Si_2Ge_2)$ in 0 magnetic fields from 60-100K.

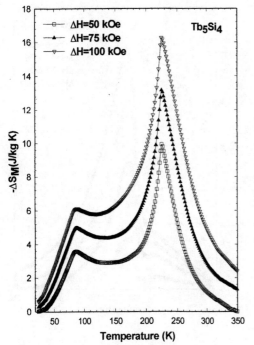

FIGURE 3A. The magnetic entropy change of Tb$_5$Si$_4$ calculated from the heat capacity data for various magnetic field changes.

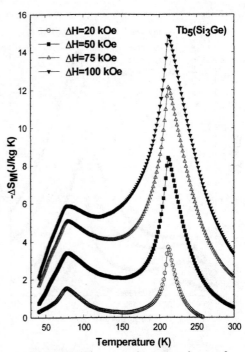

FIGURE 3B. The magnetic entropy change of Tb$_5$(Si$_3$Ge) calculated from the heat capacity data for various magnetic field changes.

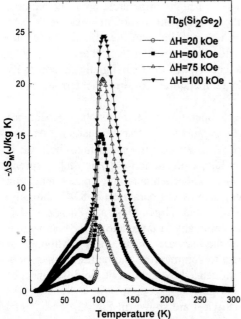

FIGURE 3C. The magnetic entropy change of Tb$_5$(Si$_2$Ge$_2$) calculated from the heat capacity data for various field changes.

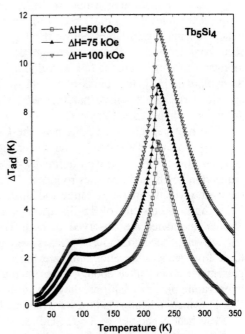

FIGURE 4A. The adiabatic temperature change of Tb$_5$Si$_4$ calculated from the heat capacity data for various magnetic field changes.

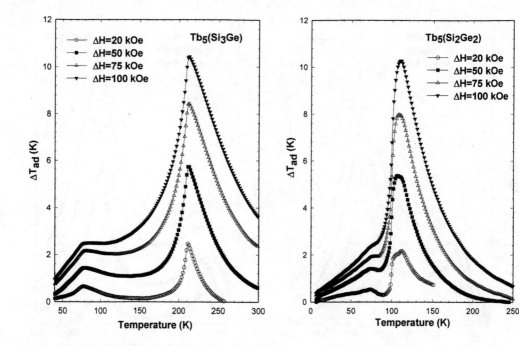

FIGURE 4B. The adiabatic temperature change of Tb$_5$(Si$_3$Ge) calculated from the heat capacity data for various magnetic field changes.

FIGURE 4C. The adiabatic temperature change of Tb$_5$(Si$_2$Ge$_2$) calculated from the heat capacity data for various magnetic field changes.

larger than the theoretical value of $9.72\mu_B$ expected for a free Tb^{+3} ion. The higher than theoretical values of p$_{eff}$ may be due to the presence of high temperature spin fluctuations, which remained significant below the high temperature limit of our magnetometer. As seen from Table 2, both the paramagnetic Curie temperatures and the effective magnetic moments of Tb$_5$Si$_4$ are in fair agreement with those reported by Holtzberg *et al.* [11], Serdyuk *et al.* [12] and Spichkin *et al.* [13]. The behavior of the dc magnetic susceptibility and the large and positive paramagnetic Curie temperatures confirm the ferromagnetic ground state in all three alloys.

The heat capacity of Tb$_5$Si$_4$ near the Curie temperature (Fig. 1) displays a behavior which is typical for a second order ferromagnetic-paramagnetic phase transition. At zero magnetic field, there is a well-defined λ-type peak. In 20 kOe magnetic field the peak is reduced in height and becomes rounded. Then in higher magnetic fields the peak is spread out over a wide temperature range and shifts towards higher temperature. In zero magnetic field, the heat capacity of Tb$_5$Si$_4$ displays a second, low temperature (T=85K) anomaly. Similar anomalies are observed in both Tb$_5$(Si$_3$Ge) and Tb$_5$(Si$_2$Ge$_2$) at ~80 and ~74K, respectively. Since no impurity phases were detected in any of three alloys, we believe that these low temperature anomalies are intrinsic to the materials under study. Although at present we do not have enough experimental data to comment on the nature of these low temperature phase transitions, they are strongly influenced by a magnetic field (see inset of Fig. 1), which indicates that they are magnetic in origin.

The heat capacities of Tb$_5$(Si$_3$Ge) and Tb$_5$(Si$_2$Ge$_2$) are shown in Fig. 2. As one can see, the heat capacity of Tb$_5$(Si$_3$Ge) is indicative of a second order ferromagnetic-paramagnetic transition, similar to that of Tb$_5$Si$_4$. For Tb$_5$(Si$_2$Ge$_2$) the heat capacity peak is no longer a λ-shape anomaly, but is representative of a first order phase transformation, which agrees well with the conclusion made by Morellon *et al.* [14].

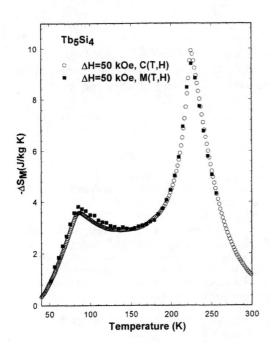

FIGURE 5. The comparison of the magnetocaloric effect, $-\Delta S_M$, of Tb_5Si_4 calculated from both magnetization and heat capacity data.

The isothermal magnetic entropy change and the adiabatic temperature change calculated from the heat capacity data, using equation (3) and (4), are shown in Figs. 3, 4. Both the magnetic entropy change and the adiabatic temperature change increase with the magnetic field strength. The magnetic entropy change for Tb_5Si_4, $Tb_5(Si_3Ge)$ and $Tb_5(Si_2Ge_2)$ reaches 16.3, 14.8, and 24.6 J/kg K, respectively when the magnetic field change is 100 kOe. The corresponding adiabatic temperature changes for the same conditions are 11.4, 10.4, and 10.3 K, respectively. The magnetic entropy change of $Tb_5(Si_2Ge_2)$ is considerably larger when compared with the other two alloys. For a moderate magnetic field change (ΔH=50 kOe), the magnetic entropy change for Tb_5Si_4, $Tb_5(Si_3Ge)$ and $Tb_5(Si_2Ge_2)$ are 9.9, 8.4, and 15.1 J/kg K, respectively. These are slightly lower than the values reported for the $Gd_5(Si_xGe_{4-x})$, but are still much larger than these for many other magnetic materials, which indicates that $Tb_5(Si_xGe_{4-x})$ alloys may be used as possible magnetic refrigerants. Furthermore, similarly to the $Gd_5(Si_xGe_{4-x})$ systems [3], the title alloys may be used between about 100 and 250K by varying their chemical composition between $2 \leq x \leq 4$.

The isothermal magnetic entropy change as a function of temperature calculated from magnetization data using equation (1) is compared with the magnetic entropy change calculated from heat capacity data for Tb_5Si_4 in Fig. 5. The two results are in excellent agreement. Both techniques confirm the presence of the low temperature anomaly of the magnetocaloric effect, as was noted from heat capacity data (Fig. 1).

CONCLUSIONS

The x-ray powder diffraction study of the $Tb_5(Si_xGe_{4-x})$ alloys shows that the alloys have the orthorhombic Gd_5Si_4-type crystal structure when x = 4 and 3 and the monoclinic crystal structure when x = 2. All three alloys order ferromagnetically between 100 and 225 K depending on their composition. The phase transformations near Curie temperatures are second order for x = 4, 3 and first order for x = 2. Additional low temperature magnetic phase transitions are found in all of the three alloys. The magnetocaloric effect calculated from both the magnetization and heat capacity data shows good agreement between the two experimental techniques. The MCE values of the title alloys are much larger than those found in many other magnetic materials, which suggests that they are promising materials for magnetic refrigeration applications.

ACKNOWLEDGMENT

The Ames Laboratory is operated by Iowa State University for the U.S. Department of Energy (DOE) under Contract No. W-7405-ENG-82. This work was supported by the Laboratory Technology Research Program, Office of Computational and Technology Research, of the U.S. DOE.

REFERENCES

1. Pecharsky, V.K., Gschneidner, K.A., Jr., *Phys. Rev. Lett.* **78**, pp. 4494-4497 (1997).
2. Pecharsky, V.K., Gschneidner, K.A., Jr., *J. Magn. Magn. Mater.* **167**, pp. L179-L184 (1997).
3. Pecharsky, V.K., Gschneidner, K.A., Jr., *Appl. Phys. Lett.* **70**, pp. 3299-3301 (1997).
4. Gschneidner, K.A., Jr. and Pecharsky, V.K., U.S. Patent No. 5743095 (1998).
5. Zimm, C.B., Jastrab, A., Sternberg, A., Pecharsky, V.K., Gschneidner, , K.A., Jr., Osborne, M.G., and Anderson, I.E., "Description and Performance of a Near-room Temperature Magnetic Refrigerator," in *Advances in Cryogenic Engineering* 43, edited by P.Kittel, Plenum, New York, 1998, pp. 1759-1766.
6. Gschneidner, K.A., Jr., Pecharsky, V.K. and Zimm, C.B., "Magnetic Cooling for Appliances," in *Proceedings of the 50th Annual International Appliance Technical Conference, Purdue University, West Lafayette, IN, May 10-12, 1999 (International Appliance Technical Conference, Inc., Largo Florida)*, pp. 144-154.
7. Gschneidner, K.A., Jr., Pecharsky, V.K. and Zimm, C.B., *Mater. Techn.* **12**, pp. 145-149 (1997).
8. Pecharsky, V.K. and Gschneidner, K.A, Jr., *J. Magn. Mag. Mater.* **200**, pp. 44-56 (1999).
9. Tishin, A.M., in "Handbook of Magnetic Materials," K.H.J. Buschow, Ed., Elsevier Science B.V., 1999, vol.12, pp. 395.
10. Gschneidner, K.A., Jr. and Pecharsky, V.K., *Annu. Rev. Mater. Sci.* **30**, pp. 387-429 (2000).
11. Holtzberg, F., Gambino, R.J. and McGuire, T.R., *J. Phys. Chem. Solids* **28**, pp. 2283-2289 (1967).
12. Serdyuk, Y.V., Krentsis, R.P. and Gel'd, P.V., *Sov. Phys. Solid State* **23**, pp. 1592-1594 (1981).
13. Spichkin, Y.I., Pecharsky, V.K., Gschneidner, K.A., Jr., *J. Appl. Phys.* **89**, pp. 1738-1745(2001).
14. Morellon, L., Magen, C., Algarabel, P.A., and Ibarra, M.R., private communication.
15. Pecharsky, V.K., Moorman, J.O. and Gschneidner, K.A., Jr., *Rev. Sci. Instum.* **68**, pp. 4196- 4207 (1997).
16. Pecharsky, V.K. and Gschneidner, K.A., Jr., *J. Appl. Phys.* **86**, 565-575 (1999).
17. Pecharsky, V.K. and Gschneidner, K.A., Jr., *J. Alloys Compounds* **260**, pp. 98-106 (1997).
18. Gschneidner, K.A., Jr., Pecharsky, V.K., Pecharsky, A.O., Ivtchenko, V.V., Levin, E.M., *J. Alloys Compounds* **303-304**, pp. 214-222 (2000).

MAGNETOTHERMAL PROPERTIES OF SINGLE CRYSTAL DYSPROSIUM

A.S. Chernyshov[1], A.M. Tishin[1], K.A. Gschneidner, Jr.[2,3], A.O. Pecharsky[2], V.K. Pecharsky[2,3], and T.A. Lograsso[2,3]

[1]Faculty of Physics, M.V. Lomonosov Moscow State University, 119899, Moscow, Russia.

[2]Ames Laboratory and [3]Department of Materials Science and Engineering, Iowa State University, Ames, IA 50011-3020, USA.

ABSTRACT

The magnetocaloric properties (the adiabatic temperature change) of the high purity single crystalline dysprosium have been measured directly over the temperature range from 78 to 220 K in magnetic fields from 0 to 14 kOe applied along the easy magnetization direction (a-axis). These results are in good to excellent agreement, except for two regions (105 to 127 K, and 179 to 182 K), with the previous magnetocaloric effect data reported on lower purity dysprosium samples. The magnetic phase diagram of Dy has been refined based on the results of these measurements and two new high magnetic field phases have been identified.

INTRODUCTION

The rare-earth metal dysprosium, Dy, has one of the largest magnetic moments in the lanthanide series, i.e., the experimentally determined value of the effective magnetic moment of a free Dy^{3+} ion is 10.83 μ_B [1]. In zero magnetic field Dy is paramagnetic (PM) above ~180 K, and its ordered magnetic structures include helical antiferromagnet (AFM) below T_N = ~180 K and a ferromagnet (FM) below T_C = ~90 K with the easy magnetization direction coinciding with the crystallographic a-axis [1,2]. At elevated magnetic fields, an intermediate fan magnetic structure is found in the temperature range between ~130 and ~180 K [2]. There is a tricritical point on the H-T phase diagram of Dy near 165 K [3] in a magnetic field of approximately 5 kOe, where the first-order AFM-fan transition becomes a second-order phase transformation. It is commonly accepted that in zero magnetic field the transition between the AFM and FM phases of Dy is a first order phase transformation, which is accompanied by a hexagonal-orthorhombic lattice

CP614, *Advances in Cryogenic Engineering:*
Proceedings of the International Cryogenic Materials Conference - ICMC, Vol. 48,
edited by B. Balachandran et al.
© 2002 American Institute of Physics 0-7354-0060-1/02/$19.00

distortion, while the transition between the AFM and PM phases is a second order phase transformation.

The zero magnetic field AFM structure of dysprosium is a modulated spiral with the wave vector oriented along the hard magnetizing direction. The investigation of thermal expansion and sound propagation [4] shows the presence of commensurability points, where the wave vector of the helix structure becomes commensurate with the c-lattice parameter. Most of the anomalies of the magnetic structure of Dy were confirmed by neutron scattering [5], and are associated with the peculiar temperature dependence of the helix turn-angle.

Experimental results available to date show that Dy also behaves unusually in the vicinity of its Néel temperature [6,7]. Neutron scattering studies [8] confirmed the absence of a discontinuity near the AFM-PM transition. Concurrently other results [9] revealed the existence of a magnetic satellite peak near the Neel point in the direction of the c-axis. This observation was explained by either the presence of an intermediate vortex state or by the commensurability of both magnetic and crystal structures. The suggestion that the AFM – PM magnetic phase transition in Dy is a second order phase transformation has been confirmed by many experimental results [10]. The AFM-fan transition takes place over a broad range of magnetic fields and, thus, has a continuous character. The absence of a magnetic field hysteresis in the temperature range extending from the tricritical point to the Néel temperature also proves that this transition is a second order phase transformation. On the other hand, the presence of the temperature hysteresis of the magnetization near T_N (see [7]) and a number of other experimental data, collected in the vicinity of Néel temperature (e.g., see Ref. 11), provide some evidence for the first-order nature of this phase transition, and also resulted in claims that the AFM – PM transition has the characteristic of a mixed order phase transition.

The presence of the extended temperature hysteresis in the paramagnetic region suggests the possibility that AFM clusters exist in the bulk of the PM phase well above Néel temperature. Both the amount and size of these clusters decrease with increasing temperature, and Dy becomes truly paramagnetic and homogeneous only at temperatures 30 to 50 K above T_N (e.g., see Ref. 11). Since impurities may have strong effect on the nature of the phase transitions and, in principle, lead to the appearance of hysteresis in the vicinity of T_N, we carried out a detailed investigation of the phase diagram of Dy using high purity single crystals. Furthermore, it appears that first order phase transition materials, where the magnetism and the crystallography change simultaneously, hold a great promise as advanced magnetic refrigerant materials [12-14]. Therefore, this work is also an attempt to gain a better understanding of the behavior of the magnetocaloric effect (MCE) during a first-order phase transformations and the influence of the impurities on the MCE values.

EXPERIMENTAL DETAILS

The high-purity single crystals of dysprosium were prepared by the Materials Preparation Center of the Ames Laboratory. The overall purity of the samples was 99.98 wt %. The sample of Dy used in this study was cut in the form of a parallelepiped from a large grain. The approximate dimensions of the parallelepiped were $1 \times 3 \times 12$ mm^3 with the longest side parallel to the a crystallographic axis. The orientation of the sample was established by using a back-reflection Laue technique. The direct measurements of the MCE were performed with the magnetic field applied along the long side of the parallelepiped. The combined accuracy of the alignment of the *a* crystallographic direction with the direction of the applied magnetic field was of the order of ±5°. The magnetocaloric effect was measured in the temperature interval from 78 K to 350 K in a quasistatic magnetic field. The magnetic field, which ranged from 0 to 14 kOe, was created around a thermally insulated specimen by charging and discharging the coil of an electromagnet. Due to the relatively large magnetic induction, the field sweeps from 0 to 14 kOe were performed during about 5 s. The quality of the thermal insulation of the sample was high enough to ensure nearly adiabatic conditions during the measurements. The equilibrium temperature of the sample was measured by a copper-constantan thermocouple before and after the magnetic field sweep. The MCE was determined as the difference between these two equilibrium temperatures. The relative experimental errors were estimated to be on the order of 7%. Most of the MCE results presented below were corrected for the demagnetizing field using the values of the demagnetizing factor for an ellipsoid with the same dimensions (N = 0.17). In some cases the magnetocaloric effect was measured both during heating or cooling of the sample. Measurements were carried out on cooling when each temperature (in zero magnetic field) was reached during cooling the sample from ~230 K to the target temperature, while in the case of heating, the target temperatures were reached during heating the sample, which had been initially cooled to ~77 K.

RESULTS

The isothermal dependencies of the MCE in the vicinity of the FM-AFM transition with the magnetic field applied along the easy magnetizing direction are shown in Fig.1. In

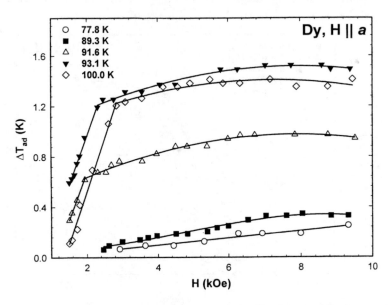

FIGURE 1. The isothermal dependencies of the magnetocaloric effect (corrected for demagnetization) in the single crystal Dy with the magnetic field applied along the easy magnetizing direction in the temperature interval from 77 to 100 K. The lines drawn through the data points are the guides for an eye.

the ferromagnetic region, i.e. at T = 77.6 and 89.3 K, the MCE is small and it increases nearly linearly with increasing magnetic field between ~2 and ~10 kOe. At higher temperatures, when Dy becomes antiferromagnetic in the zero magnetic field, the magnetocaloric effect shows quite different behavior. First, a rapid increase of the MCE is observed in low magnetic fields (2 to 3 kOe), and then the MCE is nearly saturated in magnetic fields exceeding ~4 kOe between 91.6 and 100 K. This behavior can be understood if one recalls that the magnetic field induces a first-order AFM → FM transition above ~90 K in Dy. The first step-like increase and the saturation of the MCE is observed at 91.6 K; and this temperature agrees well with all previous data confirming that Dy is antiferromagnetic in zero magnetic field above ~90 K.

Typical isothermal dependencies of the MCE in the region from 110 to 127 K with the magnetic field applied parallel to the easy magnetizing direction are shown in Fig. 2. Each MCE function shows two step-like anomalies. The first anomaly is due to a magnetic field induced AFM ↔ FM transition. The second step appears only above ~105 K. The difference between the values of the critical magnetic field corresponding to the first step as measured during the heating and cooling of the sample indicates the presence of a temperature hysteresis, which is expected during the metamagnetic AFM ↔ FM transitions. However, the values of the critical field corresponding to the second step remain essentially the same when measured during both the heating and cooling of the sample. We conclude, therefore, that first critical fields correspond to a first order phase transition, and that second critical fields are indicative of a second order phase transition. The magnetocaloric effect was practically completely reversible in the range of magnetic fields where the second step was observed.

Figure 3 illustrates the magnetocaloric effect in Dy as a function of temperature between 77 and 210 K in magnetic field varying from 0 to 10 kOe shown together with the previously reported data obtained using different purity single crystals of Dy. The general

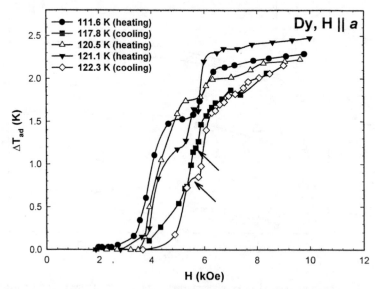

FIGURE 2. The isothermal dependencies of the magnetocaloric effect (corrected for demagnetization) in the single crystal Dy with the magnetic field applied along the easy magnetizing direction in the temperature interval from 110 to 125 K. The lines drawn through the data points are the guides for an eye.

behavior of the MCE is in a fair agreement with the earlier measurements except for the much sharper increase of the MCE around 90 K, and the much higher values of the MCE between ~180 and 210 K where they exceed those reported earlier [3,11] by about a factor of two. Note, that all data presented in Fig. 3 have not been corrected for the demagnetizing factor.

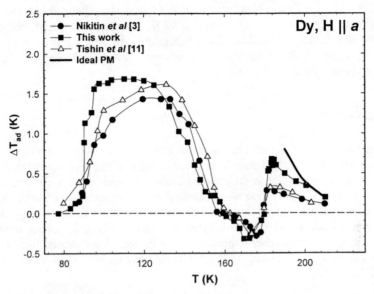

FIGURE 3. The magnetocaloric effect in the single crystal Dy in the 10 kOe magnetic field applied along the easy magnetizing direction compared with the earlier measurements [3,11]. All data are shown without correcting for demagnetization effects.

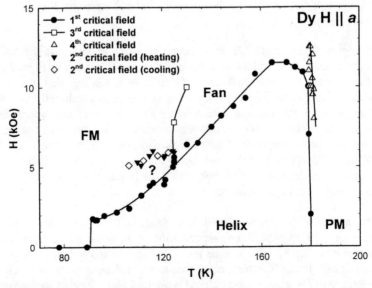

FIGURE 4. The H-T magnetic phase diagram of the single crystal of Dy along the easy *a*-axis constructed using the magnetocaloric effect data.

The magnetic phase diagram of Dy with the magnetic field applied along the easy magnetization direction based entirely on the magnetocaloric effect data is shown in Fig. 4. We note that the diagram includes points in temperature-magnetic field coordinates based on the anomalous MCE(T,H) behavior above ~120 K, which are not shown due to space limitations of this communication. The proposed phase diagram is similar to the earlier diagrams [2, 3, 5] with respect to the four major magnetic phases, i.e. PM, helix-AFM, fan, and FM structures, but some new and significant features on this diagram have been added. These include: (1) the first critical magnetic field separating helix-AFM and FM phase fields remains far above zero, at least down to 92 K, and then it is sharply reduced to zero between ~90 and 92 K; (2) additional phase fields delineating an unknown magnetic phases between ~105 and ~127 K in magnetic fields between ~ 3 and 6 kOe and between ~179 and ~182 K in magnetic fields 8-12 kOe.

DISCUSSION

In general, most of the results reported above are in good agreement with those published earlier. Dy clearly behaves as a ferromagnetic material below its Curie point, and as an antiferromagnetic material between Curie and Néel points in zero magnetic field. There is an intermediate fan phase which forms during the transition from the AFM to FM phase in the temperature region above 127 K and in magnetic fields above ~5 kOe. From magnetocaloric effect we were able to detect the helix-AFM – fan phase transition only below ~130 K, because at higher temperatures the critical fields exceed the ~14 kOe limit of our electromagnet. According to the MCE data Dy clearly behaves as a paramagnet above ~182 K.

Contrary to earlier data, the MCE results obtained using high purity single crystal Dy indicate that there is a sharp increase of the critical magnetic field required to cause the helix-AFM to FM metamagnetic transition in the immediate vicinity of the Curie temperature, i.e. between 90 and 92 K (see Fig.4). It is feasible that this can be associated with the coexistence of both the helix-AFM and FM phases just above the Curie temperature.

In the region from 110 to 125 K a second step-like anomaly of the MCE is clearly distinguishable (see Figs. 2 and 4). We believe that results shown in Figs. 2 and 4 can be interpreted in terms of commensurability effects. Assuming that the magnetic and crystallographic lattices become commensurate at a certain combination of the magnetic field and temperature, this changes (increases) the symmetry of the system and thus may change the nature of the magnetic phase transition and may result in the splitting of the phase boundary line corresponding to a single first order phase transition between the helix-AFM and FM states into the two lines: the lower magnetic field phase boundary remains a first order phase transformation, while the higher magnetic field phase boundary becomes a second order phase transformation.

The anomalies of the magnetic field dependence of the MCE found in the temperature interval between 179 and 182 K correspond to the additional phase boundary line in the phase diagram (see Fig. 4). We assume that it reflects an intermediate "vortex" state that, as noted previously in the literature, may exist in this temperature region. Hence, the true Néel temperature of Dy when magnetic field is applied along its easy magnetizing direction is 182 K where all MCE anomalies, and therefore, critical fields no longer exist.

We also found a significant difference of the measured values of the MCE in the paramagnetic state in our high purity single crystal as compared with the previously reported data [3,11], see Fig. 3. The difference is well beyond the accuracy of the experimental measurements, i.e. the new results exceed the old results by about a factor of two. First, it is important to note that the biggest difference is observed at temperatures close to Néel temperature, i.e. where short range correlations (AFM clustering) are significant, and this difference diminishes when the temperature increases and the system approaches a truly paramagnetic state. The obtained results could be understood as follows.

To consider the influence of the antiferromagnetic clusters on the MCE values, one can calculate the magnetocaloric effect of dysprosium assuming an ideal paramagnetic behavior. It is well known that the magnetization, $M(H,T)$, of an ideal paramagnetic material follows Curie-Weiss law:

$$M(H,T) = \frac{Np_{eff}^2}{3k} \frac{H}{T-T_N},$$ (1)

where N is the number of paramagnetic atoms, k is the Boltzmann constant, and $p_{eff} = g^2 J(J+1) \approx 10\mu_B$ is the effective magnetic moment for dysprosium. Our experimental temperature dependence of the heat capacity in the temperature interval above 190 K and in the range of magnetic fields 0 - 10 kOe has been approximated by the following equation [12]:

$$C(H,T) = 10^7 \left(25 + \frac{125}{T-172} \right).$$ (2)

After integration of Maxwell thermodynamic relation for MCE, using equations (1) and (2) and neglecting the quantities proportional to $(\Delta T_{ad})^2$, which is valid for small magnetic field change when the MCE is small, one can obtain the numerical expression for the magnetocaloric effect in a paramagnetic Dy:

$$\Delta T_{ad}(T) = 1.8 \frac{T}{(T-172)(T-167)}.$$ (3)

The results obtained using Eq. 3 are shown in Fig. 3 as a heavy solid line. As one can see, the calculated MCE values are in an excellent agreement with the results obtained in this work in the range of temperatures above 200 K. The earlier MCE data continue to deviate from the calculated ones over a much larger temperature range. Consider a model, where the magnetic field first suppresses AFM structure of each cluster and aligns their magnetic moments with the direction of the field. After that, the magnetism inside the clusters should become quite similar to that in the PM phase and, most likely, clusters no longer behave as a separated magnetic sub-phase. It is therefore expected that initial process of the transformation of AFM clusters results in negative ΔT and the total MCE value at low magnetic fields is thus reduced. Obviously, in the case of low purity samples there are more AFM clusters and the initial magnetization process leads to a larger reduction of the resulting MCE. Fundamentally, this difference is expected to be less pronounced in higher magnetic fields and it is reasonable to conclude that in our (higher purity) sample, AFM clusters exist over a narrower temperature interval, than in case of previous studies. Unfortunately, the purity of the Dy sample investigated in [3,11] is unknown and it is difficult to carry out a more careful analysis of the observed differences.

CONCLUSIONS

The magnetocaloric effect in a single crystal Dy have been measured from 78 to 220 K in magnetic fields ranging from 0 to 14 kOe applied along the easy magnetizing direction. In the maximum magnetic field of 14 kOe, the value of the MCE varies from -1.2 K at 175.5 K to 2.5 K at 115 – 120 K. The results of our measurements are in fair agreement with earlier data. Considerable disagreement with previous measurements is observed in the immediate vicinity of Néel temperature in the paramagnetic state, which is most likely associated with the higher purity of the specimen used in this study. The H-T phase diagram of Dy have been revised to reflect the presence of two unknown magnetic phases existing between ~105 and ~127 K in magnetic fields between ~ 3 and 6 kOe and between ~179 and ~182 K in magnetic fields 8-12 kOe, respectively.

ACKNOWLEDGMENT

The Ames Laboratory is operated by Iowa State University for the U.S. Department of Energy (DOE) under contract No. W-7405-ENG-82. Different aspects of this work were supported by the Office of Basic Energy Sciences, Materials Sciences Division of the U.S. DOE (KAG, VKP, AOP, and TAL), by NATO Linkage Grant 974570 (all authors), and by RFBF grant 01-02-17703 (ASC and AMT). We thank to M.I. Ilyn, A.S. Mischenko and Yu.I. Spichkin for their assistance with some experimental measurements and useful discussions.

REFERENCES

1. Jensen, J. and Mackintosh, A.R., *Rare earth magnetism: structures and excitations*, Clarendon Press, Oxford, 1991, p. 45.
2. Herz, R. and Kronmüller, H., J. Magn. Magn. Mater. **9**, 273 (1978).
3. Nikitin, S.A., Andreenko, A.S. and Pronin, V.A., Fiz. Tverd. Tela **21**, 2808 (1979).
4. Greenough, R.D., Blackie, G.N. and Palmer, S.B., J. Phys. C: Solid State Phys. **14**, 9 (1981).
5. Wilkinson, M.K., Koehler, W.C., Wollan, E.O. and Cable, J.W., J. Appl. Phys. **32**, 48 (1961).
6. Dudáš, J. and Feher, A., Acta Phys. Polonica, **A76**, 195 (1989).
7. Dan'kov, S.Y., Popov, Y.F. and Tishin, A.M., Vestn. Mosk.. Univ., Ser. 3: Fiz., Astron. **35**, 98 (1994).
8. Brits, G.H.F. and De V. Plessis, P., J. Phys.: F. Met. **15**, 239 (1985).
9. Bessergenev, V.G., Gogava, V.V., Kovalevskaya, Yu.A., Mandzhavidze, A.G., Fedorov, V.M. and Shilo, S.I., Pis'ma Zh. Eksp. Teor. Fiz., **42**, 412 (1985).
10. Tishin, A.M., *The investigation of magnetic, magnetothermal and magnetoelastic properties of heavy rare-earth metals and their alloys in the region of magnetic phase transitions*, Dr. Sc. Thesis, Moscow State University, Moscow, 1994 (in Russian).
11. Tishin, A.M. and Martynenko, O.P., *Physics of rare earths near magnetic phase transitions*, Nauka, Moscow, 1995, pp. 106 – 114 (in Russian).
12. Pecharsky, V.K. and Gschneidner, K.A., Jr., J. Magn. Magn. Mater., **200**, 44 (1999).
13. Tishin, A.M., in *Handbook of Magnetic Materials*, vol.12, ed. K.H.J. Buschow, (Elsevier Science B.V., The Netherlands, 1999), p.395.
14. Gschneidner, K.A., Jr. and Pecharsky, V.K., Annu. Rev. Mater. Sci., **30**, 387 (2000).

RELATIONSHIP OF ADIABATIC, ISOTHERMAL AND FIELD CONSTANT CHANGES OF A MAGNETIC ENTROPY

A.M. Tishin[1] and Yu.I. Spichkin[2]

[1]Faculty of Physics, M.V. Lomonosov Moscow State University
Moscow, 119899, Russia

[2]N. S. Kurnakov Institute of General and Inorganic Chemistry RAN,
Moscow, 117091, Russia

ABSTRACT

The character of a magnetic entropy change under the influence of the magnetic field for different isobaric thermodynamic processes is discussed. The general equations describing the relationship of adiabatic, isothermal and field constant ways of changing of magnetic entropy are presented. It is shown that a value of total magnetic entropy adiabatic change of can be close to isothermal one (for, example, at elevated temperature and/or a small value of magnetocaloric effect) as well as to constant field one (in the case of large magnetocaloric effect and/or a small value of magnetization and low temperature).

INTRODUCTION

Up to present a question about different contributions to magnetic entropy change caused by the magnetic field change and their relation is unclear. At the same time it is very important from the point of view of application of different magnetic materials in magnetic refrigerators and cryocoolers and estimation of their efficiency for this purposes. The main goal of this short communication is to clarify the question concerning different contributions to magnetic entropy change and give a push to future experimental and theoretical investigations of this question. We show that total value of magnetic entropy change under the effect of magnetic field during, for example, isobaric – adiabatic process consists of isothermal and isofield parts. The influence of both these parts on the total MCE value can be significantly changed depending from the range of magnetic field and temperature under investigation.

CP614, *Advances in Cryogenic Engineering:*
Proceedings of the International Cryogenic Materials Conference - ICMC, Vol. 48,
edited by B. Balachandran et al.

GENERAL THERMODYNAMIC CONSIDERATION

Consider a total entropy S of magnetic material at isobaric conditions as a function of temperature, T, and magnetic field H: $S = S(T,H)$. The total differential of the S has a form:

$$dS(T,H) = (\frac{\partial S(T,H)}{\partial T})_H dT + (\frac{\partial S(T,H)}{\partial H})_T dH . \quad (1)$$

With the help of Maxwell equation:

$$(\frac{\partial S(T,H)}{\partial H})_T = (\frac{\partial M(T,H)}{\partial T})_H \quad (2)$$

and the general formula for heat capacity at constant pressure and field:

$$C_H(T,H) = T(\frac{\partial S(T,H)}{\partial T})_H , \quad (3)$$

one can derive from Eq.(1) that:

$$dS(T,H) = \frac{C_H(T,H)}{T} dT + (\frac{\partial M(T,H)}{\partial T})_H dH , \quad (4)$$

where $M(T,H)$ is a magnetization. It should be noted that $C_H(T,H)$ contains contributions from lattice, electronic and magnetic subsystems.

The finite total entropy change due to the change of magnetic field from H_1 to H_2 can be calculated by Maxwell equation (2) and is the isothermal entropy change:

$$\Delta S(T,H) = S(T,H_2) - S(T,H_1) = \int_{H_1}^{H_2} (\frac{\partial M(T,H)}{\partial T})_H dH . \quad (5)$$

At certain conditions the total entropy can also be presented as a sum of three main terms (see, for example, [1]):

$$S(T,H) = S_{mag}(T,H) + S_l(T,H) + S_e(T,H), \quad (6)$$

where $S_l(T,H)$ is the lattice contribution, $S_e(T,H)$ is the electron contribution and $S_{mag}(T,H)$ is the contribution from magnetic subsystem (magnetic part of entropy or magnetic entropy). In general case all three contributions depend on temperature and magnetic field and cannot be clearly separated. The situation is especially difficult in the case, for example, of low temperature region where the value of electronic coefficient, γ, can change in few times under the influence of magnetic field or in the case of coexisting of magnetic, structure and electronic phase transitions (see Ref. [2]). However, in the first approximation we can believe that lattice and electronic parts of entropy depending only on temperature and all contributions depending on magnetic field (from any changes of magnetic subsystem) are presented in total value of entropy in Eq. (6) by $S_{mag}(T,H)$. Using Eqs. (2)-(4),(6) one can obtain the following equation:

$$dS(T,H) = \frac{C_l(T)}{T}dT + \frac{C_e(T)}{T}dT + \frac{C_{mag}(T,H)}{T}dT + (\frac{\partial S_{mag}(T,H)}{\partial H})_T dH , \quad (7)$$

where $C_l(T)$ is a lattice contribution to heat capacity, $C_e(T)$ is a electronic contribution to heat capacity and $C_{mag}(T,H)$ is a heat capacity of magnetic subsystem $(C_{mag} = C_{mag}(T,H))$:

$$C_{mag}(T,H) = T(\frac{\partial S_{mag}(T,H)}{\partial T})_H . \quad (8)$$

The first two terms in Eq. (7) are the changes of lattice and electronic entropy and the last two terms represent the total differential of the magnetic part of the entropy:

$$dS_{mag}(T,H) = \frac{C_{mag}(T,H)}{T}dT + (\frac{\partial S_{mag}(T,H)}{\partial H})_T dH . \quad (9)$$

As one can see in the considering case, dS_{mag} consists of two parts – one connects with the change of temperature (field constant part) – it is the first term in Eq. (9), let's call it $dS_{mag\,H}$ and another – due to change of a magnetic field (isothermal part) – it is the second term in Eq. (9), let's call it $dS_{mag\,T}$.

ISOTHERMAL CHANGE OF MAGNETIC ENTROPY

According to Eq. (6) and Maxwell equation (2) a finite isothermal change of S_{mag} caused by the magnetic field change (denoted us as $\Delta S_{mag\,T}$) is equal to the total isothermal entropy change:

$$\Delta S_{mag\,T}(T,H) = S_{mag}(T,H_2) - S_{mag}(T,H_1) = \int_{H_1}^{H_2}(\frac{\partial S_{mag}(T,H)}{\partial H})_T dH = \int_{H_1}^{H_2}(\frac{\partial M(T,H)}{\partial T})_H dH =$$
$$= S(T,H_2) - S(T,H_1) = \Delta S(T,H) . \quad (10)$$

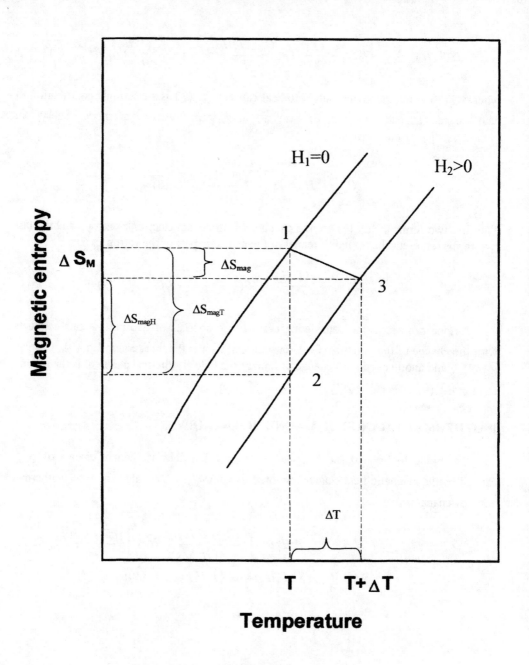

FIGURE 1. An example of temperature dependencies of magnetic entropy S_{mag} at two different fields H_1 and H_2 ($H_2 > H_1$) in a ferromagnetic material.

In this case temperature is constant, and lattice and electronic entropies $S_l(T)$ and $S_e(T)$ are constant (contributions of their changes to $\Delta S_l(T)$ and $\Delta S_l(T)$ to $\Delta S(T,H)$ are equal to zero) and we do not take into account magnetic entropy change caused by any change of temperature (as well as caused by the magnetocaloric effect itself).

ADIABATIC CHANGE OF MAGNETIC ENTROPY AND MAGNETOCALORIC EFFECT

Consider the magnetic entropy change under isobaric-adiabatic (not at isothermal) conditions. It is well known that as an external magnetic field is applied to a magnetic material under adiabatic conditions (under the conditions of constant total entropy of the body, $S(T,H)$=const) the initial temperature of the material T may vary due to magnetocaloric effect by the value ΔT. Under adiabatic conditions the left part of Eq. (7) is equal to zero and the equation can be rewritten as:

$$\frac{C_l(T)}{T}dT + \frac{C_e(T)}{T}dT = -((\frac{\partial S_{mag}(T,H)}{\partial H})_T dH + \frac{C_{mag}(T,H)}{T}dT). \quad (11)$$

The right-hand part of Eq. (11) represents magnetic entropy change under adiabatic process (adiabatic magnetic entropy change) – it is described by Eq. (9). It should be noted, that we consider equilibrium adiabatic process at which the temperature of electronic, lattice and magnetic subsystems vary simultaneously under magnetic field changing and the temperature of magnetic subsystem is permanently equal to the temperature of lattice. Usually exactly temperature of the lattice is measured in MCE experiments. However, in some special cases under nonequilibrium conditions (for example, due to the absence of possibility of energy exchange between magnetic and electronic and lattice subsystems), the temperatures of electronic and lattice subsystems may differ, which can lead to the hysteresis on the magnetocaloric effect field and temperature dependencies.

According to Eq. (9), the total finite change of magnetic entropy caused by the field change ΔH and temperature change ΔT in adiabatic process can be written as follows:

$$\Delta S_{mag}(T,H) = S_{mag}(T+\Delta T, H+\Delta H) - S_{mag}(T,H) =$$

$$= \int_H^{H+\Delta H} (\frac{\partial M(H,T+\Delta T)}{\partial T})_H dH + \int_T^{T+\Delta T} \frac{C_{mag}(H,T)}{T}dT. \quad (12)$$

The first integral in Eq. (12) represents the finite isothermal magnetic entropy change ΔS_{magT} and the second – the finite isofield entropy change ΔS_{magH}. These contributions are illustrated by FIG. 1, where the temperature dependencies of the magnetic entropy S_{mag} at the magnetic field $H_1 = 0$ and $H_2>0$ for an arbitrary magnetic material are shown.

The result of adiabatic magnetization process can be imagined as a sum of two sequential processes of entropy change: isothermal T=const change of magnetic field $H_1 \rightarrow H_1 + \Delta H = H_2$ (process $1 \rightarrow 2$ in the FIG. 1., it corresponds to the isothermal entropy change ΔS_{magT}) and temperature change ($T \rightarrow T+\Delta T$) under the constant magnetic field $H=$

const (process 2→3 in FIG. 1, it corresponds to the isofield entropy change ΔS_{magT}). In accord with above-mentioned, ΔT here is a finite value of magnetocaloric effect caused by the finite field change $\Delta H = H_2 - H_1$. The magnetization process also can by considered as isofield on the first stage (at $H_1=0$) and as constant magnetic entropy on the second stage. This consideration will be done in further work.

Thus, both (isothermal and isofield) parts give contribution to a total change of magnetic entropy under adiabatic magnetization (or demagnetization). The increase of the isofield contribution $\Delta S_{magH}(T,H)$ leads to reduce the total value of $\Delta S_{mag}(T,H)$ and to increase the value of magnetocaloric effect (large distance between points 2 and 3). In this case a total adiabatic change of magnetic entropy, $\Delta S_{mag}(T,H)$, and adiabatic isothermal change of magnetic entropy, $\Delta S_{magT}(T,H)$, may differ significantly (see FIG. 1) and cannot be assumed to be equal. Values of ΔS_{mag} and ΔS_{magT} may be close to each other only at small value of C_{mag}/T (i.e. small value of magnetic contribution, C_{mag}, to total value of heat capacity and/or high temperatures) and/or small value of ΔT (when point 2 and 3 are close, see FIG.1). The value of MCE (the distance between points 2 and 3 on T axis) is determined by the value of ΔS_{magH} and the value of ΔS_{magT} influence ΔT by displacing initial $S_{mag}(T)$ curve.

For isobaric-adiabatic process ($dS = 0$) Eqs. (4) and (2), can give the well known expression for the magnitude of MCE [1]:

$$dT(T,H) = -\frac{T}{C_H(T,H)}(\frac{\partial S(T,H)}{\partial H})_T dH \qquad (13)$$

or

$$dT(T,H) = -\frac{T}{C_H(T,H)}(\frac{\partial M(T,H)}{\partial T})_H dH . \qquad (13a)$$

Using Eq. (11) one can obtain for adiabatic process:

$$dT(T,H) = -\frac{T}{C_H(T,H)}(\frac{\partial S_{mag}(T,H)}{\partial H})_T dH = -\frac{T}{C_H(T,H)}dS_{magT}(T,H) , \qquad (14)$$

where $C_H = C_l + C_e + C_{mag}$ is the total heat capacity. From Eq. (14) it can be concluded that the value of dT is proportional to the isothermal change of S_{mag}, which can be calculated by Maxwell relation (2). However, the value of MCE is determined by isofield adiabatic change ΔS_{magH} – see FIG. 1. To reach maximum MCE value one should provide such conditions at which ΔS_{mag} should be minimal (in this case ΔS_{magH} and ΔS_{magT} are close to each other). In Eq. (14) magnetic entropy change due to a temperature change affects magnetocaloric effect indirectly through additional term (C_{mag}) in the total heat capacity.

CONCLUSIONS

Thus, we show here that isothermal change of magnetic entropy caused by the magnetic field change is equal to isothermal change of total entropy, which can be calculated on the base of magnetization data using Maxwell relation. The magnetic entropy change under adiabatic process ΔS_{mag} can be represented as a sum of two contributions: isofield magnetic entropy change ΔS_{magH} and isothermal magnetic entropy change ΔS_{magT}. Values of total adiabatic change of magnetic entropy ΔS_{mag} and isothermal part of magnetic entropy change ΔS_{magT} may be close to each other only at small values of C_{mag}/T (i.e. small value of magnetic contribution, C_{mag}, to total value of heat capacity and/or high temperatures) and/or small value ΔT of MCE. In general case the value of MCE is determined by interrelation between ΔS_{magT} and ΔS_{magH}. It should be noted that this interrelation in different range of magnetic field and temperature has not been experimentally studied yet.

ACKNOWLEDGEMENTS

This work is supported by grant Russian Fund of Basic Research No. 01-02-17703 and NATO Linkage Grant No. 974570. A.M.T. thanks Profs. K.A. Gschneidner, Jr. and V.K. Pecharsky for useful discussions.

REFERENCES

1. Tishin, A.M., *"Magnetocaloric effect in the vicinity of phase transitions,"* in Handbook of Magnetic Materials, edited by K.H.J. Buschow, North-Holland, Amsterdam, 1999, v.12, ch. 4, p. 395-524.
2. Pecharsky, V.K., and Gschneidner, Jr., K.A. *Advanced. Materials* 13, 683 (2001).

COOLING PERFORMANCE OF CERAMIC MAGNETIC REGENERATOR MATERIAL USED IN A GM CRYOCOOLER

T. Numazawa[1], T. Satoh[2], T. Yanagitani[3] and A. Sato[1]

[1]Tsukuba Magnet Laboratory, National Institute for Materials Science
3-13 Sakura, Tsukuba 305-0003, Japan
[2]R&D Center, Sumitomo Heavy Industries, Ltd.
19 Natsushima, Yokosuka, Kanagawa 237-8555, Japan
[3]Ceramics Division, Konoshima Chemical Co. Ltd.
80 Koda, Takuma-Cho, Mitoyo-Gun, Kagawa 769-1103, Japan

ABSTRACT

A ceramic magnetic regenerator material GAP (GdAlO$_3$) has been developed for cryocoolers below 4 K region. A new fabrication method has achieved Vickers hardness of ~900, relative density of 99.5 % to that of the single crystal, and smooth surface on the spherical particles in diameters between 133 μm and 500 μm. The cooling tests with a GM cycle cryocooler have been done for ^{4}He and ^{3}He as working fluids individually. The 2nd regenerator constitution was adjusted by partially replacing the HoCu$_2$ with the GAP. The cooling capacity increases up to 30 % around 3.4 K and the minimum temperature reached 2.2 K for the ^{4}He test. ^{3}He operation with the GM cycle was investigated to provide temperatures below 2 K without the superfluid transition appearing with ^{4}He. The test results showed a remarkable improvement on the cooling characteristics; the cooling capacity increased 1.8 times at 2.0 K and 2.4 times at 1.8 K, compared to those without the GAP. The minimum temperature without heat load was lowered from 1.67 K to 1.57 K at the cycle speed of 60 rpm.

INTRODUCTION

Since magnetic regenerator materials were developed, cryocoolers have provided temperatures below 10 K easily, and currently a cooling temperature of 4.2 K is obtainable without liquid helium [1-4]. Thus, those materials have contributed much to realize the various "cryogen free" applications. But further investigation is still needed to develop new regenerator materials for sub-4 K cryocoolers for cooling sensor devices such as TES or STJ [5]. In

CP614, *Advances in Cryogenic Engineering:*
Proceedings of the International Cryogenic Materials Conference - ICMC, Vol. 48,
edited by B. Balachandran et al.
© 2002 American Institute of Physics 0-7354-0060-1/02/$19.00

order to provide temperatures lower than 4 K, a regenerator material with a large heat capacity per unit volume is essentially needed. For this purpose, a conventional magnetic regenerator material $HoCu_2$ has been used, but its heat capacity decreases largely below 4 K, then the cooling capacity of the cryocooler also comes to be small. We have developed a new ceramic regenerator material GAP ($GdAlO_3$), which has a considerably higher heat capacity than that of $HoCu_2$ below 3.9 K [6]. Previous cooling test with 4 K cryocoolers showed a clear improvement in both the cooling capacity and the minimum cooling temperature [6-7].

This paper will show further development results on GAP as a regenerator material and an experimental test with a GM (Gifford–McMahon) cycle cryocooler. Since the GAP keeps a large heat capacity below 2 K, it is interesting to see how much of a minimum temperature can be obtainable by the GAP. Therefore, we have done the cooling test by using ^4He and ^3He gas as working fluids individually.

THERMAL AND MAGNETIC PROPERTIES OF GAP

GAP is an oxide rare earth magnetic material with the peroveskite structure, and its magnetic behavior is antiferromagnetic [8]. Since a localized magnetic interaction in Gd site does not make any magnetic transition above 4 K, the Neel temperature is observed around 3.8 K with a sharp λ-type anomaly in the specific heat. It is noticeable that the magnetic transition with folding the high degeneracy, i.e., high level of magnetic entropy contributes to make a large heat capacity around its magnetic transition temperature. We have proposed to use oxide magnetic materials instead of rare earth compounds especially below 4 K [6]. However, growing GAP single crystals is very difficult because of its high melting temperature of ~2050 °C. Therefore, we have fabricated the GAP into a ceramic form, which was most identical to the single crystal on the magnetic properties. The following experimental data are for the ceramic GAP.

FIGURE 1 shows the volumetric heat capacity of GAP and other conventional magnetic regenerator materials used in the 4 K cryocoolers. GAP has a considerably larger heat capacity below 4 K than those of other materials, but the heat capacity decreases sharply

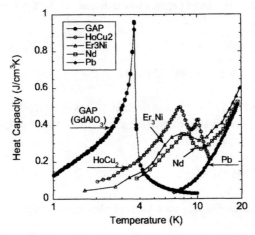

FIGURE 1. Volumetric heat capacity of the GAP and other conventional regenerator materials used in 4 K cryocoolers.

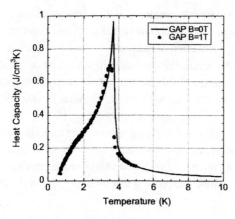
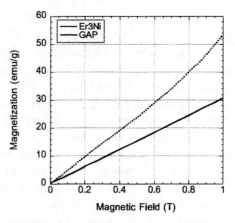

FIGURE 2. Heat capacity of the GAP with the magnetic fields of 0 and 1 T.

FIGURE 3. Magnetization of magnetic field dependence at 5 K for the GAP and Er₃Ni.

above 4 K coming from a typical antiferromagnetic behavior. At the magnetic transition temperature around 3.8 K, the peak value of the heat capacity in the GAP is about 4 times larger than that of $HoCu_2$ and the heat capacities of both materials decrease when decreasing the temperature, but the GAP still has 2.5 times larger heat capacity than that of $HoCu_2$ at 2 K.

Since the 4 K cryocoolers are often used with superconducting magnets, the magnetic field dependence of the regenerator material must be examined. FIGURE 2 shows the heat capacity of the GAP with the magnetic fields of 0 and 1 T. Since GAP is a typical antiferromagnet, the magnetic field of only 1 T does not change the shape of the heat capacity largely as seen in FIG 2, closed circles.

When pressurized helium gas passes through the regenerator material, there is a certain amount of motion in the regenerator material which consists of fine particles with an average diameter of ~300 μm. This often makes magnetic noise which decreases resolution in sensors cooled by the cryocooler. FIGURE 3 shows the magnetization vs. the magnetic field curves at 5 K on the GAP in comparison with Er_3Ni. It is clear that the magnetization of the GAP is about 40 % smaller than that of the Er_3Ni. This comes from the antiferromagnet behavior of the GAP, where the antiparallel interacting spins tend to weaken the total magnetic moment. Thus, the GAP is a much preferable regenerator material for the devices with high sensitivity on magnetic noise.

FABRICATION OF CERAMIC FORM GAP

The fabrication method of a ceramic form GAP has been reported in our previous paper [6], but we have investigated improving the characteristics of GAP used as the regenerator material.

Preparation of GAP Fine Powder

- Mix the Al_2O_3 and the Gd_2O_3 in the form of ultra fine powders less than 1mm size with ethyl alcohol.
- Dry the mixed powder by heating.

- Calcine for ~3 hours at 1200 °C. GAP single phase will be obtainable at this stage.
- Grind the GAP powder with a ball mill adding ethyl alcohol, and then dry by heating.

Fabrication of GAP Sphere Form Particles

- Adding GAP fine powder with a small amount of water into a granulator; fine spherical particles can be grown. This process is shown in FIG 4 (A).
- Calcine the granules at 1700 °C.
- Filter the proper sizes of GAP granules between 100 μm and 500 μm by using screen meshes. Irregular sized granules can be recycled.

The relative density, which is defined by the ratio of the density of fabricated poly crystal to the theoretical one (7.4 g/cm³), has been increased from 99 % to 99.5 % for the GAP granules. Also Vickers hardness has been obtainable up to ~900, but several cooling tests with GM cryocoolers showed that some amount of fine powders were found in the regenerator. This is because the GAP particle consists of two layers and the outside thin layer is relatively brittle and easy to be removed. Thus, we developed a method to remove this layer by using a polishing process schematically shown in FIG 4 (B). FIGURE 5 shows the photograph of the polished GAP particles. Those particles are not exactly spherical, but it is enough to use as the regenerator materials because of the smooth contact surface. After we introduced the new method, there have not been any the broken powders found during the cooling test so far.

EXPERIMENTAL RESULTS ON A GM CYCLE CRYOCOOLER BY USING GAP

Cooling tests with a pulse tube and a GM cycle cryocooler has been done by using the GAP at the lowest temperature portion of the 2nd regenerator [6-7]. The cooling power was increased about 40 % at 4.2 K and the minimum temperature without heat load was lowered from 2.9 K to 2.5 K in case of the pulse tube cryocooler [6]. For the GM cryocooler, only the preliminary test results have been shown, thus we will describe the further test results on the GM cycle.

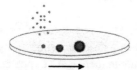

(A) Fabrication of sphere particles with granulator.

(B) Removing the thin surface layer of GAP.

FIGURE 4. Schematic fabrication process for GAP sphere form particles. The thin and weak surface layer can be removed by polishing.

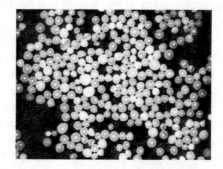

FIGURE 5. Photograph of the polished GAP particles with average diameter of ~300 μm.

	Case 1	Case 2	Case 3
Pb	150g	150g	150g
HoCu$_2$	150g	120g	90g
GAP	0g	24.5g	47.3g

(A) ^{4}He operation

	Case 4	Case 5	Case 6
Pb	150g	150g	150g
HoCu$_2$	145g	90g	60g
GAP	0g	47.3g	70.1g

(B) ^{3}He operation

TABLE 1. Constitution of 2nd regenerator used for ^{4}He and ^{3}He operations.

A two-stage GM cycle cold head unit, SRDK-205DW of Sumitomo Heavy Industries, Ltd. was used in the present study. This unit was originally designed to provide a cooling capacity of 0.5 W at 4.2 K with the compressor unit whose rated input power was 3.4 kW at a 60 Hz driving frequency, but we changed the compressor unit to a smaller one with the rated input power of 2.6 kW to decrease the volume of helium gas. The overall volume of helium gas was about 230 liters at standard condition with a charging pressure of ~15 bar.

Satoh et al. have showed the clear improvement on the cooling characteristics with the GM cryocooler replacing ^{4}He by ^{3}He gas used as a working fluid, where ^{3}He gas behaves much like ideal gas and it does not have a superfluid transition above ~3 mK [9]. The cooling capacity increased about 20 % at 4.2 K and the minimum temperature went down to 1.65 K by using the magnetic regenerator materials of HoCu$_2$ or HoCu$_2$+NdInCu$_2$ at the lowest temperature portion of the 2nd regenerator [9].

Since the heat capacity of the regenerator material contributes to the cooling efficiency significantly, it is expected to increase the cooling capacity replacing those regenerator materials by the GAP. As similar to Satho's experiment, we replaced the HoCu$_2$ by the GAP partially in the 2nd regenerator. The constitution of the regenerator material is shown in TABLE 1 (A) and (B) for the cooling test using the ^{4}He gas and ^{3}He gas, respectively.

Cooling Test By ^{4}He Gas

FIGURE 6 shows the measurement results of the cooling capacity at 4.2 K as a function of 2nd regenerator constitution, i.e., the mass ratio of HoCu$_2$ to GAP. It is clear that using the GAP increases the cooling capacity and there is an optimum value around the constitution of case 2 (HoCu$_2$/GAP=120g/24.5g) given in TABLE 1 (A). Since the heat capacity of GAP decreases steeply above 4 K, HoCu$_2$ will be needed to compensate for a lack of the heat capacity in GAP.

FIGURE 7 shows the 1st stage and 2nd stage temperatures vs. the cycle speed curves without the heat load. The 2nd stage temperature goes down with decreasing the cycle speed. On the other hand, the 1st stage temperature has an opposite tendency. As expected from FIG 6, case 2 gives the lower temperatures at any cycle speed than the other cases.

FIGURE 8 shows the cooling capacity at the 2nd stage as a function of 2nd stage temperature at the cycle speed of 60 rpm. The GAP increases the cooling capacity at any temperatures below 4.2 K. Case 2 gives the increasing cooling capacity of ~14 % compared with case 1 (HoCu$_2$/GAP=150g/0g), while case 3 (HoCu$_2$/GAP=90g/47.3g) provides that of only ~3 %. However, case 3 gives the higher cooling capacity around 3.4 K, where the increased value of ~30 % is obtainable. The cooling test clearly showed the remarkable effect by the

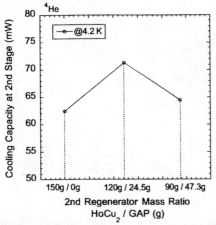

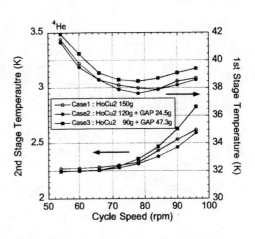

FIGURE 6. Cooling capacity at 4.2 K vs. 2nd regenerator constitution for ⁴He operation.

FIGURE 7. Cycle speed dependence of cooling temperatures without heat load for ⁴He operation.

GAP on the cooling capacity, but the minimum temperature was not lowered much. This is because the thermal expansion of ⁴He comes to be zero at the superfluid transition near 2.2 K.

Cooling Test By ³He Gas

FIGURE 9 shows the measurement results for the cooling capacity at 1.8 K and 2.0 K as a function of 2nd regenerator constitution. There is a similar tendency to the ⁴He operation as seen in FIG 6, but the increasing rate of the cooling capacity is significantly different with increasing the ratio of GAP. For the cooling capacity at 1.8 K, it became ~3 times larger at case 5 (HoCu$_2$/GAP=90g/47.3g) than that at case 4 (HoCu$_2$/GAP=145g/0g). For 2.0 K, the cooling capacity increases ~1.7 times larger with changing the mass ratio from case 4 to case 5.

FIGURE 10 shows the 1st stage and 2nd stage temperatures vs. the cycle speed curves without the heat load. The 1st stage temperatures are lowered about 3~4 K compared with the

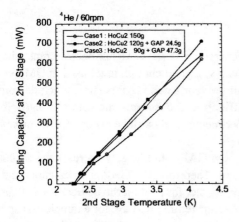

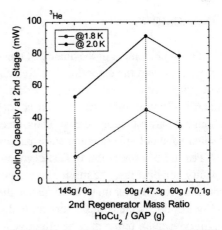

FIGURE 8. Cooling capacity as a function of 2nd stage temperature for ⁴He operation.

FIGURE 9. Cooling capacity at 1.8 K and 2.0 K vs. 2nd regenerator constitution for ³He operation.

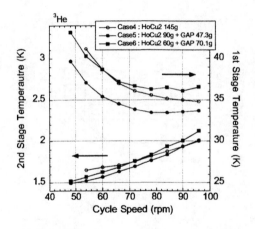
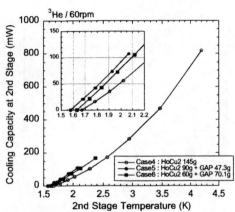

FIGURE 10. Cycle speed dependence of cooling temperatures without heat load for ^{3}He operation.

FIGURE 11. Cooling capacity as a function of 2nd stage temperature for ^{3}He operation.

^{4}He operation and case 5 gives the lowest value than those at case 4 and case 6 (HoCu$_2$/GAP=60g/70.1g). The 2nd stage temperature decreases with decreasing the cycle speed, but the temperature does not reach to a constant value even at 48 rpm, while the temperature reaches to be constant in the case of the ^{4}He operation. The lowest 2nd stage temperature of 1.49 K is observed at 48 rpm in case 5.

Measurement data for the cooling capacity on the 2nd stage temperature at 60 rpm are shown in FIG 11. It is clear that the ^{3}He operation increases the cooling capacity and lowers the minimum temperature at the 2nd stage. It is expected that the cooling capacity will be increased by using the GAP for the ^{3}He operation as well as for that of the ^{4}He, but the experiments for case 5 and 6 have not been done more than 2.5 K so far. The inset of FIG 11 shows the details below 2.2 K. The effect of the GAP apparently appears in the temperature region; the cooling capacity increases from 52 mW to 92 mW at 2 K and from 19 mW to 46 mW at 1.8 K with changing the mass ratio from case 4 to case 5. It is also noticeable that the minimum temperature is lowered from 1.67 K to 1.57 K by using the GAP.

CONCLUSIONS

The improvement of the quality of the regenerator material GAP has been accomplished. The hardness and the smoothness were achieved to a good enough level used in GM cycle cryocoolers without any broken powders. But the fabricated GAP particles include some amounts of out of spherical ones. It is not difficult to choose only the spherical particles, although it may increase the cost. Thus, an experimental study will be needed for the mixture of sphere and out-sphere regenerator particles.

For the ^{3}He operation in GM cryocooler, the GAP could largely increase the cooling capacity and decrease the minimum temperature. Therefore, the GAP is an essential and a promising regenerator material to realize the cryocooler for cooling the devices lower than 2 K. If we optimize the compressor unit to decrease the return pressure, for example, to use an auxiliary vacuum pump may be effective to provide lower temperatures less than 1 K theoretically.

REFERENCES

1. T. Hashimoto et al., "Recent Progress on Rare Earth Magnetic Regenerator Material," in *Advances in Cryogenic Engineering* 37, Plenum, New York, 1992, pp. 859.
2. K. H. J. Buschow, et al., *Cryogenics* **15**, pp. 261 (1975).
3. J. Bischof, M. Divis, P. Svoboda and Z. Smetana, *Phys. Stat. Sol.* **(a)114**, pp. 229 (1989).
4. M. Nagao, T. Inaguchi, H. Yoshimura, S. Nakamura, T. Yamada and M. Iwamoto, "*Generation of Superfluid Helium by a Gifford-McMahon Cycle Cryocooler*," ICC6 vol. II, 1991, pp. 37-47.
5. K. D. Irwin, G. C. Hilton, D. A. Wollman, and John M. Martinis, *J. Appl. Phys.* **83 (8)**, pp. 3978-3985 (1998).
6. T. Numazawa, O. Arai, S. Fujimoto, T. Oodo, and Y. M. Kang, "New Regenerator Material for sub-4K Cryocoolers," *Cryocoolers* 11, 2001, pp. 465.
7. S. Fujimoto, T. Kurihara, T. Oodo, Y. M. Kang, T. Numazawa and Y. Matsubara, "Experimental Study of a 4K Pulse Tube Cryocooler," *Cryocoolers* 11, 2001, pp. 301.
8. D. Cashion et al., *J. Appl. Phys.* **39**, pp. 1360 (1968).
9. T. Satoh, A. Onishi, I. Umehara, Y. Adachi and K. Sato, "A Gifford-McMahon Cycle Cryocooler below 2 K," *Cryocoolers* 11, 2001, pp. 381.

CRYOGENIC MATERIALS TESTING

A DEVICE TO TEST CRITICAL CURRENT SENSITIVITY OF NB₃SN CABLES TO PRESSURE

E. Barzi [1], M. Fratini[1], and A. V. Zlobin[1]

[1] Fermi National Accelerator Laboratory
Batavia, Illinois 60510, USA

ABSTRACT

Testing the critical current of superconducting cables under compression is a means to assess the performance of the final magnet. However, these cable tests are complicated and expensive. A fixture to assess the superconducting performance of a Nb₃Sn strand within a reacted and impregnated cable under compression was designed and built. It is currently being commissioned. This device was designed to operate in liquid helium at 4.2K and in high magnetic fields. A cable sample is compressed between two plates. A hydraulic cylinder mounted on the top flange allows applying a pressure up to 200 MPa to the cable sample. The copper current leads to the sample were designed to carry 2000 A. This paper illustrates the fixture assembly, describes the analyses performed, details the cable sample preparation, and shows preliminary test results.

INTRODUCTION

The critical current, I_c, of a Nb₃Sn virgin strand is reduced during magnet fabrication and operation. In the wind and react technique, both strand deformation during cabling (before reaction) and cable compression in the coil (after reaction, due to coil precompression and Lorenz force) decrease the original I_c. This latter factor is due to I_c sensitivity of Nb₃Sn to strain. In the react and wind method, the bending strain introduced during winding causes further degradation.

The Short Sample Test Facility (SSTF) at Fermilab has allowed measuring the I_c degradation due to cabling of Nb₃Sn strands made with different technologies. These strands were extracted from Rutherford cables with different packing factors, heat treated and subsequently tested [1]. A method was also developed to measure I_c degradation due to bending strains of ±0.2 and ±0.4% in strands of various diameters [2]. Recent I_c data obtained in cable tests have shown an excellent correlation with strand measurements both in the case of unbent -representative of cabling degradation only- and bent samples - inclusive of both cabling and bending strain degradation [3].

The device described in this paper was designed to provide quantitative information on the I_c degradation occurring under stress in a Nb₃Sn superconducting magnet in the

CP614, *Advances in Cryogenic Engineering:*
Proceedings of the International Cryogenic Materials Conference - ICMC, Vol. 48,
edited by B. Balachandran et al.
© 2002 American Institute of Physics 0-7354-0060-1/02/$19.00

simplest and cheapest way, *i.e.* with strand tests as opposed to cable tests, and by using an existent facility. A similar setup was successfully built in 1981 at IHEP, Protvino, USSR, to measure the inter-strand resistance of NbTi cables under pressure [4]. To reproduce the real conditions in which the superconductor will operate in the magnet, the pressure was applied to the strand without removing it from the original cable. An impregnation fixture that is used also for reaction was designed. After reaction, cable samples of Nb_3Sn are impregnated and stacked together in order to improve pressure distribution.

DESIGN

Fig. 1 shows the whole assembly. The Nb_3Sn cable (1) is compressed between two Inconel plates. The bottom plate (2) is driven up by an Inconel rod assembly (3), which is pulled up by a 20 ton hydraulic cylinder (4) placed on a stainless steel support on the top flange of the device (5). The top plate (6) is welded to an inconel tube (7), which is itself welded to the top flange. A helium venting pipe (8) and a voltage tap connector (not shown in the figure) are also present. The assembly is immersed in boiling He at 4.2 K within the 64 mm bore of a superconducting solenoid. The device was designed such as to center the cable sample within the solenoid. After insertion in the cryostat, the top flange is fastened with screws to the existent facility. A current of up to 2000 A can be carried by the copper leads (9), which include a pair of Nb_3Sn strand splices. After reaction and impregnation, the cable sample is carefully mounted at the bottom of the device by soldering the ends of a strand to the current leads. The strand ends are long enough to ensure current transfer. To allow the differential thermal contraction between copper and Inconel, the current leads are free to move within a bellow at the top of the device.

(a)

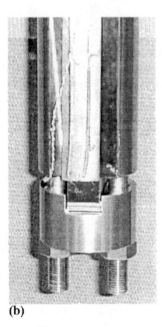
(b)

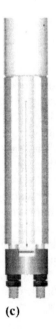
(c)

FIGURE 1. Fixture assembly.

Mechanical Analysis

The mechanical analysis was performed in two steps. First, each component was sized analytically, then the final assembly was verified with ANSYS. This was done to check possible stress concentrations in the device. The ANSYS model that was used is shown in Fig. 2. Thanks to the axial symmetry of the structure and of the loads a quarter section only of the whole assembly was modeled. This included the tube, the top plate, the cable sample and the bottom plate. A quarter of the 20 ton load was applied as a uniform pressure on the bottom plate at the contact area with the nuts. The first analysis of the assembly gave a singularity (*i.e.* a very high stress concentration) in just one element of the tube at the junction with the top plate. Therefore, the tube had to be analyzed individually. As a long beam under compression, it was verified that the tube would not buckle. The load distribution and mesh for the single tube are shown in Fig. 3A. A detail of the stress distribution is shown in Fig. 3B. A maximum equivalent stress of about 800 MPa was found close to the top plate.

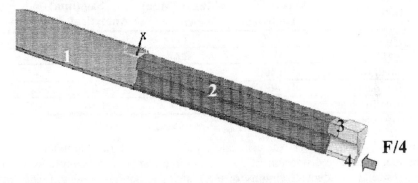

FIGURE 2. ANSYS model representing a quarter of the whole assembly. From left to right, the tube (1), the top plate (2), the cable (3), and the bottom plate (4) are shown. A quarter of the 20 ton load, F/4 in Fig., was applied to the bottom plate.

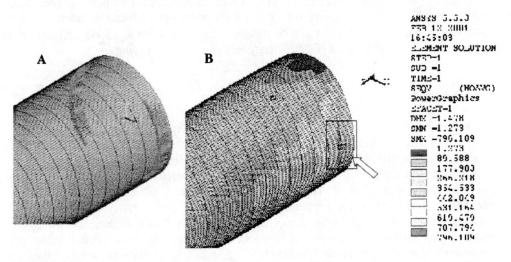

FIGURE 3A. Load distribution and mesh used for the tube. The load was applied only at the nodes shared with the superior plate, **B.** Equivalent stress in the tube. The darkest area pointed by the arrow bears a stress of about 800 MPa.

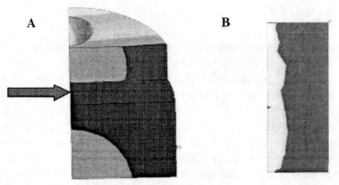

FIGURE 4A. Axial displacements of the bottom plate (a quarter is shown). The arrow indicates the sample location. The uniformity in the area of the sample location shows adequate rigidity of the plate, **B.** Axial stress on the sample (a quarter is shown). Note the nearly uniform compression.

TABLE 1. Comparison of the analytical results with ANSYS output.

Item	Max σ (MPa) Analytical	Max σ (MPa) Ansys	Sag (mm) Analytical	Ansys
Tube [1]	606	799	-	-
Large rod [1]	387	296	-	-
Small rod [1]	1400	1600	-	-
Top plate	460	398	~ 0	~ 0
Bottom plate	846	617	0.046	0.03

[1] CJP welds.

The sag of the two plates was also calculated. Fig. 4A shows the axial displacements of the bottom plate. The displacement distribution in the area of the sample location is very uniform, ensuring an adequate rigidity of the plate. The bottom plate sag calculated with ANSYS was only 0.03mm. Fig. 4B shows the pressure distribution on the sample.

The results of the analytical calculations are compared with ANSYS output in Table 1. Due to space restraints in the design, the highest stress was found in the small rods of the rod assembly, which however are designed to always operate in liquid helium. The large stress in the rods and the tube imposed (with a 1.1 safety factor) a structural material with a tensile yield strength of 1190 MPa at room temperature (1600 MPa at 83 K), and required complete joint penetration welds (CJP). Therefore, Inconel 718 was chosen.

Thermal Analysis

The results from two different thermal codes were compared to determine the size of the current leads: LEAD [5], and CCLAMP [6]. CCLAMP was run by fixing the temperature at the top of the leads, by deactivating the mass flow controller, and by considering a large difference between gas and copper temperature. The parameters used in both cases are listed in Table 2. The maximum temperature of the leads and the He mass flow rate as a function of the copper RRR are shown in Fig. 5 for both codes. LEAD appeared to be more conservative than CCLAMP, and the leads were sized using the former. With a RRR of 100, the loss of He coolant was calculated to be about 0.3 liter per measurement for a measurement duration of 5 minutes. However, due to thermal conduction of the device that weights about 20 kg, a significantly larger loss is expected.

TABLE 2. Parameter (for both leads) used in the two codes.

Parameters (for the two leads)	
Length (m)	0.884
Perimeter (m)	0.13
Cu area (m^2)	0.0002
Gas area (m^2)	0.0007
Hydraulic diameter (m)	0.046
Warm end temperature (K)	300
Cold end temperature (K)	4.2
Current (A)	4000

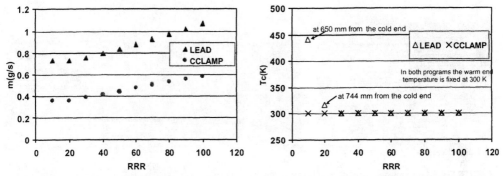

FIGURE 5. Mass flow rate (left) and maximum temperature (right) reached by the leads as a function of the copper RRR calculated with LEAD and CCLAMP.

CABLE SAMPLE PREPARATION

Reaction and impregnation of Nb$_3$Sn cables are very delicate procedures due to the brittleness of Nb$_3$Sn. Cable samples were prepared by cutting a piece length of cable, paying attention to leave within the cable a longer strand. This strand would be used for critical current testing after applying pressure to the whole cable area. The ends of the selected strand had to be long enough to ensure current transfer once they would be soldered to the current leads. The procedure was first to cut the cable at a length of twice the current transfer length plus the cable sample length, then to cut it again leaving out the strand tails. The wires in the cable, except the selected strand, are then welded together to prevent tin leaks during heat treatment.

A single fixture was designed for both reaction and impregnation in order to limit handling of the sample after reaction. This fixture is shown in Fig. 6. It is composed of four stainless steel parts: the body with grooves that support the strand tails along their length; the clamp plate with holes for pouring the epoxy; two cover plates with the double function of connecting the body to the clamp plate and of containing the epoxy bath.

After reaction in argon atmosphere, the sample is gently removed from the fixture in order to spray the latter with Easy Release 300, manufactured by Mann Formulated Products Incorporated. Boron Nitride Spray high temperature mold release, by Advanced Ceramics Corporation, to use before reaction was also tried. This would have avoided the need to remove the sample after heat treatment. However, that method did not prove effective. The sample is then very carefully replaced in the fixture. Two small copper plates, sitting in the clamp plate at the cable ends, are soldered to the strand to reinforce the bends. The fixture is then completely assembled, before sealing it with General Electric

RTV 21 for impregnation. Several studied were performed to determine the best impregnation procedure. To avoid bubbles in the epoxy, the fixture is vacuumed overnight and CTD 101 K epoxy mix is degassed for 45 minutes at 60°C. A very slow potting (drop by drop) is also used. Reaction and curing cycles are given in Tables 4 and 5.

Several cable samples were impregnated and are ready to be tested with this device. A few of them are described in Table 3, where MJR stands for Modified Jelly-Roll, IT for Internal Tin, and PIT for Powder-in-Tube technologies by Oxford Superconducting Tecnology (OST), Intermagnetic General Corporation (IGC), and Shape Metal Innovation respectively. The 28-strand keystoned cables had width, lay angle and keystone angle within 14.24±0.025 mm, 14.5±0.1 degree, and 0.9±0.1 degree respectively. For rectangular cables, the keystone angle was 0.0±0.1 degree [1].

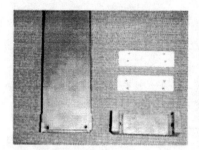

FIGURE 6. Main components of the impregnation fixture.

TABLE 4. Heat treatment schedule.

Parameters	
Ramp Rate, °C/h	6
Temperature, °C	700
Duration, h	40

TABLE 5. Impregnation parameters.

Parameters	
Curing temperature	110°C for 5 hr
Post-curing temperature	125°C for 16 hr

TABLE 3. Cable parameters

Cable ID	Strand no.	Strand Diameter, mm	Strand Type	Manufacturer	Core	Type	Thickness, mm	PF, %
746-I	28	1.0	MJR	OST	Y	KS	1.808	89.4
745-C	28	1.0	MJR	OST	N	R	1.839	86.6
1	28	1.0	PIT	SMI	Y	KS	1.860	88.3
2	28	1.0	IT (ITER)	IGC	Y	KS	1.779	90.6

PRELIMINARY TEST RESULTS

To verify the mechanical resistance of the device, it was first tested at room temperature. Using the Power Team hydraulic cylinder supplied by a Enerpack pump, a force up to 13 tons (*i.e.* the maximum force tolerated by the small rods at room temperature) was applied in steps of 2 tons. At each step, the strain of the tube assembly due to compression was also measured in order to check the clearance for the movement of the leads.

To test the LabView DAQ software, a small current was supplied at room temperature to a dummy cable sample. The sample had been impregnated and its strand tails soldered to the current leads. A pair of voltage taps were placed 53 mm apart along the selected strand, just outside the cable length. An additional pair of voltage taps was placed further apart to ensure sample protection. The sample was then compressed up to 3 tons. The voltage-current (VI) characteristics was measured, obtaining the expected resistive curve.

A sample of OST 745-C cable was used for the first test in boiling helium at 4.2 K. The device was inserted warm within the superconducting solenoid of the Teslatron (Oxford Instruments) cryostat, and fastened with screws. Nitrogen pre-cooling was then started, followed by helium cooling. During this process, a leak in the seal at the top of the current leads was noticed. The helium evaporation rate was measured to be about 20 l/h. At 4.2 K, the magnetic field was brought up to 10 T, and tests were carried out at pressures up to 200 MPa. However, the sample appeared to be broken, no current went through it, with a measured voltage of 3.3V, *i.e.* the power supply voltage limit. After extraction of the device, one could see that the sample was broken at both bends. The cylinder was dismounted and it was determined that frosting of the dynamic O'ring that sealed the rod, pulling the rod dragged the current leads too, which pulled and broke the sample ends.

The seal was promptly repaired and the dynamic O'ring tested at room temperature. This showed that the O'ring is ineffective even warm, so that the rod is still connected to the leads. At present, the O'ring is being replaced with a bellow.

SUMMARY

A new device to test critical current degradation in Nb_3Sn cables under pressure was designed and built, and is currently being commissioned at Fermilab. Preliminary tests have further improved the design of this device. A sample impregnation procedure has been established. Reacted and impregnated cable samples are ready to be tested as soon as the system is commissioned.

ACKNOWLEDGMENT

The authors would like to thank Jay Hoffman and Cristian Boffo for their kind help.

REFERENCES

1. E. Barzi et al., "Strand critical current degradation in Nb_3Sn Rutherford cables", ASC '00, Virginia Beach, VA, Sept. 17-22, 2000.
2. G. Ambrosio at al., "Study of the react and wind technique for a Nb_3Sn common coil dipole", MT-16, IEEE Transaction on Applied Superconductivity, Vol.10, no.1, pag.338, March 2000.

3. P. Bauer et al., "Fabrication and testing of Rutherford cables for react and wind accelerator magnets", ASC '00, Virginia Beach, VA, Sept. 17-22, 2000.
4. A. Zloblin, G. Enderlein, "Transverse resistance measurements of flat transposed superconducting cables under pressure", Preprint IHEP 81-57, Serpuskhov, 1981.
5. M. Wake, "Current lead design for VLHC magnet development", TD-00-071, October 27, 2000.
6. R. Carcagno et al., "A technical description and users manual for cryogenic current leads analysis model program", SSC Cryo note 94-31 version 1.0, February 1994.

TENSILE TESTING METHODS OF HIGH STRENGTH AND HIGH MAGNETIC FIELD COMPOSITE WIRES

V. J. Toplosky and R. P. Walsh

National High Magnetic Field Laboratory
Tallahassee, FL 32310

ABSTRACT

Tensile tests to measure the ultimate tensile strength of full-section high strength composite wire have been improved by using tabs that are bonded directly to the metal. Reduced section specimens are not always desirable due to the non-uniform properties of the composite wire. The use of friction grips on tensile tests of full-section wire without tabs creates combined stresses in the material in the grip region usually resulting in premature failure. For a tabbed specimen, the stress concentration is decreased making failure within the gage section possible. Analysis is included to show that, by using tabs, direct contact of the wire to serrated friction grips is eliminated, reducing the high stress-concentration factor. The net results are higher ultimate strength values that better approximate the wire's actual strength.

INTRODUCTION

The goal of a tensile test is to apply uniaxial tensile stress to a control load volume of the test material. This is usually done through the application of an external force that is transmitted to the test section via shear stress. For isotropic materials, machining the test sample with a reduced section and gripping the material in a larger cross-sectioned area is possible. A common problem encountered when tensile testing composite materials is premature failure of the sample caused by stress concentrations in the grip region. High strength conductors for high field magnet applications are often composite materials that require full cross-sectional tests to realize the conductor's actual strength. Traditional gripping methods of a constant cross-section wire, such as spools, tapered wedge grips and serrated friction grips introduce stress concentrations that influence mechanical data such as elongation and ultimate strength. The problem is further compounded for high strength conductors where the high tensile forces necessary to break the sample result in wire slippage in the grip region, especially at cryogenic temperatures. To reduce the effect of grip stress concentration, one can bond tabs to the material with a high strength adhesive, effectively increasing the cross-sectional area of the gripped region.

CP614, *Advances in Cryogenic Engineering:*
Proceedings of the International Cryogenic Materials Conference - ICMC, Vol. 48,
edited by B. Balachandran et al.
© 2002 American Institute of Physics 0-7354-0060-1/02/$19.00

BACKGROUND

The need to accurately measure a composite wire's ultimate strength and elongation is important for the design of high-field pulsed magnets. Pulsed magnets have extreme materials requirements since they typically operate with the conductor's stress above the material's elastic stress range [1]. They are also cooled to liquid nitrogen temperature (77 K) before the magnet is energized and then warmed (due to ohmic heating) to temperatures greater than 300 K during a pulse.

The traditional wire gripping methods cannot always be employed due to restrictions in the material's composition and makeup. Typical materials used for pulse magnet conductors are metal matrix composites that have some of the following geometric properties that make them difficult to tensile test.

1). Fiber-reinforced metal matrix micro-composites have gripping problems similar to those encountered with fiber-reinforced polymer composites. By machining the conductor into a reduced-section specimen, the grip tab is weakened by the cutting of the axial reinforcing fibers.

2). In metal matrix macro-composites where the conductor has a sheath, the sheath is used as a reinforcement material or low strength material to aid the wire drawing process. If the sheath is removed to make a reduced section sample the test results will be altered compared to a full section properties.

3). The third case that can occur in both micro-composites and alloys where non-uniform strain-hardening through the cross-section results in a strength gradient through the thickness of the material. The outermost material is the hardest region and its removal will result in decreased tensile strength.

Wedge grips are cumbersome for 77 K testing. Spool grips are effective but require longer specimens and introduce bending stress due to the wire being wrapped around the spool. This bend radius can lead to peeling and premature failure. Another method of transmitting stress to a composite wire is the use of serrated bolt-together grips. Disadvantages of using this method are the stress concentrations introduced by the grip digging into specimen surface, as well as the combined stress in the grip region caused by the normal clamping force. Slippage within the grip is also a problem for large cross-section wire. Increasing the clamping force leads to increased stress concentration. Tapered wedge grips exert a normal force that is proportional to the tensile force due the resolution of forces the taper provides. This helps to alleviate slippage but stress concentrations within the grip still exist.

Since specimen slippage and premature failure due to stress concentrations lead to lower than expected results, an alternative method has been developed. Bonding metal tabs to the ends of the specimen in conjunction with the bolt together friction grips has helped to reduce the stress concentration within the gage length and has led to reduced slippage in large cross-sectional specimens. Figure 1 shows this method and the associated forces; F_N (normal clamping force) and F_T (applied tensile force).

DESIGN AND ANALYSIS

For a conductor within friction grips, the wire can be modeled as having a normal load (clamping) and a tensile load. Interfacial friction helps to further increase bonding. In order to obtain failure, the following equation must be satisfied:

$$F_N\mu > F_T \qquad (1)$$

where μ is the coefficient of friction between the inside grip surface and the bare wire.

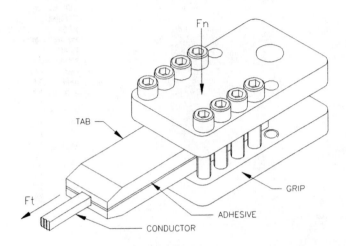

FIGURE 1. Drawing showing components and forces.

For the case of a constant cross-section wire, the application of a normal force creates a combined stress situation in the grip region. The von Mises stress will exceed the materials strength before the axial tensile stress does. Figure 2a shows a finite element analysis (ANSYS) of the non-tabbed metal wire and the bolt-together friction grip. By tabbing the sample (see Figure 2b), the normal stress decreases for the same applied normal force and a situation can be obtained where the combined stress is lower than the tensile stress allowing for failure in the gage section. The normal force is obtained by clamping the serrated friction grips onto the sample creating high stress concentration regions. The coefficient of friction for this metal grip-on-wire situation is in the value of 0.15 – 0.60 [2]. Tabs with beveled edges help to smoothly transition the combined stress in the grip region into uniaxial tension in the test section.

The other aspect in preventing premature failure of the conductor are the stress concentrations within the grip region. Using the tabs removes any direct contact of the wire material to the serrated grip thus reducing the stress concentration. In Figure 2a, the interface between the non-tabbed wire and grip results in a stress concentration ratio of approximately 4.2. Figure 2b, shows the tab and wire interface, that effectively reduces the stress concentration ratio to 3.2, a decrease of 24%.

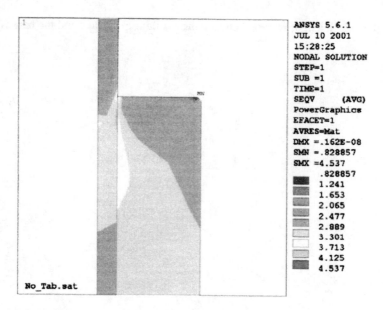

(a). Non-tabbed wire

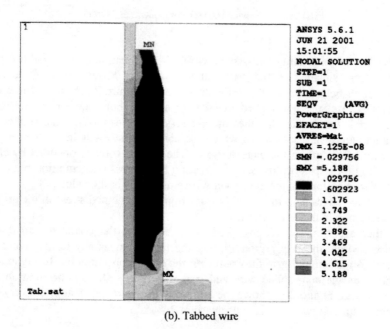

(b). Tabbed wire

FIGURE 2. Finite element analysis showing stress concentrations of non-tabbed and tabbed wire.

Test Program

The test program's objective is twofold. The first objective is to evaluate and establish a tab method for composite wire specimens. The second objective is to use the

56

established method on composite wires to evaluate its effectiveness. The test program is divided into two parts:

a). Tab materials selection: this consisted of investigating a number of parameters (2 tab materials, 2 adhesives and 3 bond preparation procedures). Tab materials were: NEMA G-10, a high-pressure glass epoxy laminate and aluminum alloy 6061-T6. Two bonding agents were used: EA 9394, an aluminum filled epoxy paste that exhibited good shear strength [3] and Stycast 2850 FT, a two part, thermally conductive epoxy that has a low coefficient of expansion [4]. First, two tab materials were evaluated along with 2 adhesives. Tensile tests were then conducted on copper niobium wire to evaluate the test variables. The results of these tests allowed the selection of the tab material (6061-T6) and adhesive (Stycast). Separate sets of tests were conducted to evaluate the effect of surface preparation variables. This consisted of using extended tabs that were pin connected to the tensile machine rather than using friction grips which eliminate the variable of normal force.

Three different methods of tab/sample preparation were evaluated. The first consisted of cleaning the tabbed surfaces and sample using a prescribed strain gage cleaning method. The specimens are degreased then rinsed with mild phosphoric acid solution (conditioner) and dried [5]. Then the surfaces are rinsed with ammonia-based solution (neutralizer). For the second method, the tab/sample interface regions were abraded with high-pressure gritblasting of aluminum-oxide/glass-bead mix. The samples were then rinsed with methanol before applying the adhesive. The third method consisted of an acid dip. Copper niobium (CuNb) wire samples were dipped in a 10% nitric acid-90% ethanol solution for 30 seconds. Copper stainless steel (CuSS) wire was etched in a 12.5% hydrochloric acid-4.5% nitric acid-83% distilled water for 5 minutes [6].

b). Material Strength Measurements: once the tab material, adhesives and sample preparation method were established, tensile tests were conducted on copper niobium and copper stainless steel wire – two commonly used pulse magnet conductors. Tests were conducted as per ASTM E 8M. The specimens were loaded until failure occurred and the ultimate strength was determined. The specimen failure region was noted as were any other discrepancies such as slippage. Testing was conducted at both 295 and 77 K. For comparative purposes, non-tabbed wire was tested in wedge grips at 295 K and bolt-together friction grips at 77 K.

RESULTS AND DISCUSSION

Tab materials selection

The initial phase of testing consisted of tensile testing a 4.70 x 5.65 mm CuNb rectangular wire. The conditioner-neutralizer method of sample preparation was used for cleaning every sample. The variables consisted of using different tab materials (G-10 and Aluminum 6061-T6) and different adhesives (EA 9394 and Stycast 2850 FT). For reference purposes, non-tabbed CuNb wire was also tested using wedge grips. Table 1 summarizes results for this initial testing.

TABLE 1. 295 K results using different tab materials and adhesives w/CuNb conductor.

Tab Matl	Epoxy	Ult. (MPa)	Comment
6061-T6	Stycast 2850 FT	1048	failed gage length
6061-T6	EA 9394	768	slipped in tab
G-10	Stycast 2850 FT	768	slipped in tab
G-10	EA 9394	558	slipped in tab
untabbed	n/a	1010	failed in grip

One conclusion to draw from this table is that the bonding of G-10 to copper niobium was not effective. A second observation is that Stycast works better that EA9394 to bond aluminum to copper niobium. The comparative differences between the Stycast/6061-T6 combination and the non-tabbed wire tested using the wedge grip shows that the tabbed wire is approximately 3% stronger than the wedge grip ultimate strength. Equally important is the failure mode: the tab sample failed within the gage length while the wedge grip sample failed within the grip.

The results of sample preparation method tests direct pullout of pin connect tabs shown in Figure 3. Both CuNb and CuSS were tested. All testing was performed at 295 K. The acid etch method shows superior strength for both CuSS (a 63% increase) and CuNb (46% increase) over the conventional conditioner-neutralizer method previously used.

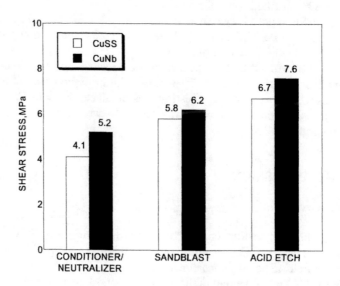

FIGURE 3. Shear strengths of aluminum tabs to wire.

Materials Strength Tests

Aluminum tabs and Stycast 2850 FT were used to test a series of CuSS and CuNb wires. Testing was conducted at 295 K and 77 K. The average tensile stress of two samples are presented in Figures 4a (CuSS) and 4b (CuNb).

For both materials and temperatures, it clearly shows that there is an increase in the ultimate strength with the bonded tabs compared to the conventional wedge grips. The effect appears to be more dramatic with the CuNb where there is an increase of 5.1% at 77 K and 4.0% at 295 K. Increases were also observed for the CuSS – 1.6% at 77 K and 1.1% at 295K.

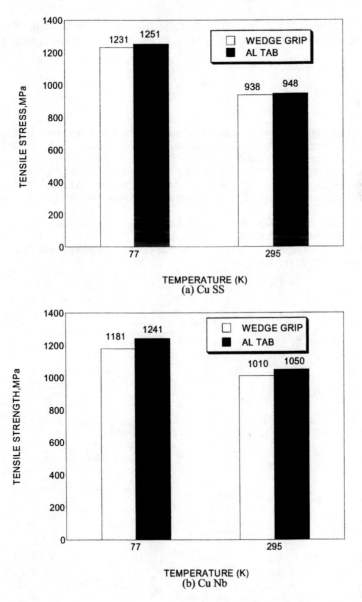

FIGURE 4. Temperature versus tensile strength for tabbed and non-tabbed samples.

CONCLUSION

Finite element analysis of a non-tabbed wire and tabbed wire has shown the large difference in combined stress and stress concentrations between the two scenarios. Higher ultimate strengths have been achieved by optimizing the tab design, testing different tab materials with different adhesives and tensile testing several sample preparation methods for shear strength. Using aluminum alloy 6061-T6 tabs in conjunction with Stycast 2850 FT adhesive has consistently produced higher ultimate strength values over non-tabbed metal wire tests. Increases of about 5% are observed at both 295 and 77 K. This will help to better approximate the wires engineering ultimate strength for the case of pulsed magnet design.

While a 5% difference in strength may not seem significant, the improved measurement of the conductors properties helps in the design of these cutting edge pulsed magnets.

REFERENCES

1. Li, L. et al., "High Performance Pulsed Magnets with High Strength Conductors and High Modulus Internal Reinforcement", IEEE Transactions on Applied Superconductivity, **10** (1), pp. 542-545 (2000).
2. Beer, F.P. and Johnston, E.R., Vector Mechanics for Engineers, 5th edition, McGraw-Hill, 1988, p. 319.
3. Goeders, D.C. and Perry, J.L., "Adhesive Bonding PEEK/IM-6 composite for cryogenic applications", Society for the Advancement of Material and Process Engineering, Vol. 36 (1), pp. 348-361 (1991).
4. Technical Data Sheet, Stycast 2850 FT Thermally Conductive Epoxy Encapsulant, Emerson & Cuming Inc., 1999.
5. M-Line Accessories, Strain Gage Applications with M-Bond AE-10, AE-15 and GA-2 Adhesive Systems, Instruction Bulletin B-137-16, Measurements Group Inc., 1979.
6. Metals Handbook Ninth Edition, Volume 5, Surface Cleaning, Finishing and Coating, American Society for Metals, 1982, pp. 59-67.

INTERNATIONAL STANDARDIZATION THROUGH VAMAS ACTIVITIES ON MECHANICAL PROPERTIES EVALUATION AT 4K

T. Ogata and Participants of VAMAS TWA 17

National Institute for Materials Science
Tsukuba, Ibaraki 305-0047, Japan

ABSTRACT

Since 1986, a series of international interlaboratory comparisons on the evaluation of mechanical properties of cryogenic structural materials have been performed among the participants of VAMAS (the Versailles Project on Advanced Materials and Standards) TWA (Technical Working Area) 17 in order to establish unified test methods. Through these international collaborations and Round-Robin Tests (RRTs), we have accumulated knowledge about mechanical tests at 4 K, and have prepared a draft of an international standard for tensile testing in liquid helium. The draft was submitted to ISO TC164 /SC1 as a new work item, and testing conditions, strain measurements, and other technical points have been discussed. The outline, development, and discussion of the document so far, with the results of RRTs, are presented. Objectives of other international standardization programs on mechanical testing in high magnetic field, J-Evaluation on Tensile Test, and for composite material are summarized also in this paper.

INTRODUCTION

For large-scale cryogenic applications, such as a nuclear-fusion reactor, a superconducting power generator, a superconducting magnetic levitated train, and so on, it is very important to construct intellectual infrastructure, which can support the practical use of advanced cryogenic structural materials, high-strength stainless steels, high-Mn steels, and large-thickness weld joints. This research has been carried out with a close contact with VAMAS.

VAMAS was formed in 1982 at the Economic Summit of the G7 Heads of State as a way to stimulate trade in technologies that depend on advanced materials. The signatories of the governing Memorandum of Understanding are Canada, France, Germany, Italy, Japan, UK, USA, and the European Community. Through its technical programs, VAMAS emphasizes international collaborative pre-standards-measurement research.

CP614, *Advances in Cryogenic Engineering:*
Proceedings of the International Cryogenic Materials Conference - ICMC, Vol. 48,
edited by B. Balachandran et al.

Working together, scientists from these and many other countries carry out the technical activities necessary to form the basis for subsequent national and international standards. VAMAS has formal cooperative agreements with ISO and IEC that include publishing and distributing worldwide the outputs of its pre-standardization activities as Technology Trends Assessments. VAMAS has brought an ordered, non-bureaucratic structure to international collaboration in activities leading to harmonization of materials measurements. The mission of VAMAS is "to Support World Trade in Products Dependent on Advanced Materials Technologies by Providing the Technical Basis for Harmonized Measurements, Testing, Specifications, and Standards."

VAMAS has more than ten Technical Working Areas (TWA) aimed at the development in the evaluation of advanced materials and its standardization through international collaboration. TWA 17, cryogenic structural materials, which has been organized in the VAMAS to promote the prestandardization program on mechanical-properties tests of composite materials and alloys at liquid-helium temperature. The goal is to develop an understanding of mechanical-property determinations at liquid-helium temperature (4 K) and establish a unified method. A series of international interlaboratory comparisons of both tensile and fracture-toughness tests for high-strength stainless steels, a titanium alloy, and a aluminum alloy, and compression and shear tests for composite material G-10CR were performed so far [1-6]. Figure 1 presents the schedule of RRTs and other activities of cryogenic structural group.

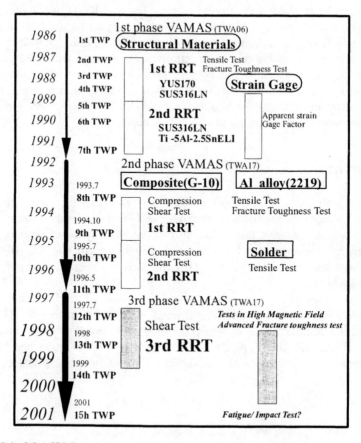

FIGURE 1. Schedule of RRTs and other activities of cryogenic structural working area.

TABLE 1. The institutes and contact persons who participate in this project.

Institute	Nation	Contact Person	Composite	JETT	HMF
University of Tokyo	Japan	K.Shibata	●	●	●
Tohoku University	Japan	Y. Shindo	●	●	●
Osaka University	Japan	S. Nishijima	●		
Natio. Inst. Mater. Sci.	Japan	T. Ogata	●	●	●
NIFS	Japan	A.Nishimura		●	●
NHMFL	USA	R. P. Walsh	●	●	●
Compo. Tech. Dev. Inc.	USA	P. Fabian	●		
FZK	FRG	A.Nyilas	●	●	●
Tech. Univ. Wien	Austria	K. Humer	●		
RAL	UK	D.Evans	●		
EMPA	Switzerland	R. Huwiler	●		
UCLA	USA	J.W.Morris		●	
SEP	France	J-P Lecornu			
LINDE	FRG	Mitterbacher		●	
CAS	China	Z. Zhang	●	●	

Since 1997, we have promoted projects to establish and report pre-standard testing methods of a) tensile test in high magnetic field (HMF), b) J-evaluation on tensile test (JETT) by round bar with circumferrential notch, and c) interlaminar shear test of G-10CR, glass fiber reinforced plastic (Composite), at cryogenic temperature.

The institutes and contact persons who participate in the projects are listed in Table 1. Fifteen research institutes from eight nations have participated in the projects of TWA 17.

According to the agreement of the 14th TWP meeting, a draft of the international standards "Metallic Materials - Tensile Testing in Liquid Helium" was discussed in TWA 17 and submitted to ISO TC164/SC1 the end of 1999 as a New Work Item for the SC1.

This paper presents the outline and the results of RRT of each project, and discussion so far of the ISO proposal.

PROJECTS AND RESULTS

Tensile Test at 4K in High Magnetic Field

Prof. Shibata leads this project and the objectives of this project are to clarify the effect of high magnetic field on mechanical properties and establish reliable methods of evaluating Young's modulus, yield strength, tensile strength at 4K in high magnetic field.

In a preliminary test, small effect of magnetic field on load-cell was confirmed in proposed testing system [7]. The specimens of titanium (CP-Ti) for the first RRT were distributed to the participants and results have been reported from four participants so far.

Values of Young's modulus obtained by all participants are shown in Figure 2 [8]. The mean values at 0 and a magnetic field are 120.3 and 118.5 GPa, respectively. That is to say, the effect of magnetic fields on the value of Young's modulus seems to be negligibly small. However, the standard deviation seems to be fairly large comparing with the results of the RRTs under VAMAS project [4].

After further discussion of these results, a technical report for ISO/TC164/SC1 (Uni-axial testing) as a TTA document will be prepared.

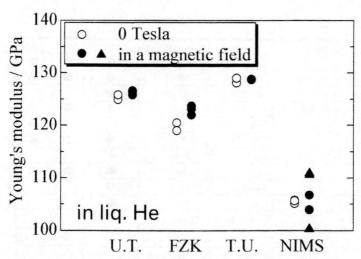

FIGURE 2. Comparison of the values of Young's modulus measured by each participant at zero and magnetic fields. U.T.: the University of Tokyo, FZK: Forschungszentrum Karlsruhe, T.U.: Tohoku University, NIMS : National Institute for Materials Science

FIGURE 3. Mechanical testing system with liquid helium free 6T magnet in NIMS.

Figure 3 shows a mechanical testing system with 6 T magnet installed in National Institute of Materials Science to carry out mechanical tests in high magnetic field without using liquid helium. The Nb-Ti superconducting magnet is cooled by refrigerator. The bore diameter is 101 mm and the inside diameter of a vacuum-insulated cryostat is 80 mm. The loading capacity is 50 kN. The magnet is cooled to 6 K by refrigerator in 40 hours and excited to 6 T in 20 min. With this testing system, remaining problems will be resolved and testing procedures in high magnetic field will be refined especially in viewpoints of simplicity and safety.

J-Evaluation on Tensile Test (JETT)

Fracture toughness is conventionally evaluated according to ASTM E 813 or JIS Z 2284, which typically require a one-inch compact-tension specimen that is not good for the test in liquid helium, and gives difficulty for obtaining valid fracture-toughness in the case of weld joints and thin plate. A new fracture-toughness test method was proposed by Rice et al.[9], developed by Nyilas [10,11], and examined by Nishimura [12].

Prof. Nishimura leads this project with the objective of developing and refining the new fracture-toughness test through a series of RRTs. First RRTs on SUS316LN were carried out and five institutes reported the results. Figure 4 [13] presents the results of RRTs on SUS316LN which show the possibility of a new test method in the evaluation of fracture-toughness of small sections. A second set of RRTs were also performed on 9% Ni steel and five institutes reported the results. The results of these RRTs have been analyzed; including the effect of the test machine stiffness, the agreement of the JETT test method with the conventional method, and the applicability of the JETT method for measuring fracture toughness [14]. After further discussion of these results, a technical report for ISO/TC164/SC4 (Toughness testing) as a TTA document will be prepared.

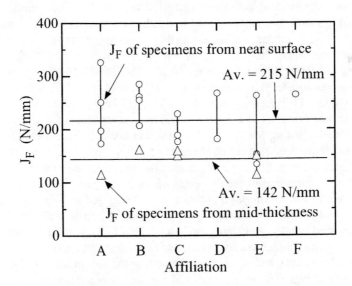

FIGURE 4. The results of RRTs of JETT on SUS316LN.

Interlaminar Shear Test on G-10 CR

Prof. Shindo leads this project and the objectives of this project are to refine and establish unified and reliable testing procedures of interlaminar shear test on G-10CR at cryogenic temperature through a series of RRT. A double-notch and short-beam shear test have been examined and analyzed using a three-dimensional finite element analysis and a standard specimen geometry and size for both double-notch and short-beam shear tests were proposed to obtain the true shear strength of G-10 CR [15].

According to this proposal, specimens were machined and distributed to the participants as the third RRTs on interlaminar shear test. Five institutes reported the results so far [16]. Parameters for the maximum shear stress of both testing methods have been determined through analysis and good agreement among the data was obtained [17].

After further discussion of these results, a technical report for ISO/TC61 (Plastic) as a TTA document will be prepared.

STANDARDIZATION OF TESTING METHODS

Standardization of methods for cryogenic mechanical testing is very important to develop and disseminate because of its difficulty.

In the ISO, there is no discussion on the international standards of mechanical testing at cryogenic temperatures under 77 K, and to solve the difficulties, we will promote our activities and output better results for the standards. Our results toward the ISO will be utilized for not only research but also international trades of these materials.

Proposal of Testing Methods to ISO

Through the international collaboration and RRTs, we have accumulated knowledge and understandings of the variables related to mechanical tests at 4 K, and prepared a draft of international testing standards of tensile testing in liquid helium as a first attempt.

A draft of New Work Item (NWI) of ISO/TC164/SC1 titled "Metallic Materials - Tensile Testing in Liquid Helium" was submitted to the ISO/TC164/SC1 at the end of 1999 as a NWI Proposal in the SC1 according to a request of the SC1 in June 1999 and the agreement of the last TWP meeting of VAMAS TWA 17 held in July 1997.

The draft is based on ASTM E 1450 and added a recent view of VAMAS activities and JIS Z 2277 (Tensile Testing Method for Metallic Materials in Liquid Helium); a diameter of a standard specimen is 7 mm but alternatives are allowed, main strain measurement methods use extensometers but strain gages are also allowed and so on.

The draft was distributed to the delegates of the SC1 from nineteen various countries for voting. A hurdle for the proposal is that it must be agreed to by more than half of the participating countries. It must to be approved by more than five countries to support a working group for discussing the new proposal. The difficulty in this process is that most countries in those committees are not familiar with cryogenic technology.

Our proposal was accepted by six countries under the recognition of the importance of cryogenic structural materials in the field, and the draft was submitted to ISO TC164/SC1 as a NWI in June 2000. Then, testing conditions, strain measurements, and other technical points have been discussed in the working group consisting of the experts from the positively agreed countries, the United States, Germany, UK, Japan, China, Belgium. Hot

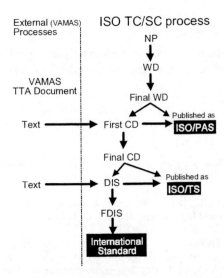

FIGURE 5. ISO TC/SC process for standardization and a liaison between ISO and VAMAS.
NP: New Proposal, WD: Working draft, CD: Committee Draft, PAS: Publicly Available Specification
DIS:Draft of International Standards, TS: Technical Specification, FDIS: Final Draft of International Standards

discussions on the document were made in the working group of the SC1 including the VAMAS participants last year. The Working Document ISO/AWI 19819, accompanied by a bibliography of VAMAS activities for references, was submitted to ISO TC 164/SC 1 in January 2001 as a result of discussion and approval among the experts of the working group and was distributed to the SC1 members for further comments as a Committee Draft. According to the resolutions of the last SC1 meeting, expected time schedule are as follows:

> 2001. 9 -- Committee Draft
> 2002. 4 -- Draft of International Standard
> 2003. 6 -- Final Draft of International Standard
> 2003.12 -- International Standard.

Future Proposal using Liaison between ISO and VAMAS

Figure 5 presents the conventional Technical Committee and Sub Committee process for standardization and the A-liaison between ISO and VAMAS, which enables mutual transfer Technical Transfer Assessment (TTA) document. This liaison could lower the hurdle for our proposal and shorten the time to the IS.

SUMMARY

Standardization of testing methods at cryogenic temperatures is very important to develop and disseminate cryogenic technology and application, and we would like to continue this activity as a result of international collaborations to establish the reasonable standard and not be controlled in future by any other standard. Any companies and any proposals are welcome to our activities.

ACKNOWLEDGMENT

Authors greatly appreciate the collaboration of participant institutes and their key persons. Research supported by the Special Coordination of Science and Technology Agency, Japan.

REFERENCES

1. Ogata, T., Nagai, K., Ishikawa, K., Shibata, K., and Fukushima, E., "VAMAS Interlaboratory Fracture Toughness Test at Liquid Helium Temperature," in *Advances in Cryogenic Engineering* 36, 1990, pp.1053-1060
2. Ogata, T., Nagai, K., Ishikawa, K., Shibata, K., and Fukushima, E., "VAMAS Second Round Robin Test of Structural Materials at Liquid Helium Temperature," in *Advances in Cryogenic Engineering* 38, 1992, pp.69-76
3. Ogata, T., Nagai, K., Ishikawa, K., "VAMAS Tests of Structural Materials at Liquid Helium Temperature," in *Advances in Cryogenic Engineering* 40, 1994, pp.1191-1198
4. Ogata, T., and Evans, D., "VAMAS Test of Structural Materials on Aluminum Alloy and Composite Material at Cryogenic Temperatures," in *Advances in Cryogenic Engineering* 42, 1996, pp.277-284
5. Ogata, T., Evans, D., and Nyilas, A., "VAMAS Round Robin Tests on Composite Material and Solder at Liquid Helium Temperature," in *Advances in Cryogenic Engineering* 44, 1998, pp.269-276
6. Ogata, T. and Participants of VAMAS TWA 17, "Results of VAMAS Activities on Pre-standardization of Mechanical Properties Evaluation at 4K," in *Advances in Cryogenic Engineering* 46, 2000, pp.427-434.
7. Shibata, K., Kadota, T., Kohno, Y., Nyilas, A., and Ogata, T., "Mechanical Properties of A Boron Added Superalloy at 4K and Magnetic Effect," in *Advances in Cryogenic Engineering* 46, 2000, pp.73-80
8. Shibata, K., Nyilas, A., Shindo, Y., and Ogata, T. "Magnetic Effect on Young's Modulus Measurement of CP-Ti at 4K (Result of Round Robin Test)," in *Advances in Cryogenic Engineering* 48, (to be published)
9. Rice, J.R., Paris, P.C., and Merkle, J.G., "Some further results of J-integral analysis and estimates," in *ASTM Special Technical Publications* 536, (1973), pp.231
10. Nyilas, A. and Obst, B., "A new test method for characterizing low-temperature structural materials, " in *Advances in Cryogenic engineering (materials)* 42A, 1996, pp.353-360
11. Nyilas, A., Obst, B., and Nishimura, A., "Fracture Mechanics Investigations at 7 K of Structural Materials with EDM Notched Round and Double Edged-Bars, " in *Advances in Cryogenic Engineering* 44, 1998, pp153-160.
12. Nishimura, A., Yamamoto, J., and Nyilas, A., "Fracture toughness evaluation at cryogenic temperature by round bar with circumferential notch," in *Advances in Cryogenic engineering (materials)* 44A, 1998, pp.145-152
13. Nishimura, A., Ogata, T., Shindo, Y., Shibata, K., Nyilas, A., Walsh, R.P., Chan, J.W., and Mitterbacher, H., "Local Fracture Toughness Evaluation of 316LN Plate at Cryogenic Temperature," in *Advances in Cryogenic Engineering* 46, 2000, pp.33-40.
14. Nishimura, A., Nyilas, A., Ogata, T., Shibata, K., Chan, J.W., Walsh, R.P., and Mitterbacher, H., "International Round Robin Test Results of J-Evaluation on Tensile Test at Cryogenic Temperature within the Framework of VAMAS Activity," in *Advances in Cryogenic Engineering* 48, (to be published)
15. Shindo, Y., Sanada, K.., and Horiguchi, K., "Fracture behavior of G-10 woven glass-epoxy laminates at low temperatures," in *Advances in Cryogenic Engineering* 42, 1996, pp.129-136
16. Shindo, Y., Horiguchi, K., and Wang, R., "Cryomechanics and Short-beam Interlaminar Shear Strength of G-10CR Glass-cloth / Epoxy Laminates," in *Advances in Cryogenic Engineering* 46, 2000, pp.167-174.
17. Shindo, Y., Horiguchi, K., Ogata, T., Nyilas, A., Zhang, Z. and Humer, K., "Results of a VAMAS Round Robin on the Cryogenic Interlaminar Shear Strength Determination of Glass-cloth / Epoxy Laminates," in *Advances in Cryogenic Engineering* 48, (to be published)

PHYSICAL AND MECHANICAL PROPERTIES AT CRYOGENIC TEMPERATURES

LOW TEMPERATURE HYDROGEN EMBRITTLEMENT BEHAVIOR OF Zr-4 ALLOY

Y.F. XIONG [a], L.F. LI [a], W.H. WANG [a], L.Z. ZHAO [a], and L.J. RONG [b]

[a] Cryogenic Laboratory, Technical Institute of Physics and Chemistry, Chinese Academy of Sciences, Beijing 100080, P. R. China

[b] Institute of Metal Research, Chinese Academy of Sciences, Shenyang, 110013, P. R. China

ABSTRACT

The mechanical properties and hydrogen embrittlement behaviors (HE) of Zr-4 alloy in the temperature range of 40-4.2 K have been investigated. The hydrogen free Zr-4 alloy sample exhibits a very good tensile strength and superior ductility as high as a 21% increase at liquid hydrogen temperature (20 K) over that at room temperature. It also shows the typical features that appeared in austenitic stainless steel, such as unstable plastic flow, strengthening and multi-necking behaviors. The difference is that the temperature of these behaviors occurred much lower than those in austenitic stainless steel. The experiment on hydrogen charged sample shows a heavy hydrogen embrittlement behavior at low temperatures. The hydride cracking mechanism of HE has been discussed in the paper.

INTRODUCTION

Alloys based on zirconium, having good mechanical properties and small neutron absorption cross section, are widely used as the container materials in a nuclear reactor. However, this material has a tendency to be embrittled by hydrogen. Many fundamental investigations have been performed on the HE of Zirconium alloys[1-2]. Recently, Zr alloy has been used as the liquid hydrogen container material in cold neutron sources. But the mechanical properties of the Zr alloys are not very clear. In this situation the authors focus

CP614, *Advances in Cryogenic Engineering:*
Proceedings of the International Cryogenic Materials Conference - ICMC, Vol. 48,
edited by B. Balachandran et al.
© 2002 American Institute of Physics 0-7354-0060-1/02/$19.00

on the low temperature mechanical properties of Zr-4 alloys, their hydrogen embrittlement, and also special behavior compared to other cryogenic materials. The scanning electron microscopy (SEM) is used as a method to investigate the mechanism of HE of the alloy.

MATERIAL AND EXPERIMENTAL PROCEDURE

Commercial Zr-4 alloy was chosen as the object of investigation. The chemical composition is 1.45 wt. %Sn, 0.22 wt.% Fe, 0.11 wt% Cr and balance Zr. The alloy was annealed in vacuum at 813K for 2 h and cooled in air. Specimens 3mm in dia and 20mm in gauge length were tested on a tensile testing machine at the rate $\varepsilon = 1.67 \times 10^{-2}$ mm/s in the temperature range 4.2 to 40 K. Hydrogen charging of the tensile samples was performed in a hydrogen pressure cell of 100 MPa at 573K for 240 h. The specimens were covered with a deep blue surface film after charging. Fractographic investigations were accomplished by the method of scanning electron microscopy (SEM) at an accelerating voltage of 15 kV in the second-electron regime.

RESULTS AND DISCUSSION

Figure 1 shows the stress-strain curves for the typical Zr-4 alloy at different test temperatures. At 4.2 K deformation is discontinuous, the unstable plastic flow (serrated yielding), showing a series of suddenly load drops accompanied by well-audible click appears in the curve, and the jump amplitude increasing with the strain. The multi-necking behavior is also observed in this sample. But the curves for 20 and 40 K are seen to have a parabolic monotonic character with a pronounced loading maximum achieved before fracture and corresponding to the tensile strength σ_B. The jump-like deformation and multi-necking behaviors at low temperature seem like the common characteristic of cryogenic materials[3,4]. The occurrence of these behaviors of the Zr-4 alloy are strongly dependent on temperature, they appear at still low temperature. The mechanism of this interesting phenomenon needs to be still investigated. Table 1 shows the dependence of the mechanical characteristics (plasticity and strength) on the temperature. The tabulated values are averaged over three samples. As we see, the tensile strength and percentage elongation after fracture of this alloy are good at cryogenic temperature, respectively, 1090MPa and 21% at liquid hydrogen temperature (20 K). Although its strength level is relatively lower than those of cryogenic titanium alloys, its cryogenic toughness is much better. The results revealed that Zr-4 alloy has sufficiently good mechanical properties for the container at cryogenic temperature.

Table 1. Mechanical properties of the Zr-4 alloy

Temperature, K	Ultimate strength, MPa	Elongation, %
4.2	1059	16
20	1090	21
40	974	22

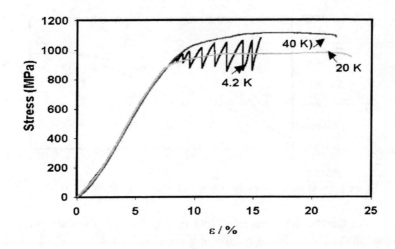

Figure 1. Stress-strain curves for Zr-4 alloys at different temperatures.

Since Zr-4 alloy is used as a the liquid hydrogen container material in a cold neutron source, the mechanism of hydrogen embrittlement must be studied thoroughly. Figure 2 shows the stress-strain curves at different test temperatures for the Zr-4 alloy after hydrogen charging as mentioned above. It is shows the Zr-4 alloy will lose its toughness and ductility after the saturation hydrogen charging (256ppm). The results of the tensile tests conducted on charged specimens are presented in table 2 which also shows the comparative percentage loss in plastic elongation on hydrogen charging. Comparative percentage loss (CPL) in plastic elongation is defined as

$$CPL = \frac{\delta_{uncharged} - \delta_{charged}}{\delta_{uncharged}} \times 100 \quad \dots\dots\dots\dots\dots\dots\dots\dots(1)$$

The CPL has been used as the measure of hydrogen embrittlement previously by other investigations[5]. The CPL numbers are over 80 in our experiments, this implies that the Zr-4 alloy is very susceptible to the hydrogen embrittlement. But this result does not mean that Zr-4 alloy cannot be used as the liquid hydrogen container material in cold neutron source. Huang et al.[1] point out the Zr alloys will have high HE resistance when they are charged at lower temperature. So the results of saturation hydrogen charged specimens cannot represent the HE behavior of materials working in a cold neutron source. More experiments are required. In Figure 2, the stress-strain curve of the alloy at 4.2 K shows very small jump-like deformation before fracture. This implies that the serrated character even exist in the hydrogen charged specimen.

Table 2 Mechanical properties of the hydrogen charged Zr-4 alloy

Temperature, K	Ultimate strength, MPa	Elongation, %	CPL, %
4.2	923	2.8	82
20	967	2.8	86

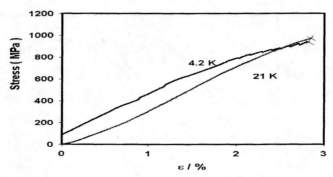

Figure 2, Stress-strain curves for hydrogen charged Zr-4 alloy at different temperature.

Studies of the fracture surfaces of Zr-4 alloy deformed at low temperatures show that all samples fracture with the formation of a pronounced macro-neck. Loss of the sample stability during straining due to neck formation and subsequent continuous decrease in cross-section stipulates the existence of a maximum and the next load decrease before fracture in the stress-strain curves (See Figure 1). In addition to this the fracture region is of the "cup-and-cone" type of appearance which is typical for ductile materials. These reveal the Zr-4 alloy has good cryogenic mechanical properties.

Fracture surfaces at 40K, 20 K and 4.2 K are shown in figure 3a, 3b and 3c. Coarse wavy slip bands, crimps and large broken cracks following the slip band bending and crimp peaks are observed in all the samples. A detail analysis of the fractured surfaces will appear elsewhere.

Fracture surfaces of hydrogen charged Zr-4 alloy differ totally from the above, as show in Figure 4a. It exhibits flat fractographic features, and many cracks can be seen. A higher magnification view of the flat feature is presented in Figure 4b. The flat fractographic features of the hydrogen charged sample proves that hydrogen is responsible for these features. We attribute the cracks of the fracture surfaces of embrittled Zr-4 alloy to the presence of hydride under HE conditions.

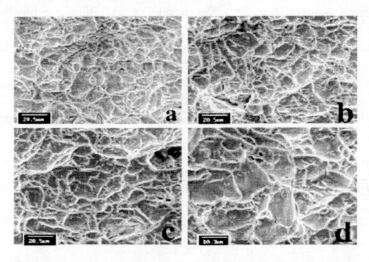

Figure 3, Fracture surface of Zr-4 alloy at 40K(a), 20 K(b) and 4.2 K (c and d).

CONCLUSIONS

The mechanical properties and fracture character of Zr-4 and its hydrogen embrittlement alloys in the temperature range 40-4.2 K have been investigated. Our results show that Zr-4 alloy has very good tensile strength and superior ductility at cryogenic temperature. Zr-4 alloy is very susceptible to the hydrogen embrittlement, hydride cracking mechanism has been proposed to explain the fracture surfaces of hydrogen charged Zr-4 alloy.

Figure 4, Fracture surface of hydrogen charged Zr-4 alloy 4.2 K(a) and its higher magnification view(b).

REFERENCES

1. Huang, J.H. and Yeh, M.S. Met. Trans. **29A**, pp. 1047(1998).
2. Kawamura *et al.*, Proceedings of the 1st International Conference on Processing Materials for Properties, pp. 1285(1993).
3. Lavrentev, F.F. *et al.*, Mat. Sci. Eng. **58**, pp. 157(1992).
4. Lavrentev, F.F. et al., Cryogenics. **3**, pp. 170(1983).
5. Kimm, S.S. et al., Scripta Metall. **22**, pp. 1831(1988).

CHARACTERIZATION OF TITANIUM ALLOYS FOR CRYOGENIC APPLICATIONS

M. Reytier, F. Kircher, B. Levesy

CEA Saclay, DSM / DAPNIA / STCM
Gif sur Yvette, 91191, France

ABSTRACT

Titanium alloys are employed in the design of superconducting magnet support systems for their high mechanical strength associated with their low thermal conductivity. But their use requires a careful attention to their crack tolerance at cryogenic temperature. Measurements have been performed on two extra low interstitial materials (Ti-5Al-2.5Sn ELI and Ti-6Al-4V ELI) with different thickness and manufacturing process. The investigation includes the tensile properties at room and liquid helium temperatures using smooth and notched samples. Moreover, the fracture toughness has been determined at 4.2K using Compact Tension specimens. The microstructure of the different alloys and the various fracture surfaces have also been studied. After a detailed description of the experimental procedures, practical engineering characteristics are given and a comparison of the different titanium alloys is proposed for cryogenic applications.

INTRODUCTION

The Compact Muon Solenoid (CMS) is one of the detectors to be built for the LHC project at CERN. The magnet system consists of a 4 Tesla solenoidal superconducting coil enclosed in a steel return yoke. To ensure the suspension of the cold mass inside the vacuum tank, a set of tie rods has been designed. The goal of this support system is to sustain the 225 tons of the superconducting coil and any magnetic forces due to a coil misalignment inside the yoke. Moreover, these tie rods are attached at one end on the cold mass at 4.5 K, and on the other side on the vacuum tank at room temperature.
Titanium alloy appears as a candidate for this application thanks to its high mechanical strength associated with a low thermal conductivity and low magnetic permeability. But, toughness is also an important requirement for this structural device. That is why to find the best compromise between the mechanical strength and the toughness, a characterization campaign has been realized on three alloys. A particular attention has been paid to an eventual notch sensitivity since the tie rods are threaded at both ends.

CP614, *Advances in Cryogenic Engineering:*
Proceedings of the International Cryogenic Materials Conference - ICMC, Vol. 48,
edited by B. Balachandran et al.
© 2002 American Institute of Physics 0-7354-0060-1/02/$19.00

TABLE 1. Chemical analysis (% wt)

	Al	Sn	V	C	Fe	O	N	H
Alloy A	5.60	2.61	-	0.011	0.05	0.072	0.036	0.0012
Alloy B1	6.38	-	4.48	0.022	0.10	0.121	0.013	0.0030
Alloy B2	6.26	-	4.27	0.014	0.13	0.116	0.0077	0.0026

MATERIALS

In order to choose the proper material for the suspension tie rods of CMS, two extra low interstitial types have been melted and forged by two different suppliers. The alloy A consists of a Ti-5Al-2.5Sn ELI bar with a diameter of 110 mm, whereas the alloys B1 and B2 are of Ti-6Al-4V ELI type with a diameter of 80 mm. Alloys A and B1 come from the same producer. The control of the chemical analysis is shown in Table 1. Here, due to the cryogenic application, lowering the permissible impurity content is the first important requirement [1]. For the three alloys, the standards [2] and [3] are satisfied. The microstructure of titanium alloys also strongly depends on the thermo-mechanical conditions of the manufacturing process [4,5]. The controls realized for each alloy are presented on Figure 1. The alloy A is composed of a mix between equiaxed alpha grains and an acidular microstructure. It should also be stressed that the same microstructure is observed at different thickness in the raw bar, which reveals a good homogeneity. The alloy B1 and B2 present, near the tensile samples, the microstructure presented on Figure 1. It is characterized by equiaxed alpha grains with intragranular beta phase. But these microstructures are not homogeneous in the whole bars. The alloy B1 presents also platelike alpha grains and intergranular beta phase in the heart of the raw bar, whereas, for the alloy B2, some alignments in the rolling direction can be noticed.

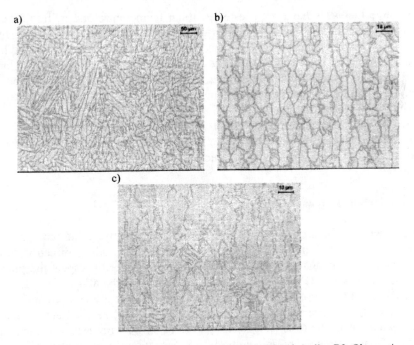

FIGURE 1. Optical micrographs of each alloy. a) alloy A, b) alloy B1 and c) alloy B2. Observations containing the longitudinal direction of the raw bars and close to the tensile samples.

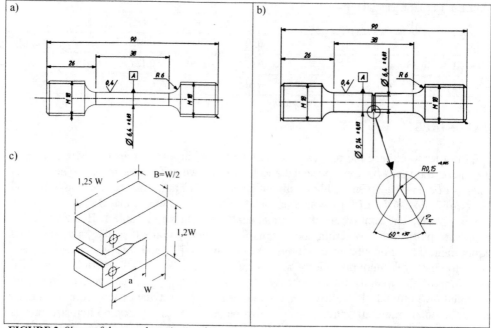

FIGURE 2. Shape of the samples : a) smooth tensile specimen b) notched tensile specimen c) compact tension specimen

EXPERIMENTAL PROCEDURE

Tensile tests

The three materials have been tested at room temperature and in liquid helium at 4.2 K. These tensile tests have been carried out at a constant cross head speed of 0.3 mm/min on both smooth and notched specimens shown on Figure 2. The load has been measured with a 150 kN cell fixed outside the cryostat and the displacement has been followed with a regular extensometer over a gauge length of 25 mm for the smooth samples and 12.5 mm for the notched specimens. Its signal is obviously calibrated at room temperature and in liquid helium at 4.2 K.

The notch shape leads to a stress concentration factor of 6.5. The ratio of the tensile strength on notched specimens (NTS) by the tensile strength on smooth specimens (TS) is used to study the notch sensitivity of each alloy and should be above one at any temperature.

Fracture Toughness Measurements

The fracture toughness (K_{IC}) has been determined by using compact tension specimens in the LT direction (the load is applied in the longitudinal direction of the raw bars and the crack propagates in the radial direction). In order to respect the plane strain conditions, the thickness (B) of the samples is of 15 mm for the alloy A and 12.5 mm for the alloy B1 and B2 (Figure 2c). Each specimen is first fatigue pre-cracked at 77 K to start the fracture toughness test with a thin crack whose length (a) reached among half of the specimen width (W). The tests have been realized at 4.2 K with a displacement speed of 1 mm/min. To determine the mean length of the propagation, five points equally located in the thickness, have been measured after optical observations of the fracture surface.

TABLE 2. Tensile characteristics at room temperature

	Yield Stress (MPa)	Tensile Strength (MPa)	Elongation (%)	NTS / TS
Alloy A	790 ± 19	875 ± 22	16.2 ± 2.1	1.47
Alloy B1	881 ± 12	957 ± 7	15.6 ± 2.0	1.49
Alloy B2	827 ± 7	897 ± 1	15.2 ± 1.6	1.46

TABLE 3. Tensile characteristics in liquid helium at 4.2 K

	Yield Stress (MPa)	Tensile Strength (MPa)	Elongation (%)	NTS / TS
Alloy A	1348 ± 12	1475 ± 1	9.9 ± 0.5	1.12
Alloy B1	1650 ± 13	1698 ± 18	4.8 ± 2.3	0.94
Alloy B2	1643 ± 22	1673 ± 13	5.1 ± 2.5	0.94

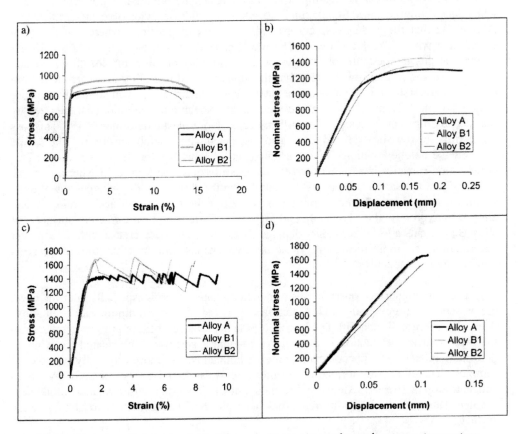

FIGURE 3. Typical tensile behavior obtained for each alloy on a) smooth samples at room temperature, b) notched samples at room temperature, c) smooth samples at 4.2 K and d) notched samples at 4.2 K.

RESULTS

Tensile characteristics at room temperature

The results of the tensile tests at room temperature are presented in the Table 2 on Figure 3a and 3b. No unexpected result is revealed at room temperature. The mechanical strength of the alloy A is lower than the two others, associated with the same order of ductility. The engineering properties of the three materials are in agreement with the standards [2] and [3] and also with other typical values [6,8]. No notch sensitivity is revealed at room temperature and the fracture surfaces are covered with equiaxed dimples, sign of a classical ductile rupture mode.

Tensile characteristics at 4.2 K

The main differences between the three alloys are revealed at cryogenic temperature. The results are shown in Table 3 and on Figure 3c and 3d. First, the mechanical behavior in plasticity is serrated for the three materials. But alloys B1 and B2 present few load drops of high amplitude, whereas alloy A shows many little drops as presented on Figure 3c. This marked discontinuous yielding, known as adiabatic deformation [9], is still the same when decreasing the speed to 0.05 mm/min. Moreover, this behavior is so sudden for alloy B1 and B2 that the yield stress corresponds to 0.15% of plasticity whereas it has been determined with 0.2% for alloy A. The mechanical characteristics present a significant increase at this temperature, but it is associated with a loss of ductility for alloys B1 and B2. This seems to confirm the fact that the alpha type alloys (alloy A) suffer less from ductility reduction at low temperature than other titanium alloys containing beta phase (alloys B1 and B2) and that a serrated yielding at cryogenic temperature characterized by many load drops can be a sign of significant capacity for plastic deformation [9]. This loss of ductility for the alloy B1 and B2 is also associated with a notch sensitivity. The tensile tests on the notched samples reveal a notch ratio below one for these two alloys. This deformation capacity lack on a notch structure is illustrated on Figure 3d where it can be observed that the samples B1 and B2 break during the elastic loading whereas the alloy A sample sustains some plastic deformation. Moreover the fracture surface of alloy A and alloy B2 are presented on Figure 4a and Figure 5a. There is no brittle fracture surface, but alloy B2 presents a lot of secondary damage characterized by deep cracks with no apparent deformation or dimple formation. These observations also confirm that alloy A presents higher ductility capacities.

The fracture surface observations of notched samples reveal, especially near the edge, the presence of large among of elongated facets instead of classical dimples for both alloys A and B2 (Figure 4b and 5b). For Nagai [6], the appearance of these grooves would seem to correspond to a reduction in ductility or toughness and would be related to twinning deformation. Here, this phenomenon is met in both alloys but more especially on notched samples. Therefore the formation of these grooves seems to be favored by stress concentration or stress triaxiality and reveals a modification in the deformation mode near a notch. But no significant difference has been observed between the two alloy fracture surfaces.

a)

b)

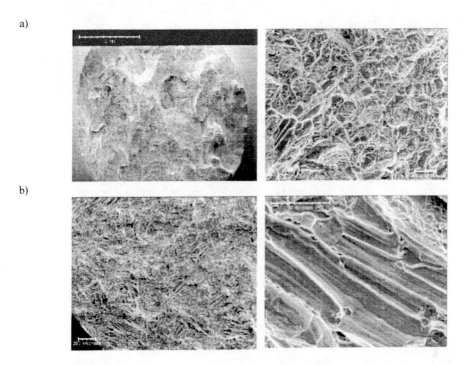

FIGURE 4. Scanning electron micrographs of fracture surface for alloy A at 4.2K on a) smooth samples and b) notched samples.

a)

b)

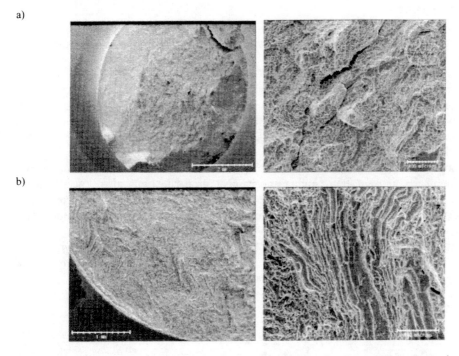

FIGURE 5. Scanning electron micrographs of fracture surfaces for alloy B2 at 4.2K on a) smooth samples and b) notched samples.

Fracture Toughness Measurements

Finally, the fracture toughness tests presented in Table 4 confirm that the alloy A can keep deformation capacities at cryogenic temperature, even in the presence of a thin crack. It shows a significant higher toughness than the other two alloys. Our results are comparable with other values [6,8] and stresses once more that significant differences are still present between extra low interstitial titanium alloy families. The alpha type alloy (alloy A) appears softer even at cryogenic temperature, with therefore a better toughness.

TABLE 4. Fracture Toughness at 4.2 K

	K_{IC} (MPa.m$^{1/2}$)		K_Q (MPa.m$^{1/2}$)
Alloy A	77 84 74	78 ± 5	
Alloy B1	58 61 63	61 ± 2	
Alloy B2	67 50	58 ± 12	60

CONCLUSIONS

In order to choose the proper titanium alloy for the CMS tie rods, a characterization campaign has been realized on three extra low interstitial titanium alloys produced by two different suppliers.

The chemical analysis are in agreement with the standards but the study of the microstructure reveals also some lack of homogeneity for alloys B1 and B2.

No unexpected result has been found by tensile tests at room temperature. The alpha type alloy A presents lower mechanical strength associated with the same order of ductility than the two other alloys.

The major differences are revealed at cryogenic temperature. Alloys B1 and B2, containing some beta phase, present both higher mechanical properties but also lower fracture toughness and a significant notch sensitivity. Moreover, their behavior is characterized by few sudden and important load drops at the very beginning of yielding and their ductility is significantly reduced.

The microstructure obtained for the alloy A (mix between equiaxed alpha grains and acidular shape) seems to reach a good compromise between the necessary mechanical strength and the toughness for our cryogenic application.

ACKNOWLEDGEMENTS

The authors wish to thank the CMS collaboration for its financial support as well as the technical staff of the DAPNIA / STCM involved in these tests, and more particularly S. Cazaux and G. Lemierre.

REFERENCES

1. Carman C. M. and Katlin, J. M., *ASTM STP*, **432**, pp. 124-144, (1968)
2. Aerospace Materials Specifications 4924 D
3. ASTM B 381 –97
4. Shannon, J.L. and Brown, W. F, *ASTM STP*, **432**, pp. 33-63, (1968)
5. Campbell, J. E., *ASTM STP*, **556**, pp. 3-25, (1974)
6. Nagai, K., *Cryogenics*, **vol. 26**, pp. 19-23, (1986)
7. Van Stone, R. H. and Shannon, J. L., *ASTM STP*, **651**, pp. 154-179, (1978)
8. Ito Y. and Nishimura T., "cryogenic properties of extra low oxygen Ti-Al-4V alloy", *sixth conference on titanium*, France, 1988, pp. 87-92.
9. Reed R. P., Clark A. F., *Materials at low temperature*, ASM, Metals Park Ohio, 1983, pp.237-267

FRACTURE AND DEFORMATION POTENTIAL OF MAGNESIUM ALLOYS AT LOW TEMPERATURES

A. Bussiba[1], M. Kupiec[1], S. Ifergane[1], A. Stechman[2] and A. Ben-Artzi[2]

[1]NRCN P. O. Box 9001, [2]ROTEM Industries, LTD P.O.Box 9046
Beer-Sheva 84190, ISRAEL

ABSTRACT

Thermal dependency of uni-axial mechanical properties in ZK60 and fracture resistance of AZ31 magnesium alloys were determined at low temperatures regime. In ZK60, this mechanical behavior was characterized in terms of stress, strain, deformation and fracture modes, followed by Acoustic Emission (AE) tracking. In AZ31, fracture toughness was assessed in relation to crack orientations, accompanied by fracture modes classification and AE monitoring. As for the former alloy, a moderate decrease in ductility was found, followed by a considerable increase in stresses, while decreasing test temperature. In addition, regardless of low temperatures, slip is still a predominant deformation mode, which controls the early stages of plastic deformation. The transition to mechanical twinning propagation mode controls the linear strain hardening stage. This transition was found to be temperature-dependent, expressed by a moderate decrease in transition strain, whereas the transition stress remains almost constant. In AZ31 alloy, a significant decrease in fracture toughness was found in the L-T orientation at low temperatures. This transition was accompanied by a mixed fracture mode, a pop-in phenomenon in the load-cod curves and by changes in AE parameters. The unlike mechanical behavior at low temperature of ZK60 compared to AZ31 was attributed to the occurrence of additional slip deformation mechanisms, based on literature findings concerning the addition of elements to pure magnesium.

INTRODUCTION

Persistent efforts are being devoted in weight reducing in several structural applications such as weight/cost ratio and improving performances viewpoint. These can be fulfilled thanks to low-density materials or advanced processing methods. For the latter, lamella composite layered -structures (blanks covers metal foam) are being designed, on one hand, and rapid solidification, nano-structural materials, thermo-mechanical processing and metal matrix composite which lead to improved properties, on the other hand [1]. However, the development of low-density alloys is the more conventional way of reducing weight, as reflected by massive research in developing Mg based alloys (the lightest structural alloy) [2-3]. As such, Mg alloys have been lately considered as structural

CP614, *Advances in Cryogenic Engineering:*
Proceedings of the International Cryogenic Materials Conference - ICMC, Vol. 48,
edited by B. Balachandran et al.
© 2002 American Institute of Physics 0-7354-0060-1/02/$19.00

materials for components in automotive [4], aerospace and space applications [5-6]. However, some Mg alloys show poor mechanical properties (strength/ductility, fracture toughness, fatigue and creep) compared to Al alloys, besides the fact that limited data is available, especially concerning low temperature regimes [7]. This temperature range imposes various constraints on structural material performances, particularly in materials with limited deformation capability, as in Mg.

Magnesium is characterized by a HCP crystal structure, which plastic deformation is controlled by slip and twinning mechanisms. The twinning deformation mode compensates, at high temperatures, the insufficient independent slip system and provides strain compatibility conditions at grain boundaries. However, at low temperatures, twins may act as a trigger in initiating brittle fracture and enhance micro-cracking occurrences, events that are undesirable in the behavior of structural materials.

Consequently, in order to integrate Mg based alloys safely in aerospace applications, more data is requested in fundamental and practical levels at low temperatures. As for the former, effects of mutual interaction of slip/twinning, heterogeneous and homogeneous twinning nucleation on plastic flow and the temperature dependency of twinning stress have to be clarified. These complex issues arise from the lack of basic understanding in mechanical twinning mechanisms. For the latter, a possible transition from ductile to brittle fracture, at uniform and triaxial state of stress and strain rate effects has to be determined.

Generally, mechanical characterization in conjunction with micro-feature observations provides a successful method in elucidating micro-mechanisms of plastic deformation. Nowadays, applications of in-situ techniques have become very attractive in visualizing plastic deformation features related to micro-fracture processes. Among others, just as TEM and STEM, AE provides an important complementary data related to mechanisms of different deformation modes [8]. The use of AE technique has increased lately, due to the great advance in AE data processing. Hence, the current study adopts the AE technique besides the use of ex-situ methods, in exploring some aspects related to slip/twinning deformation mechanisms and temperature effects on mechanical behavior.

Therefore, the objectives of the current study were twofold: First, to characterize the plastic response of ZK60 magnesium alloy in terms of slip and twinning deformation mechanisms, twinning stress as well as their mutual interaction during plastic flow as a function of test temperatures (123K-296K). Secondly, to determine the tensile mechanical properties of ZK60 and fracture toughness of AZ31 at low temperatures regime.

EXPERIMENTAL PROCEDURE

The experimental program focused on two commercial Mg alloys: ZK60 and AZ31. The ZK60 (4.9Zn, 0.49 Zr in wt.%, ratio of impurities less than 500 ppm) alloy was tested in the form of extruded bar with 10mm in diameter. Extrusion process was performed at 593K, applying a strain rate of 10^{-1} sec^{-1}. The treatment solution was conducted at a temperature of 775K/2h, revealing a grain size of about 30 μm. The AZ31 (2.5 Al, 0.092 Zn, 0.37 Mn in wt%, ratio of impurities less than 500 ppm) was examined in the form of a 8 mm rolled plate, obtained by cross-rolling process at 825K. The plate was annealed at 620K/1h, which exposed an equi-axed grain size of about 50 μm.

The uni-axial mechanical response of ZK60 was established at temperatures ranging between 123K and 296K. Tensile tests were performed with computerized machine, applying a strain rate of 10^{-3} sec^{-1} on round specimen (diameter of 5mm and a gauge length of 21mm). AE were monitored simultaneously with load trace, using resonant sensors with a typical frequency range of 50-300 kHz. The pre-amplification value was 60 dB and ambient noise was filtered using a threshold of 30 dB. A computerized AE system enabled to process from AE waveforms five parameters (counts, energy, rise time, amplitude, average frequency). Preliminary measurements determined the acquisition parameters for the current material (PDT = 50 μs; HDT = 100 μs; HLT = 300 μs).

The thermal dependency (123K-296K) of the fracture resistance behavior of AZ31 was achieved through a pre-cracked miniature three point bend specimens with dimensions of: 8mm in width, 8mm in thickness and an initial crack length of 1.5mm. Specimens were machined with different crack plane orientations (T-L, L-T, T-S and L-S) according to the notation in ASTM E399. To minimize effects due to pre-cracking temperatures, both pre-cracking and toughness tests were performed at the same temperature. COD gauge was used for crack length measurements. AE setup was mainly the same as for the tensile case.

Deformation features were characterized in two levels: At the macroscopic level, optical microscopy was used mainly in order to track twinning (primary and secondary) and slip traces of the deformed specimens. At the microscopic level, TEM was utilized for more comprehensive observations of the twinning morphology and common intersections, and their interaction nature. Finally, fracture modes examination were characterized using SEM. Special attention was given to a possible transition in the fracture mode characteristics at low temperatures.

EXPERIMENTAL RESULTS

ZK60 characterization

FIGURE 1 shows test temperature effects on the uni-axial flow properties, stresses and ductility. An increase of approximately 50% in the various stresses was obtained as the temperature diminished, followed by a moderate decrease in elongation, with minor change in strain hardening coefficient values. The true stress-strain curve profiles were characterized by two main effects: (a) Regardless of test temperatures, a discontinuous elastic to plastic deformation transition was observed, marked by a small amount of strain extension with negligible increase in stress (FIG 2). In some cases, stress relaxation was noticed; (b) Parabolic strain hardening region decreases at low test temperatures, and the transition from gradually decreasing work hardening rate to nearly constant one becomes more evident. These changes are related to a modification in the rate-controlling deformation mechanism as the temperature decreases. FIG 3 illustrates this behavior in terms of two mechanical parameters: strain ($\varepsilon_{trans.}$) and stress ($\sigma_{trans.}$) transitions, as defined in FIG 4. As shown previously, $\varepsilon_{trans.}$ decreases moderately as temperature decreases, however the $\sigma_{trans.}$ is almost insensitive to the test temperatures.

FIGURE 5 shows selected simultaneous loads: AE counts (FIG 5a, d), average frequency (FIG 5b, e) and amplitude (FIG 5c, f) vs. time at 296K and 173K, respectively. Generally, most of the AE activity in terms of counts (FIG 5a, d) occurs during the elastic regime, with a notable peak near the elastic/plastic transition. After a sharp increase in AE counts, a decay in the AE activity was detected, more emphasized at 173K (the small figure is plotted in different scales to emphasize the AE counts after the sharp transition). At 173K, the AE counts level is higher than at 296K, in a magnitude of three orders approximately. At 296K, additional peaks were observed close to the strain-hardening rate transition point. Between these two peaks a reduced amount of AE counts was generated. After the second peak, AE counts decrease gradually to discrete counts, while approaching the fracture point. The corresponding average frequencies are illustrated in FIG 5b-e with a typical sinusoidal wave at 173K at the range of 200-300KHz, with 700KHz at the transition point. At 296K mainly 1MHz frequency was noticed with some frequencies

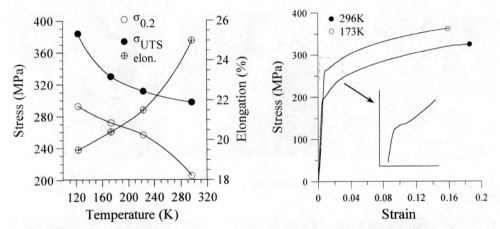

FIGURE 1. Temperature effect on uni-axial properties of ZK60.

FIGURE 2. Typical true stress-strain curve of ZK60 at various temperatures.

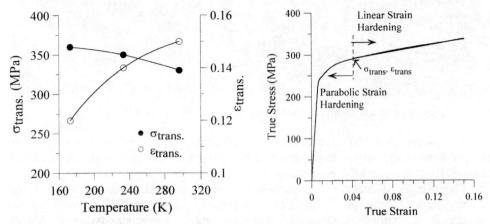

FIGURE 3. Stress, strain transition vs. temperature.

FIGURE 4. Definition of transition stress and strain.

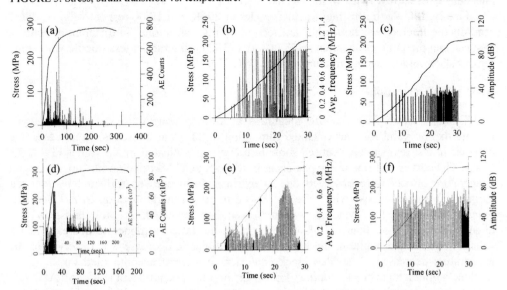

FIGURE 5. Stress, AE counts Average frequency and amplitude vs. time at: (a-c) 296K, (d-f) 173K.

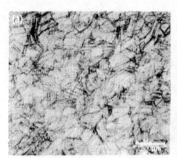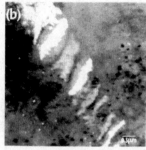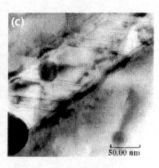

FIGURE 6. Deformation features in specimen deformed at 123K up to fracture: (a) Twinning intersection and slip traces; (b) lens type secondary twins; (c) narrow secondary twins.

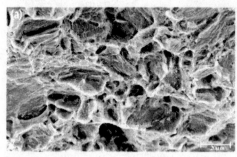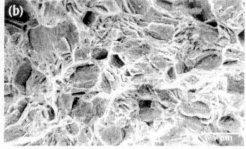

FIGURE 7. Fracture modes as a function of test temperature: (a) 296K; (b) 173K.

ranging between 50 to 500 KHz. These significant changes between both temperatures are also reflected in the amplitude parameter (FIG 5c, f). The average amplitude in 173K is about 60dB, while at 296K it is less than 40dB.

FIGURE 6a illustrates extensive twinning intersections slip traces developed after straining up to fracture at 173K. TEM deformation findings at 123K show secondary twinning in two types of morphologies within the primary twins: lens type (FIG 6b) and high density of narrow twins were detected (FIG 6c).

Finally, FIG 7a, b illustrate fracture modes at 296K and 173K respectively. At 296K semi-ductile fracture was noticed, characterized by fine "striations" in different directions, emphasizing the grain boundaries (FIG 7a). At 173K the fracture is less ductile with some regions showing low energy fracture surface (FIG 7b).

AZ31 characterization

Table 1 summarizes fracture toughness at various temperatures and orientations. Despite that no valid K_{IC} was obtained (limitation in the plate thickness) in all cases, the decrease in test temperatures caused some brittleness as indicated in K_C values (Table 1), Load-COD curves and fracture appearance. In the L-S and T-S orientations, no significant temperature effects were observed due to extrinsic influences as reflected by inclined fatigue and fracture cracks which are introduced in mixed mode loading. In L-T and T-L orientations, mode I of loading was preserved. As expected, the toughness in L-T orientation was higher than in T-L orientation. In the latter, a moderate decrease in K_C values was obtained, while in the L-T orientation dramatic decrease was observed as the temperature decreased. A sudden crack extension and arrest accompanied this trend of brittleness after a main crack blunting, known as "pop-in" phenomenon (FIG 8b). At 296K, a stable crack growth and a crack blunting were observed with no evidence of brittle crack

88

extension (FIG 8a). With regard to AE findings, at 296K, a typical accumulation profile was observed up to maximum load, followed by a moderate decrease in AE activity with load dropping (FIG 8a). At 173K, almost constant AE counts were generated up to the localized unstable crack growth (FIG 9b). This distinction in AE counts appears also in other AE features, namely: a higher rising time was obtained at 173K (90μs), compared to 30-40μs at 296K. In the energy parameter, more than two orders of magnitude were found at 173K, compared to 296K (especially in pop-in events). Finally, FIG 9 illustrates the fracture modes at both temperatures. At 296K, ductile-like fracture was observed while quasi-cleavage facets accompanied by micro-cracking (see arrows, FIG 9b) and isolated ductile regions, characterized the mixed mode fracture at 173K.

TABLE 1. Fracture toughness as a function of temperatures and orientations in MPa.m$^{1/2}$.

Temperature	Crack plane orientations			
	T-L	L-T	T-S	L-S
296K	9.5	14.3	N/A	N/A
200K	10.5	N/A	N/A	N/A
173K	10.1	15.5	8.6	9.5
123K	9.5	10.2	7	10.1

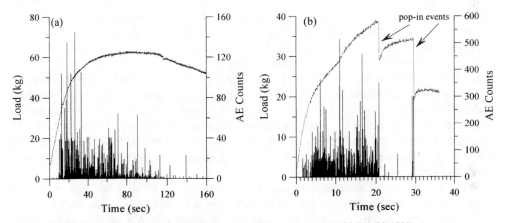

FIGURE 8. Load and AE counts vs. time at L-T orientation specimens: (a) 296K, (b) 173K.

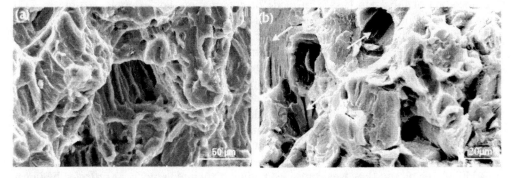

FIGURE 9. Fracture modes as a function of test temperature: (a) 296K; (b) 173K.

DISCUSSION

The evident distinction existing in the true stress-strain curve profile in ZK60 clearly indicates an inter-related transition in the deformation mechanisms according to test temperatures. The deformation behavior followed a sequential transition from initially slip mechanisms that control the parabolic strain-hardening region (decreasing strain hardening rate) to a mechanical twinning propagation, which controls the linear strain hardening stage (constant strain hardening rate). This trend is preserved regardless of test temperature. The yield stress for slip increases due to temperature since the Peierls stress is a short-range barrier. Nevertheless, the yield stress levels are still lower than the uniform twinning nucleation stress. The transition to twinning deformation mode is achieved at stresses higher than those needed for heterogeneous twinning nucleation because twinning takes place without the assistance of large numbers of pre-existing dislocations in the matrix. Therefore, at a critical strain ($\varepsilon_{trans.}$) which is temperature-dependent, the critical shear stress for twinning is achieved due to various strain hardening sources: twinning broadening, secondary twinning, dislocations emission at twinning intersection points and the increase in the twinning volume fraction, contributing to a constant strain hardening rate. As shown previously, this twinning stress ($\sigma_{trans.}$) is almost insensitive to temperature parameters, since it involves a mutual movement of a large number of atoms. The minor increase in σ_{trans} may be associated with the growth of available dislocations in twinning nucleation, generated at the elastic regime, near yield stress, as it will be explained later.

Although the twinning mode is dominated at progressive stages of the plastic deformation, twinning was observed at the elastic regime by metallographic observations at the triple point or at defect locations. These findings clarify the AE activity intensification at the elastic regime, which may be correlated to a heterogeneous twinning growth. This type of twinning nucleation occurs at low stress due to pre-existing mobile dislocation near these regions. This behavior is expressed in the equation given by Song et al. [9], which emphasizes the term related to pre-existing dislocations. These localized dislocations control the burst twinning nucleation type accompanied by a microscopic load drop, intensified at the elastic/plastic transition. Microscopic load release has been widely reported in cases of compression of refractory alloys [10] and U-Ti alloys [11]. After reaching the twinning stress, the twinning nucleation is more uniform and causes a decrease in AE counts.

Even though low temperature regime was applied, the ZK60 alloy behaves in a ductile manner, without any significant features of brittleness nor in ductility nor in fracture mode. Following Hardie results [12] on lattice spacing relationship in Mg solid solution, it was concluded that decreasing axis ratio (c/a) encouraged cross glide of non-basal slip, beside the operation of the main basal slip mode of deformation. For example, Yoshinaga [13] found that by adding Li, the value of the critical resolved shear stress for non–basal slip is dramatically diminished compared to pure Mg, as temperature decreases. Thus, the combined modes of slip deformation are responsible for delaying brittleness tendency at low temperatures. The addition of the main element Zn to Mg to form the ZK60 alloy leads to a slight decrease in the c/a ratio, which still may affect and reduce the resolved shear stress and introduce the non–basal slip.

Based on the same arguments, the addition of Al to Mg to form AZ31 results in increasing the axis ratio, which inhibits the initiation of the non-basal slip. As a result, one may expect a tendency towards brittleness at low temperatures of Mg –Al alloys. In fact, Busk [14] has shown a significant decreasing in ductility at 223K of AZ91 alloy (9% Al). In addition, the plastic flow was controlled mainly by twinning mode, as reflected by linear strain hardening from the early stage of plastic deformation. Moreover, compression of

AZ31 at 78K leads to a reduction of deformability as has shown by the true stress-strain curve [15]. This behavior has been attributed to the limited deformation caused by dislocation and twinning as well. Indeed, decreasing in toughness accompanied by pop-in phenomenon (event which characterizes semi-brittle materials), cleavage-like facets with twins traces (see arrows, FIG 9b) in the fracture surface, as presented in this work, confirm the propensity to brittleness of this alloy as reported also in the above-mentioned studies.

CONCLUSIONS

1. The rate controlled deformation mode in ZK60 changes from dominated slip at 296K, to dominated twinning mechanism at low temperatures ($\varepsilon_{trans.}$ decreases).
2. The yield stress for slip is lower than that for uniform twinning stress nucleation in the whole range of test temperatures. At εtrans., critical shear stress for uniform twinning is attained due to various strains hardening sources. The slip/twinning transition stress is insensitive to test temperatures.
3. Prior slip promotes heterogeneous twinning nucleation by providing necessary dislocations, supported by AE and metallographic findings. In contrast, yield stress for slip is dependent on temperature, since Peierls stress is a short-range barrier.
4. Axis ratio (c/a) decreases slightly by adding Zn to pure Mg, while increasing in Al alloys. These results control the operating and non-operating phases respectively of the non-basal slip beside the basal plane slip, which are responsible for the ductile and semi-brittle behavior of ZK60 and AZ31 respectively at low temperature regimes.

ACKNOWLEDGMENTS

The authors wish to express their gratitude to, Mr. A. Sobel, Mr. R. Shefi, Mr. K. Salah, Mr. D. H. David and Mr. Y. Gershon from the NRCN, for their experimental assistance.

REFERENCES

1. Ward-Close, C. M., Froes, F. H. and Cho, S. S., Synthesis and Processing of Lightweight Metallic Materials - An Overview Synthesis (Paper presented at Lightweight Metallic Materials II; TMS, Warrendale, US, 1997.
2. Das, S. K. and Davies, L. A., Mater. Sci. Eng. **98**, pp. 1-12 (1988).
3. Li, Y. and Jones, H., Mater. Sci. and Technology, **12**, 981 (1996).
4. Schumann, S. and Frederich, F., "The Use of Magnesium in Cars – Today and in Future", in *Proc. Magnesium Alloys and Their Applications*, April 23-30, Wolfsburg, Germany, 1998.
5. Agion, E., Brofin, B. and Schwartz, I., Magnesium Application in the Aerospace Industry (1996).
6. Zeumer, N., in *Proceedings of the Int. Magnesium Conf.*, McLean, VA:DMA 1998, p.125.
7. Iskander, S. K., et al., in *Proc. of Magnesium Technology 2000*, edited by H. J. Kaplan, J. N. Hryn and B. B. Clow, TMS Publications p. 325-329.
8. Yuyama, S., in: *Corrosion Monitoring in Industrial Plants Nondestructive Testing and Electrochemical Methods*, edited by G. C. Moran and P. Labine, ASTM STP 908, 1986.
9. Song, S. G. and Gray, G. T. III, Metall. and Mater. Trans. A. **26A**, pp. 2665-2675 (1995).
10. Jagannadham, K., Armstrong, R. W. and Hirth, J. P., Phil. Mag. **68 (2)**, pp. 419-451 (1993).
11. Bussiba, A, "Mechanical Behavior and Fracture Processes in Uranium Titanium Alloy", Ph.D. Thesis, July 1998.
12. Hardie, A. D. and Parkins, N., Philos. Mag., **4**, pp. 815-825 (1959).
13. Yoshnaga, H. and Horiuchi, R., Trans. JIM., **4**, pp. 134-141 (1963).
14. Busk, R. S, "Magnesium Products Design", Marcel Dekker, Inc., ISBN 0-8274-7576-7, 1986.
15. Lach, E., Kainer, K. U., Bohmann, A. and Scharf, M., in Int. Congress "Magnesium Alloys and their Applications", edited by K. U. Kainer, Wiley-Vch Press, p. 354-358, 2000.

MAGNETIC PROPERTIES OF THE AUSTENITIC STAINLESS STEELS AT CRYOGENIC TEMPERATURES

T. Kobayashi[1], S. Kobayashi, K. Itoh[2], and K. Tsuchiya[1]

[1] High Energy Accelerator Research Organization
1-1 Oho, Tsukuba, Ibaraki, 305-0801 Japan

[2] National Institute for Materials Science
3-13 Sakura, Tsukuba, Ibaraki, 305-0047 Japan

ABSTRACT

The magnetization was measured for the austenitic stainless steel of SUS304, SUS304L, SUS316, and SUS316L with the temperature from 5K to 300K and the magnetic field from 0T to 10T. The field dependences of the magnetizations changed at about 0.7T and 4T. The dependence was analyzed with ranges of 0-0.5T, 1-3T, and 5-10T. There was not so much difference between those stainless steels for the usage at small fields and 300K. The SUS316 and SUS316L samples showed large non-linearity at high fields and 5K. Therefore, SUS304 was recommended for usage at high fields and low temperatures to design superconducting magnets with the linear approximation of the field dependence of magnetization.

INTRODUCTION

The austenitic stainless steels are widely used for cryogenic apparatus because of their high strength at 4.2K.[1] The magnetic properties, were considered to be paramagnetic suitable as structural materials for superconducting magnets. The components of the steels were ferromagnetic materials and the chemical inhomogenities give small ferromagnetic and super-paramagnetic properties. This is a disadvantage for use in superconducting accelerator magnets for high energy physics because these magnets are required to have the field integral along beam line with accuracy of 10^{-4}. The magnetic investigation of these stainless steels is, therefore, important.

In this paper, the magnetization of the austenitic stainless steels, SUS304, SUS304L, SUS316, and SUS316L are reported for temperatures from 5K to 300K and the magnetic field from 0T to 10T.

CP614, *Advances in Cryogenic Engineering:*
Proceedings of the International Cryogenic Materials Conference - ICMC, Vol. 48,
edited by B. Balachandran et al.

EXPERIMENTAL METHOD

The samples were taken from plates of commercial materials of SUS304, SUS304L, SUS316, and SUS316L with the dimensions of about 3mmx3mmx1.5mm. The sample surface was polished chemically. The magnetizations were measured by using the MAGLAB VSM SYSTEM CF1200V (Vibrating Sample Magnetometer of Oxford Instruments). The calibration was performed by using Ni and Pd. The accuracy was 0.5%. The frequency was 45Hz and the vibrating amplitude was 1mm. The field was swept between ±3T at 300K and ±10T at 5K.

RESULTS AND DISCUSSIONS

FIGURE 1 shows the magnetizations of the samples at 300K and FIG 2 shows the magnetizations at 5K. The samples were non-magnetic because the magnetization was less than 1A/mkg under the external field $\mu_0 H$ of 3T while Fe, Cr, and Ni have magnetizations of 220, 22, and 60A/mkg. The field dependences of the magnetizations changed at about 0.7T and 4T. The dependence was analyzed with ranges of 0-0.5T, 1-3T, and 5-10T.

For use in the low field application, the slopes between 0.5T and -0.5T were calculated. For the SUS304 sample, the slope was 0.339A/mkgT at 300K and 0.714A/mkgT at 5K. For the SUS304L sample, the slope was 0.331A/mkgT at 300K and 0.753A/mkgT at 5K. For the SUS316 sample, the slope was 0.322A/mkgT at 300K and 1.15A/mkgT at 5K. For the 316L slope was 0.551A/mkgT at 300K and 1.70A/mkgT at 5K. At 300K, SUS304, SUS304L and SUS316 showed almost the same value of slope. The SUS316L sample showed a large value of slope. The slopes increased at 5K. The SUS 304 and SUS 304L had values 2 times larger than that at 300K. The SUS316 and the SUS316L showed 3times larger values compared to 300K. The values are summarized in TABLE 1.

Table 1. Field (B=0-0.5T) Dependence of Magnetization $M = aB + b$ at 300K and 5K

	a_{300K} (A/mkgT)	b_{300K} (A/mkg)	a_{5K} (A/mkgT)	b_{5K} (A/mkg)
SUS304	0.339	0.008	0.714	0.027
SUS304L	0.331	0.005	0.753	0.035
SUS316	0.322	0.000	1.151	0.005
SUS316L	0.551	0.009	1.700	0.008

When the field became large, the field dependence of the magnetization showed a non-linearity. To investigate this tendency, the field dependences were analyzed with field range between 1T and 3T. The SUS304 sample showed the intercept of 0.032(A/mkg) and the slope of 0.268A/mkgT at 300K. The intercept suggested the presence of magnetic materials. On the other hand, hysteresis could not be observed so the origin of the intercept was considered to be superparamagnetism.[2-5] The magnetic impurities formed magnetic clusters and each cluster behaved as magnetic moment. The cluster moments behaved as paramagnetic moments. Te cluster magnetic moments were easy to saturate at low fields. In this case, the saturation field strength was about 0.7T. The SUS304 sample showed the intercept of 0.0113A/mkg at 5K. The intercept was as large as expected from the comparison of the slopes at 300K and at 5K. As shown in FIG 2, the slope changed at about 4T. The SUS304L sample showed very similar results. The intercept was 0.024A/mkg at 300K. These values were the same as the SUS304 sample. At 5K, the intercept was 0.014A/mkg. There was not significant difference between SUS304 and SUS304L. The SUS316 sample showed linear dependence and the intercept was

0.000A/mkg. The slope was slightly larger than SUS304 and SUS304L. The 5K data showed the intercept was -0.170(A/mkg). The negative intercept was due to the positive curvature around 2T. In the case of SUS316L at 300K, the intercept was 0.135(A/mkg). At 5K, the intercept was 0.172(A/mkg). The SUS316 sample showed the change in the slope at about 4T and the SUS316L showed such a change at about 2.5T. These values are summarized in Table 2. These intercepts are considered to be due to the ferromagnetism imperfections or bcc structure. The SUS304 has an fcc structure and is nonmagnetic at room temperature. The bcc SUS304 stainless steel that is strongly ferromagnetic and cold-working or low-temperature deformation are known to partially retain this bcc state. The bcc phase of SUS304 has a magnetization of 130A/mkg at room temperature, due largely to the Fe moment.[6] For the SUS304 stainless steel sample could be considered to be with the ferromagnetic compositions of 0.025% because the intercept was 0.032(A/mkg).

Table 2. Field (B=1-3T) Dependence of Magnetization $M=aB+b$ at 300K and 5K

	a_{300K} (A/mkgT)	b_{300K} (A/mkg)	a_{5K} (A/mkgT)	b_{5K} (A/mkg)
SUS304	0.268	0.032	0.734	0.011
SUS304L	0.280	0.028	0.782	0.014
SUS316	0.320	0.000	1.355	- 0.170
SUS316L	0.310	0.135	1.583	0.172

The high field (B=5T-10T) dependences of the magnetizations were investigated. The slopes were almost same about 0.06A/mkgT. The intercept of SUS304 was 0.034A/mkgT. The intercept of SUS304L was 0.065A/mkgT. The intercept of SUS304L was about two times larger than that of SUS304. The intercept of SUS316 was 0.213 and the intercept of SUS316L was 0.330. These values are summarized in Table 3.

Table 3. Field (B=5-10T) Dependence of Magnetization $M=aB+b$ at 5K

	a_{5K} (A/mkgT)	b_{5K} (A/mkg)
SUS304	0.062	0.034
SUS304L	0.057	0.065
SUS316	0.059	0.213
SUS316L	0.065	0.330

For the usage of these stainless steels at low temperatures and high fields, the non zero intercept was the important factor. For the linear approximation design of a superconducting magnet, the small intercept was important and SUS304 and SUS304L were preferable. For the non-linear design, though, the field dependence of the magnetization should be taken into account but SUS316 and SUS316L could be used because those field dependences were obtained as reported above.

CONCLUSIONS

The field dependences of the SUS304, SUS304L, SUS316, and SUS316L magnetizations changed at about 0.7T and 5T. There is not much difference between those stainless steels for the usage at small fields and 300K. The SUS316 and SUS316L samples showed large non-linearity at high fields and 5K. Therefore, SUS304 is recommended for the usage at high fields and low temperatures for design of superconducting magnets with the linear approximation of the field dependence of magnetization or the approximated relations should be used for the stainless steels.

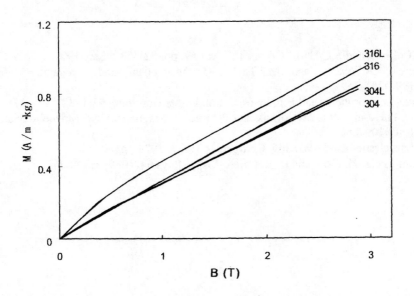

FIGURE 1. Magnetization M as a function of applied field B at 300K

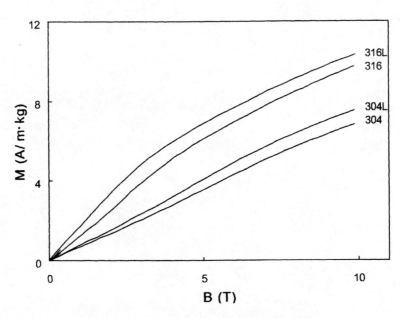

FIGURE 2. Magnetization M as a function of B at 5K

REFERENCES

[1] E.W.Collings and S.C.Hart, " *Cryogenics*, vol.**19**, pp. 521-530, Sep. 1979.

[2] J.Crangle, A.Fogarty and M.J.Tayler, J. Magnetism and Magnetic Materials **111**(1992)255-259.

[3] T.Datta, D.Thornberry and E.R.Jones,Jr, Solid State Commun. **52**(1984)515-517.

[4] J.Ding, H.Husang, P.G.McCormik and R.Street, J. Magnetism and Magnetic Materials **139**(1995)109-114.

[5] D.C.Larbalestier and H.W.King, Cryogenics March 1973 pp160.

[6] J. Childress, S. H. Liou, and C. L. Chien, J.Appl.Phys.**64**(1988)6059-6061.

EFFECTS OF δ-FERRITE AND WELDING STRUCTURE ON HIGH-CYCLE FATIGUE PROPERTIES OF AUSTENITIC STAINLESS STEELS WELD METALS

T.Yuri[1], T.Ogata[1], M.Saito[2], and Y.Hirayama[2]

[1]National Institute for Materials Science
Tsukuba, Ibaraki, 305-0047, Japan
[2]Mitsubishi Heavy Industries Ltd.
Nagasaki, Nagasaki, 851-0301, Japan

ABSTRACT

We studied the effects of δ-ferrite and welding structure on high-cycle fatigue properties for austenitic stainless steel weld metals at cryogenic temperatures. SUS304L and SUS316L weld metals contained 0% δ-ferrite (0% material) and 10% δ-ferrite (10% material) were prepared. High-cycle fatigue tests were carried out at 293 , 77 and 4 K. The S-N curves of those weld metals shifted towards higher stress levels, i.e., the longer life side, with decreasing test temperature. The ratios of 10^6-cycles fatigue strength (FS) to tensile strength (TS) of 0% material decreased from 0.8 to 0.45 and those of 10% material decreased between 0.35 to 0.65 with decreasing test temperature. Fatigue crack initiation sites of SUS304L 10% material were almost at blowholes, and those of SUS316L 10% material were at weld pass interface boundaries. On the other hand, those of 0% materials were considered to be due to the interface of the solidification structure. Although δ-ferrite reduces toughness at cryogenic temperatures in austenitic stainless steel weld metals, the effects of δ-ferrite on high-cycle fatigue properties are not significant.

INTRODUCTION

In the application of cryogenic technology for the clean energy, there are projects of constructing large-scale facilities of superconducting magnets at liquid helium temperature,

CP614, *Advances in Cryogenic Engineering:*
Proceedings of the International Cryogenic Materials Conference - ICMC, Vol. 48,
edited by B. Balachandran et al.
© 2002 American Institute of Physics 0-7354-0060-1/02/$19.00

Table 1. Chemical compositions of base and each weld metals for SUS304L and SUS316L (wt%) and volum contents of δ-ferrite (%).

Materials		C	Si	Mn	P	S	Ni	Cr	Mo	Cu	N (ppm)	O (ppm)	H (ppm)	δ-ferrite(%)
SUS304L	Base	0.017	0.56	0.87	0.031	0.002	9.05	18.33	0.17	0.24	412	28	1.7	—
	Weld (δ =0%)	0.013	0.39	1.45	0.025	0.002	11.78	17.53	0.02	—	356	26	0.6	0
	Weld (δ =10%)	0.014	0.40	1.51	0.021	0.003	9.74	19.08	0.01	—	350	40	1.3	10.5
SUS316L	Base	0.022	0.52	0.85	0.026	0.005	12.09	17.59	2.13	0.28	380	15	3.2	—
	Weld (δ =0%)	0.015	0.40	1.45	0.022	0.002	13.83	16.94	2.01	—	433	21	0.9	0
	Weld (δ =10%)	0.013	0.41	1.43	0.019	0.002	12.08	19.03	2.03	—	300	25	0.8	9.9

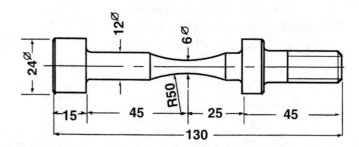

Figure 1. Dimensions of high-cycle fatigue specimen (mm).

and some projects [1, 2] examined the effective utilization of the hydrogen through its transportation. The structural materials used in liquid hydrogen huge tankers or storage tanks have been required an object of study. However, there are concerns that the structural materials used in such large-scale structures for handling liquid hydrogen may be subjected to highly rigorous service conditions that differ from the conventional conditions, such as long-term use in cryogenic temperatures, hydrogen environment, the application of various welding processes, and complicated loading conditions. In order to guarantee the long-life reliability and safety of cryogenic machinery used under harsh operating conditions such as cyclic loading and deformations, the securing of safety evaluations for fatigue is considered to be particularly important. For example, liquid hydrogen storage tanks may be cyclically stressed by temperature variations due to the amount of liquid hydrogen when it is injected or discharged, and liquid hydrogen supertankers may be cyclically stressed by the wave motion in maritime transportation. Therefore, the evaluation of fatigue properties is a critical issue for material selection and container design.

Austenitic stainless steels are widely used for cryogenic applications, however, there are very few data on fatigue properties of weld metals at cryogenic temperatures [3,4]. In austenitic stainless steels, about 5% δ-ferrite is introduced to prevent hot cracking during welding, but δ-ferrite is brittle at cryogenic temperatures [5,6]. However, the effects of δ-ferrite on the high cycle fatigue properties below the liquid hydrogen temperature have

Table 2. Tensile properties of base and each weld metals for SUS304L and SUS316L at 293, 77 and 4 K.

Materials		T (K)	Y.S(MPa)	T.S(MPa)	ε (%)	ϕ (%)
SUS304L	Base	293K	251	592	69.5	80.4
		77K	350	1557	40.0	65.7
		4K	389	1727	36.5	54.0
	Weld (δ =0%)	293K	350	512	62.4	74.8
		77K	275	275	42.8	45.4
		4K	336	1432	39.6	35.6
	Weld (δ =10%)	293K	402	577	56.4	81.3
		77K	383	1371	41.6	54.7
		4K	470	1563	34.0	36.6
SUS316L	Base	293K	256	575	61.0	82.5
		77K	403	1300	54.9	71.8
		4K	561	1536	49.0	53.5
	Weld (δ =0%)	293K	410	531	50.2	62.5
		77K	494	1145	41.5	37.9
		4K	572	1344	41.5	36.7
	Weld (δ =10%)	293K	453	582	54.6	74.1
		77K	565	1229	49.8	57.6
		4K	678	1403	34.6	25.2

T :Temperature   YS:Yield strength   TS:Tensile strength

ε :Elongation ϕ :Reduction of area

been scarcely investigated [7].

In this study, we obtained high-cycle fatigue properties of 0% and 10% materials for SUS304L and SUS316L and discussed the effects of δ -ferrite and welding structure on the fatigue properties at cryogenic temperatures.

EXPERIMENTAL PROCEDURE

Material and Specimens

The materials used in this study were the commercial austenitic stainless steels SUS304L and SUS316L. The chemical compositions of base and weld metals are listed in Table 1. The weld metals were made by joining plates (28t mm x 200w mm x 1000L mm) with tungsten-inert-gas (TIG) welding. The contents of Cr and Ni of the filler metals were regulated using the Delong [8] diagram and the filler metals have 0% and 10% δ -ferrite. The chemical compositions of weld metals and amount of ferrite (measured by ferrite meter and not only δ -ferrite) are also listed in Table 1. The configuration of the joint groove was U-shaped and the weld was completed in 17 passes with a voltage of 8~10 V, a current of 120~210 A, a speed of 8 cm/min and interpass temperature below 423 K. The weld metals also satisfy the JIS 1 grade by non-destructive inspections (screening tests and radiant rays inspections).

The dimensions of high-cycle fatigue specimen are shown in Figure 1; a hourglass type with the minimum waist diameter of 6 mm. The specimens of the base metal were cut from C direction and those of the weld metal were cut in perpendicular to welding directions. The tensile properties of base and each weld metal are summarized in Table 2 [7].

High-Cycles Fatigue Test

The testing machine is a servohydraulic model with a dynamic load capacity of ±50kN. A sinusoidal cyclic load with a stress ratio R=0.01 (minimum load / maximum load) was applied. Fatigue tests were carried out in liquid helium (4 K), in liquid nitrogen (77 K) and in room temperature (293 K). In 4 K test, a recondensation type refrigerator [9] was operated to keep the liquid helium level constant during the test. To avoid specimen heating by cyclic loading, the tests at 4 K [10] were done with a testing frequency of 4 Hz in the range of cycles less than 10^4 and at 10 Hz over 10^4 cycles. The frequency at 77 and 293 K tests was 10 Hz. The maximum measurable fatigue life in the present work was restrict to around 10^6 cycles due to various limitations concerning long term testing.

Observation of Microstructure and Fracture Surface

The microstructure of base and weld metals were observed by optical microscopy. The fatigue fracture surfaces were also examined using scanning electron microscopy (SEM). X-ray energy dispersion spectroscopy (EDS) was utilized for the chemical analysis of microstructure.

RESULTS AND DISCUSSION

High-Cycle Fatigue Properties

The S-N curves of 0% and 10% materials for SUS304L and SUS316L at 293 , 77 and 4 K are shown in Figures. 2 and 3, respectively. The S-N curves of 0% and 10% materials shifted to higher stress level or longer life side with a decrease of test temperature. The fatigue properties of 0% material are higher than those of 10% material at all test temperatures, respectively. However, 10^6- cycles fatigue strength (FS) of SUS304L 0% material at 4 K seemed nearly equal to that 77 K.

The ratios of 10^6- cycles fatigue strength to tensile strength (TS) as a function of test temperature are shown in Figure 4. Empirically it is told that the fatigue strength is proportional to tensile strength even if at cryogenic temperatures. The authors had previously reported [7,11] that the ratios of FS to TS are 0.5 to 0.8, which are almost

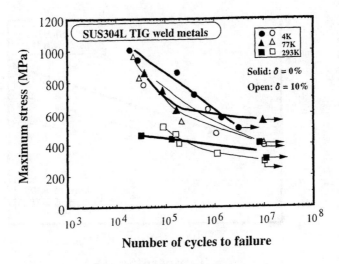

Figure 2. S-N curves of SUS304L weld metals at 293, 77 and 4 K.

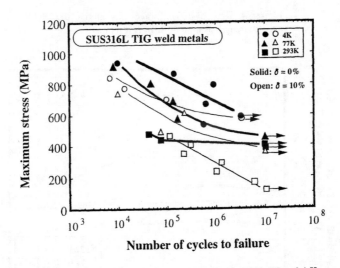

Figure 3. S-N curves of SUS316L weld metals at 293, 77 and 4 K.

constant and/or slightly increase with a decrease of test temperature. The ratios of FS to TS for SUS304L and SUS316L base metals were 0.5 to 0.8 at all test temperatures. However, the ratios of FS to TS of 0% material decreased from 0.8 to 0.45 and those of 10% material decreased between 0.35 to 0.65 with decreasing test temperature, as shown in Figure 4. At low temperatures, the ratios of FS to TS of 10% material were a little bit lower than those of 0% material.

Yield strength is an important measure for the selection of structural material. In Figure 5, the ratios of FS to yield strength (YS) for SUS304L each weld metal is higher than those of SUS316L. This is considered that SUS304L has low yield strength and due to strain-induced martensitic transformation.

Figure 6 represents the typical SEM micrographs of fracture surface of 0% material

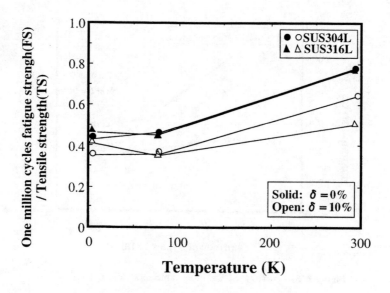

Figure 4. Ratio of one million cycles fatigue strength to tensile strength as a function of test temperature.

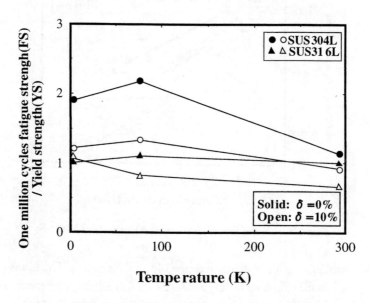

Figure 5. Ratio of one million cycles fatigue strength to yield strength as a function of test temperature.

for SUS304L and SUS316L. Fatigue crack initiation sites of 0% material initiated almost from the inside of the specimen and those were decreased the high-cycle fatigue properties. Through the EDS observation, the chemical compositions of the fatigue crack initiation sites of 0% material are almost the same as those of matrix, respectively. On the other hand, each 0% material contains no δ-ferrite and no microfissures [12,13] by optical microscopic observations, and these tensile specimens failed by a ductile fracture even at 4 K. Therefore, it is considered to be due to the interface of the solidification structure. As previously reported by the authors [12], all the fatigue crack initiation sites of 10% material were at the very small blowholes or some weak defects which could not be detected by non-destructive screening tests nor microscopic observations in advance. This means that undetectable microvoids by usual inspection procedures may act as stress concentration sites in fatigue tests and cause lower fatigue properties compared with the base metals, especially at low temperatures. On the other hand, fatigue crack initiation sites of 0% material were also almost at some weak defects. Although, high-cycle fatigue properties of 0% material at low temperatures were a little bit higher than those of 10% material, shown in Figure 5, the effects of δ-ferrite on high-cycle fatigue properties are not significant.

We accept that the reduction to a minimum of weld defects needs to be attained to further improve the fatigue properties of austenitic stainless steels at cryogenic temperatures. In general, fatigue strength increases with decreased test tempareture. Therefore, it is believed that fatigue strength at 20 K (in liquid hydrogen) would be higher than that at 77 K. However, fatigue strength in liquid hydrogen may be subject to deterioration due to hydrogen embrittlement. It is of crucial important to investigate the effect of hydrogen on δ-ferrite in the weld metal. Further studies are clearly needed to evaluate the behavior of austenitic stainless steel welds in a liquid hydrogen environment.

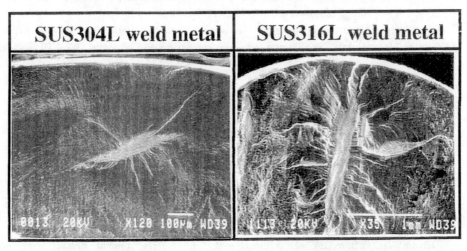

| SUS304L weld metal | SUS316L weld metal |

4 K =1002 MPa, δ = 0%   77 K = 916 MPa, δ = 0%

Figure 6. Typical scanning electron micrographs of fracture surface of high-cycle fatigue specimen of 0% material for SUS304L and SUS316L tested at low temperatures.

SUMMARY

High cycle fatigue properties of 0% (δ=0%) and 10% (δ=10%) materials for SUS304L and SUS316L and the effects of δ-ferrite and welding structure on the fatigue properties have been investigated at cryogenic temperatures.

1) The S-N curves of those weld metals shifted to higher stress levels, i.e., the longer life side, at decreased test temperature.

2) The ratios of 10^{6}-cycles fatigue strength (FS) to tensile strength (TS) of 0% material decreased from 0.8 to 0.45 and those of 10% material decreased between 0.35 to 0.65 at decreased test temperatures.

3) Fatigue crack initiation sites of SUS304L 10% material were almost at blowholes, and those of SUS316L 10% material were at weld pass interface boundaries created during welding. On the other hand, those of 0% materials were considered to be due to the interface of the solidification structure.

4) Microvoids, which are Undetectable by conventional inspection procedures, cause reduced fatigue resisitance.

5) Although δ-ferrite reduces toughness at cryogenic temperatures in austenitic stainless steel weld metals, the effects of δ-ferrite on high-cycle fatigue properties are not significant.

REFERENCES

1. Horiya,T., Yamamoto,N., Iida,T., Yamamoto,A., Okaguchi,S., Yaegashi,N., Doko,T., Saito,M., Yokogawa,K. and Ogata,T., Proceedings of 16th International Cryogenic Conference/International Cryogenic Materials Conference Part3,(1997), pp.1915-1918

2. New Energy and Industrial Technology Development Organization, 1997 Annual Summary Report on Results " International Clean Energy Network Using Hydrogen Conversion (WE-NET)" , March (1998)

3. Suzuki,K., Fukakura,J. and Kashiwaya,H., Journal of the Japan Society of Materials Science, 36, (1987), pp.1090-1096, in Japanese

4. Handbook on Materials for Superconducting Machinery, Battele Columbas Lab., Ohio ,(1997)

5. Okaguchi,S., Horiya,T., Ishige,K., Saito,M., Yaegashi,N., Yamamoto,A., Nakagawa,H., Yokogawa,K. and Ogata,T., Proceedings of Sixteenth International Cryogenic Conference/ International Cryogenic Materials Conference Part3,(1997), pp.1923-1926

6. Yushuchenko,K.A., Savchenko,V.S., Solokha,A.M. and Voronin,S.A., " Effect of Delta-Ferrite on the Properties of Welds in Austenitic Steels at Cryogenic Temperatures," in *Advances in Cryogenic Engineering (Materials)* 40,edited by R.P.Reed et al., Plenum Press,New York, (1994), pp.1263-1266

7. Delong,W.T., Ostrom,G.A. and Szumachowski,E.R., Welding Journal, Reseach Supplement, 35,(1956), pp. 521-528

8. Yuri,T., Ogata,T., Saito,M. and Hirayama, Y., Cryogenics 40,(2000), pp.251-259

9. Yuri,T., Nagai.K., Ogata,T., Umezawa,O. and Ishikawa,K., Cryogenic Engineering; Vol.26, (1991), pp.184-189, in Japanese

10. Ogata,T., Ishikawa,. K, Nagai,K., Yuri,T. and Umezawa,O., Cryogenic Engineering; Vol.26, (1991), pp.190-196,in Japanese

11. Nagai,K., Yuri,T., Umezawa,O., Ogata,T. and Ishikawa,K.,Cryogenic Engineering; Vol.26, (1991), pp.255-262, in Japanese

12. Hull, F.C., Welding Journal, Reseach Supplement, 46,(1967),pp.399-409

13. Lundin, C.D. and Spond,D.F., Welding Journal, Reseach Supplement, 56, (1976), pp.356-367

HALL CURRENTS IN INHOMOGENEOUS MAGNETIC FIELDS

J.H.Parker, Jr and O.R.Christianson

WEMD Technology Center
Mount Pleasant, PA 15666

ABSTRACT

A theoretical treatment of Hall currents in inhomogeneous magnetic fields is given, and both analytic and numerical procedures to calculate the effects are described. The conductor geometry studied is a long rectangular strip conductor with axial currents, and is subjected to a spatially varying magnetic field aligned perpendicular to the strip surface. The current density is described in terms of a T vector perpendicular to the strip surface. Current density components are derived from spatial derivatives of the T vector that depends upon the resistivity of the conductor and the Hall coefficient. Hall currents produce non-uniform current distributions due to the inhomogeneous magnetic fields that can create a significant increase in resistance in excess over that produced by magneto-resistance alone. Results are presented for aluminum at cryogenic temperatures.

INTRODUCTION

Transverse Hall currents can produce apparent anomalous effects in cryogenic conductors. An apparent anomalous magneto-resistance effect was observed at cryogenic temperatures for a square normal metal conductor made of a high purity aluminum core surrounded by a high resistance Al-Fe-Ce alloy jacket.[1] Resistance measurements at zero magnetic field were consistent with the expected resistance of the high purity core shunted by the high resistance jacket. Magneto-resistance measurements on this conductor gave results that deviated from predictions of this simple parallel circuit model. This anomalous behavior was due to transverse Hall currents in the composite which resulted in increased I^2R losses and a higher effective resistance for the composite.[2,3]

Recent studies have observed this anomalous magneto-resistive effect in Al-Cu composite conductors.[4,5,6,7] This anomaly was analyzed on the basis of transverse Hall currents using a conformal mapping model for a round conductor.

CP614, *Advances in Cryogenic Engineering:*
Proceedings of the International Cryogenic Materials Conference - ICMC, Vol. 48,
edited by B. Balachandran et al.
© 2002 American Institute of Physics 0-7354-0060-1/02/$19.00

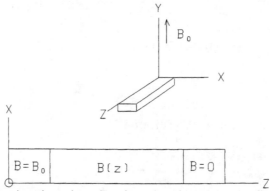

FIGURE 1. Axis orientation and top view of sample with starting and ending field sections.

The effect Hall currents produce in conductor joints at cryogenic conductors has also been analyzed.[8] Results suggest an increase in the joint resistance.

The purpose of this note is to carryout calculations for currents flowing in a long rectangular strip conductor subjected to a spatially varying magnetic field aligned perpendicular to strip surface. The geometry is illustrated in FIGURE 1 showing the strip conductor oriented relative to a xyz coordinate system. The dc magnetic field is along the y-axis and the long length of the conductor is parallel to z axis.

THEORY

The magnetic field is oriented in the y direction and varies along the z direction. The resistivity is a function of magnetic field which, since it varies along the z direction, results in a variation along the z direction. The current density consists of an x and a z component.

$$\vec{B} = B_0(z)\hat{j} \tag{1}$$

$$\rho(B_0(z)) = \rho(z) \tag{2}$$

$$\vec{\rho}_h = R_h B_0 \hat{j} \tag{3}$$

$$\vec{j} = j_x\,\hat{i} + j_z\,\hat{k} \tag{4}$$

Taking the curl of the electric field and the divergence of the current density gives

$$\vec{\nabla} \times \vec{E} = 0 \tag{5}$$

$$\nabla \cdot j = 0 \tag{6}$$

$$\vec{E} = \rho\,\vec{j} + \vec{\rho}_h \times \vec{j} \tag{7}$$

$$\vec{\nabla} \times \vec{E} = \nabla \times (\rho\,\vec{j} + \vec{\rho}_h \times \vec{j}) \tag{8}$$

$$\vec{\nabla} \times (\rho \vec{j}) = \vec{\nabla} \rho \times \vec{j} + \rho \vec{\nabla} \times \vec{j} = \frac{\partial \rho}{\partial z} j_x \hat{j} + \rho \left(\frac{\partial j_x}{\partial z} - \frac{\partial j_z}{\partial x} \right) \hat{j} \tag{9}$$

$$\vec{\nabla} \times (\rho_h \times \vec{j}) = (\vec{j} \cdot \vec{\nabla}) \vec{\rho}_h - (\vec{\rho}_h \cdot \vec{\nabla}) \vec{j} + \vec{\rho}_h \nabla \cdot \vec{j} - \vec{j} \nabla \cdot \vec{\rho}_h = j_z \frac{\partial \rho_h}{\partial z} \hat{j} \tag{10}$$

$$(\vec{\nabla} \times \vec{E})_y = \frac{\partial \rho}{\partial z} j_x + \frac{\partial \rho_h}{\partial z} j_z + \rho \left(\frac{\partial j_x}{\partial z} - \frac{\partial j_z}{\partial x} \right) \rho\rho \tag{11}$$

Now write the current density in terms of a "T" vector along the y axis,

$$\vec{j} = \vec{\nabla} \times (T_y \hat{j}) \tag{12}$$

or

$$j_x = -\frac{\partial T_y}{\partial z} \tag{13}$$

and

$$j_z = \frac{\partial T_y}{\partial x} \tag{14}$$

Using these expressions for the current density components in terms of the "T" vector yields the governing equation for T_y

$$\rho \, \nabla^2 T_y + \frac{\partial \rho}{\partial z} \frac{\partial T_y}{\partial z} - \frac{\partial \rho_h}{\partial z} \frac{\partial T_y}{\partial x} = 0 \tag{15}$$

Note that the spatial derivatives of ρ and ρ_h represent, respectively, the effect of the spatial variation of the magneto-resistance and magnetic field with z. The lower figure in FIGURE 1, shows the conductor made up of three sections. The first, starting at z=0 and extending to z=S, is a lead in section with constant magnetic field, B_0. The middle section, starting at z=S and ending at z=L, is the region of variable magnetic field. The last section, starting at z=L and ending at z=E, is the ending section with either B=0 or some constant B that is less than B_0. The boundary conditions on T for a rectangular strip are, $\frac{\partial T}{\partial z} = 0$ at each end of the strip, with the z axis being parallel to long axis of the strip. Along the length of the strip, J_z is zero on each side boundary (x axis direction). This means that $\frac{\partial T}{\partial z} = 0$ and therefore T is a constant along these two boundaries.

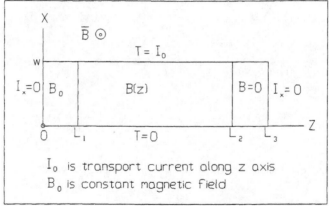

FIGURE 2. Top view of sample showing the T vector boundary conditions. Note: $I_x = -\dfrac{\partial T}{\partial z}$.

SPATIALLY LINEAR FIELD WITHOUT MAGNETO-RESISTANCE

In this case, the resistivity is independent of z and the equation for T is,

$$\rho \nabla^2 T - \alpha \frac{\partial T}{\partial x} = 0 \tag{16}$$

and $\alpha = R_h B'$, with B' the derivative of B with respect to z, a constant. If the distance between the starting and ending transition regions of FIGURE 2, is sufficiently long for linear B and no magneto-resistance, T becomes independent of z with $J_x=0$ in this middle region. The solution of the T equation is,

$$T(x) = I_0 (\exp(\alpha x) - 1) / (\exp(\alpha w) - 1) \tag{17}$$

where w is the width of the sample. Using this expression, the ratio of the resistance per unit length with the Hall effect to the resistance without is,

$$Ratio = w \alpha^2 (\exp(2\alpha w) - 1) / (\exp(\alpha w) - 1)^2 \tag{18}$$

Also, the voltage across the sample, V_x is

$$V_x = R_h B(z) I_0 \tag{19}$$

FIGURE 3 is the loss ratio as a function of the β Parameter for the middle steady flow region. FIGURE 4 shows a surface plot the J_z Ratio as a function of the β Parameter and x/w.

FINITE DIFFERENCE SOLUTIONS FOR THE GENERAL CASE

The full equation for $T(x,z)$ must be solved when $B(z)$ is non-linear and the magneto-resistance is taken into account. The method of finite difference can be used to solve this more complex equation. The solution is obtained using the method of over relaxation. The

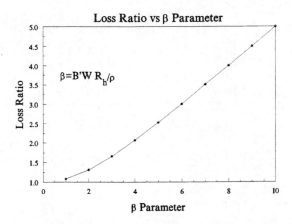

FIGURE 3. Ratio of loss with Hall to loss without Hall as function of the β parameter, $B'wR_h/\rho$.

conductor is represented in width by 20 nodes and axially the first region consists of 20 nodes, the middle region consists of 60 nodes, and the final region consists of 20 nodes. For Al, it is convenient to use the analytic form of Corruccini correlation for the magneto-resistance.[9] FIGURES 5 and 6 show the results for the J_z flow pattern for a B_0=4 T, w=1 cm and without magneto-resistance in FIGURE 5 and with magneto-resistance in FIGURE 6. FIGURE 7 shows a comparison between J_z vs x/w from the analytic formula for linear field dependence with no magneto-resistance to that obtained by the finite differences method. It is evident that except at x=0 and x=w the results are in excellent agreement. (The difference at the two sides of the sample is due to the difference in the numerical derivatives for the sides versus the middle region.) The parameters applicable to the results shown in FIGURES 5, 6 and 7 are: B_0=4 T, w=1 cm, length of middle section=6 cm, starting section=1 cm, ending section=1 cm, ρ=1x10^{-10} Ωm, and R_h=1x10^{-10} Ωm/T.

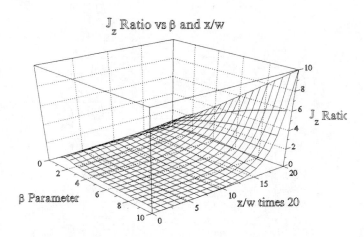

FIGURE 4. Ratio of J_z as function of x/w and the β Parameter, $B'wR_h/\rho$.

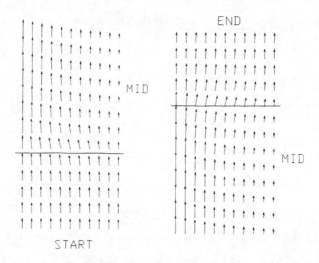

FIGURE 5. J_z current flow pattern for linear field dependence on z and without magneto-resistance.

SUMMARY

A rectangular strip of aluminum or copper at cryogenic temperatures exhibits an increased resistivity due to transverse Hall currents produced by an inhomogeneous magnetic field. The Hall currents produce a redistribution of currents across the conductor.

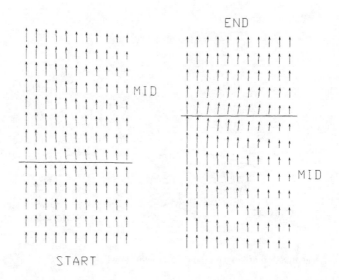

FIGURE 6. J_z current flow pattern for linear field dependence on z with magneto-resistance.

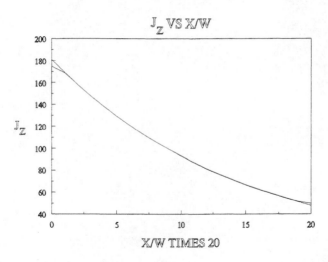

FIGURE 7. Comparison of J_z from analytic formula with finite difference result for J_z as function of x/w in the middle steady flow region with no magneto-resistance.

REFERENCES

1. Eckels, P.W., Iver, N.C., Paterson, A., and Male, A.T., "Magnetoresistance: the Hall effect in composite aluminium cryoconductors," *Cryogenics* **29(7)** p. 748 (1989).
2. Eckels, P.W. and Parker, J.H. Jr., "Magnetoresistance in composite conductors," *Advances in Cryogenic Engineering* **36**, 1990, pp. 655-662.
3. Krefta, M.P., Christianson, O.R., and Parker, J.H., Jr., "Steady-state Hall losses in composite cryogenic conductors," *Cryogenics* **36** pp. 291-301 (1996).
4. Kaneko, H. and Yangai, N., "Enhancement of magnetoresistance due to Hall current in Aluminum-copper composites," *Cryogenics* **32** p. 1114 (1992).
5. Kaneko, H., "Hall currents in aluminum-copper composites," *Advances in Cryogenic Engineering (Materials)* **40A**, 1994, pp. 451-458.
6. Yanagi, N., Mito, T., Takahata, K., and Sakamoto, M., "Experimental observation of anomalous magnetoresistivity in 10-20 kA class aluminum stabilized superconductors for the Large Helical Device," *Advances in Cryogenic Engineering* **40A**, 1994, pp. 459-468.
7. Imagawa, S., Yanagi, N., Mito, T., and Satow, T., "Analysis of anomalous resistivity in an aluminum stabilized superconductor for the Large Helical Device, "*Advances in Cryogenic Engineering* **40A**, 1994, pp. 469-477.
8. Krefta, M.P. and Parker, J.H., Jr., "The Intrinsic Magneto-Resistance of Cryogenic Joints," *IEEE Trans. on Applied Superconductivity*, 1997, p. 820.
9. Corruccini, R.J., "The Electrical Properties of Aluminum for Cryogenic Electromagnets," **NBS TN 218** (1964).

STRUCTURAL MATERIALS TESTING

INTERNATIONAL ROUND ROBIN TEST RESULTS OF J-EVALUATION ON TENSILE TEST AT CRYOGENIC TEMPERATURE WITHIN THE FRAME WORK OF VAMAS ACTIVITY

A. Nishimura[1], A. Nyilas[2], T. Ogata[3], K. Shibata[4], J. W. Chan[5], R. P. Walsh[6], and H. Mitterbacher[7]

[1]National Institute for Fusion Science
Toki, Gifu, 509-5292, Japan
[2]Forschungszentrum Karlsruhe, ITP
D-76021, Karlsruhe, Germany
[3]National Institute for Materials Science
Tsukuba, Ibaraki, 305-0047, Japan
[4]University of Tokyo
Bunkyoku, Tokyo, 113-8655 Japan
[5]Lawrence Berkeley National Laboratory
Berkeley, California 94720 USA
[6]National High Magnetic Field Laboratory
Tallahassee, Florida 32306 USA
[7]Linde AG
D-82049 Hoellriegelskreuth, Germany

ABSTRACT

A new J-evaluation method named JETT has been investigated through international round robin tests within the frame work of VAMAS activity. On the first stage, SUS316LN was tested as a common test material and it was found that JETT has potential to clarify the local fracture toughness. During last two years, 9% Ni steel was used and the effect of serrations on the toughness was investigated. As the results, the serration produces additional blunting at the notch tip resulting in delaying the crack initiation and giving higher fracture toughness. Nevertheless, it is recognized that JETT is a very conventional method to evaluate a relative toughness at the local area of the material.

INTRODUCTION

The evaluation method of the fracture toughness has been developed in 1970s to 80s and ASTM standard for R-curve determination was established at 1981 [1]. Based on this

CP614, *Advances in Cryogenic Engineering:*
Proceedings of the International Cryogenic Materials Conference - ICMC, Vol. 48,
edited by B. Balachandran et al.
© 2002 American Institute of Physics 0-7354-0060-1/02/$19.00

standard, the elastic-plastic fracture toughness measurement was developed and ASTM E813 was constituted [2]. The standard covered a stable crack growth, and since then the standard was referred in various fields. In the cryogenic structural materials field, the method described in ASTM E813 has been referred and applied to even unstable crack growth. In addition, a need for measuring the local fracture toughness arose and a certain trial was performed using a small specimen with an electro-discharge machining notch [3].

The J-calculation of the circumferentially notched round bar was proposed by J. R. Rice et al. as a conventional equation [4], and some studies on the fracture toughness tests near room temperature were published [5]. These papers showed that the crack opening displacement at stable crack initiation is almost constant in the range of 150 K to 310 K. The same concept was brought in the cryogenic field and the new method was named J-evaluation on tensile test (JETT) [6,7]. The potential to evaluate the fracture toughness was described in these papers, and one of the authors has applied this method to the local toughness evaluation of the large size conduit and the relative difference depending on the location of the sampling was investigated successfully [8].

In 1997, JETT was taken as a candidate measurement method for an international standard and the research on the characterization started. On the first stage, 1997 and 1998, SUS316LN was used as a common test material [9]. To determine the valid J_{IC}, 30 mm thick plate was prepared and the compact tension specimens (1T CT) were machined out. The test was performed in liquid helium according to ASTM E813-89. After the tests, large de-lamination or a large crevasse-like crack was observed on the fracture surface and it was recognized that the valid J_{IC} could not be obtained. The results of JETT, however, showed that the midsection of the plate had smaller toughness than near surfaces. Unexpectedly, the series of tests indicated that JETT could clarify the relative fracture toughness locally. Then the second stage was planed to clear the relation between J_{IC} and J_F which is a final J-integral on JETT. On the selection of the common material for the second stage, a lower toughness material was discussed, because the plasticity would make additional difficulties, and 9% Ni steel was taken for the test material. The 9% Ni steel is usually used for the LNG tank and never used in liquid helium environment.

In this paper, the international round robin tests results are presented and the relation between J_{IC} and J_F is discussed. Moreover, the effect of the serration on the toughness measurement is clarified.

MATERIAL AND TEST PROCEDURE

The thickness of the 9% Ni steel was 37 mm and the chemical composition of the steel in wt % was Fe-0.066C-0.22Si-0.64Mn-0.002P-0.001S-0.01Cu-0.01Cr-9.05Ni-<0.01Mo-0.001Nb-0.045Al-0.0002B. The tensile tests were performed with a round bar of 6.2 mm diameter and heavy serrations with a large load-drop were observed. Yield stress, tensile strength, elongation and reduction of area of the steel in liquid helium is 1311 MPa, 1335 MPa, 23.6 % and 62.5 %, respectively. Even in liquid helium, the steel showed necking and dimple fracture.

To obtain the critical fracture toughness according to ASTM E813-89, CT specimens were prepared together with JETT specimens. Location of the CT, JETT and tensile specimens in the plate is shown in Figure 1. Direction of 1.5 m length was the rolling direction. The CT specimens were machined out of the center of the plate thickness and the crack plane positioned to propagate along the rolling direction. Fatigue pre-crack was induced to the CT specimens at room temperature. The CT specimen code and the test result are summarized in Table 1. The fracture toughness tests were carried out in liquid

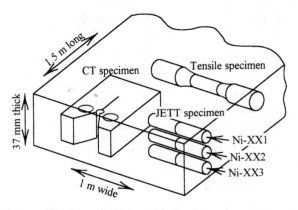

FIGURE 1. Location of CT, JETT and tensile specimens in the 9% Ni steel plate.

helium according to ASTM E399 (K test) and E813 (J test). Three JETT specimens were taken out of the plate. The longitudinal direction of the specimen was a plate width direction and the specimens were designated as NiXX1, NiXX2 and NiXX3 from the side of right surface. (see Fig. 1)

The JETT specimen is a round bar with a circumferential notch as shown in Figure 2. The shaft diameter was 6 mm and the notch of 0.1 mm width was induced by an electro-discharge machining (EDM). The net section diameter and the notch width were measured with a reading microscope of 20 times magnification. Table 2 lists the JETT specimen code and the test results. When the JETT was performed in liquid helium, two or three extensometers were attached to the specimen to cancel the bending component. Laboratories 1 and 5 used three extensometers and other laboratories used two gages. The type of the extensometer, the way of attachment and the gage length was different among the affiliations.

J - CALCULATION

For the JETT specimen, a conventional calculation formula presented by J. R. Rice et al. was applied [4]. The J-integral can be obtained by a following equation:

$$J = \frac{1}{2\pi r^2} \left[3 \int_0^{d_c} P \, d(d_c) - P \, d_c \right], \tag{1}$$

where r is a radius of a circular neck, P is a tension force and d_c is a displacement between the load points due to introducing the crack (or notch). In this study, d_c was calculated by a following equation:

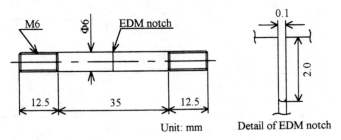

Unit: mm   Detail of EDM notch

FIGURE 2. Standard specimen for the international round robin test on JETT.

TABLE 1. Fracture Toughness Test Results with CT Specimen at 4.2K.

Specimen code	J_Q (N/mm)	K_{IC} (MPa-m$^{1/2}$)	Note	Affiliation
Ni201	16.6	(58.9)	J test	Lab. 2
LBL2	15.9	(58.8)	J test	Lab. 3
FZK1	(16.9)	60.9	K test	Lab. 4
FZK2	(17.3)	61.6	K test	Lab. 4
Ni205	13.8	(55.1)	J test	Lab. 5
Ni206	13.0	(53.5)	J test	Lab. 5

Note: Figure in parentheses is derived from $K^2 = E \times J / (1-v^2)$.

$$d_c = d_{ex} - (G.L.) \cdot P/(E\pi R^2) , \tag{2}$$

where d_{ex} is a measured displacement by two or three extensometers spanning the notch, G.L. and R are the gage length and radius of the round bar specimen, respectively.

J calculation of the CT specimen was conducted according to ASTM E813-89.

RESULTS AND DISCUSSION

Fracture Toughness of 9% Ni Steel

To measure the fracture toughness of the 9% Ni steel, J test was carried out according to ASTM E813. The steel was rather brittle at 4.2 K and showed serration or small jump during the test. When CT specimen was loaded gradually, the load went up and suddenly dropped and COD increased. Any plasticity indicating strain hardening was not observed in the load-COD diagram. The crack length measured by a compliance method increased after the load-drop happened, and it was clear that the crack extension was not stable. At Laboratory 4, the first serration led the specimen to the final fracture. It is well known that the load-drop at the serration depends on the stiffness of the mechanical testing machine and test apparatus. Since the load-load line displacement was recorded continuously during the test, P_Q was measured on the load-COD curve and K_{IC} was obtained.

One example of the R-curve is shown in Figure 3 and the results are shown in Table 1. Since the crack front was not investigated, it was not clear if those values satisfied the

TABLE 2. Summary of JETT Specimens of 9% Ni Steel.

Specimen code	Notch width (mm)	Notch depth (mm)	Gage length (mm)	J_F (N/mm)	J_M (N/mm)	Affiliation
Ni021	0.750	1.999	25.6	43.9	14.1	Lab. 1
Ni022	0.086	2.061	25.6	36.2	12.7	Lab. 1
Ni031	0.060	2.060	15.0	37.7	21.4	Lab. 2
Ni032	0.067	1.974	15.0	40.2	21.6	Lab. 2
Ni033	0.069	2.004	15.0	48.3	23.1	Lab. 2
Ni041	0.056	1.986	11.324	41.6	16.7	Lab. 3
Ni042	0.088	2.040	11.324	88.0	21.4	Lab. 3
Ni051	0.075	1.953	8.0	45.1	18.6	Lab. 4
Ni052	0.075	2.064	8.0	57.5	18.6	Lab. 4
Ni053	0.068	1.974	8.0	29.9	17.3	Lab. 4
Ni061	0.107	2.047	25.4	119.3	20.1	Lab. 5
Ni062	0.097	1.990	25.4	62.2	28.7	Lab. 5
Ni081	0.070	2.019	14.5	36.8	19.8	Lab. 6

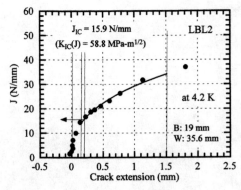

FIGURE 3. R-curve of 9% Ni steel at 4.2 K (Specimen code; LBL). J_{IC} is small but it is clearly determined by ASTM E813.

requirements described in the standard. However, K_{IC} satisfied the K test standard and the converted values from J_Q are almost the same as K_{IC}. Therefore, the critical fracture toughness (J_{IC}) is supposed to be around 14 to 17 N/mm and is very low in comparing with SUS316 and other cryogenic structural materials.

JETT Test Results

One data set of Ni041 is shown in Figure 4 (a) for an example. In the figure, the horizontal axis is a displacement due to introducing the crack/notch, which is d_c in Equation 2. The displacement does not include an elastic component. When the load applies, the notch opens and the serration occurs. Since the serration makes a jump on the deformation, the J also jumps and gets higher values. In this case, the final J is 41.6 N/mm and is rather higher than J_{IC} obtained with CT specimens shown in Table 1.

The serration is caused with adiabatic deformation [10]. The slip band would occur very quickly and the strain energy converts to heat. Since the specific heat is very small at cryogenic temperature, a large temperature rise would easily happen with a small slipping. The temperature rise makes the material soften and the material deforms until the load drops to the balance point where the material resistance becomes equal to the outer load. The deformation at notch tip during the temperature rising is not the mechanical

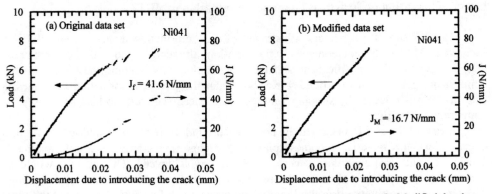

FIGURE 4. Load – displacement (d_c) curves of Ni041 specimen. (a) Original data. (b) Modified data by subtracting the deformation by serration. J_M is almost equal to or little larger than J_{IC} obtained by CT.

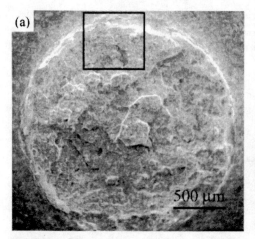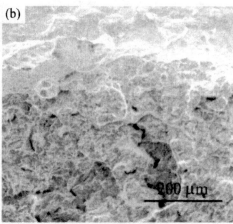

FIGURE 5. Fracture surface of Ni041. (a) Total image of the surface. (b) Enlarged image of the box in (a). No cup & cone type fracture is observed. Dimple fracture is dominant.

deformation induced by the testing machine. Therefore, the amount of the deformation by the serration was subtracted from d_c and J was calculated again. The modified results are shown in Figure 4 (b) and the final J is designated as J_M. Naturally, J_M becomes smaller depending on the magnitude of the serration.

The modified results are summarized in Table 2. The big scatter in a range of 29.9 to 119.3 N/mm reduced to a small band of 12.7 to 28.7. In comparison with J_{IC} of around 14 to 17, the modified scatter is in almost the same range or a little bit larger. The stiffness of the testing machine is different. Nevertheless, the remarkable difference among the test affiliations and the dependence of sampling location is not observed in Table 2.

From this result, it is noticed that the modification by subtracting the deformation with serration is useful.

The fracture surface of the specimen of Ni041 is shown in Figure 5. (a) shows the ground image of the fracture surface. It is relatively flat and not cup & cone type fracture. The enlarged image of the box region in (a) is shown in (b). The edge part of the fracture surface is also covered with dimples and it is hard to find shear fracture. Therefore, it is considered that the crack initiated at the notch tip and the final fracture happened at the same time.

Effect of Serration on Fracture Toughness Evaluation by JETT

Figure 6 shows a schematic illustration of the deformation at net section of JETT specimen. Most of the materials show the serration, although Incoloy 908 does not show [7]. In general, the specific heat and the heat conductivity become very small as the temperature goes down to around 4 K. So, when the plastic deformation continues, a certain strain is released by slipping, and the strain energy converts to heat and raises the temperature strongly.

In case of the JETT specimen, the heat stays around the net section and would warm it over. Since the material becomes tougher at higher temperature, the deformation is accelerated and continues until a certain balance has achieved. This balance would be performed when the load drops to the resistance force generated by the work hardening. The net section is elongated in the longitudinal direction and necking happens at the same time resulting in loosing the strain restraint. When the material has enough ability to endure the deformation and necking during the serration, the temperature recovers and the

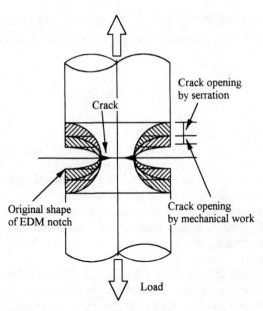

FIGURE 6. Schematic illustration of deformation on circumferential notch tip.

mechanical deformation continues. On the other hand, when the material is in critical situation, the serration causes the final fracture and the specimen becomes broken.

The fracture toughness is defined as the toughness at the final fracture point in the JETT. So, when the crack initiated at the notch tip or voids are formed at the center of the net section, the final fracture would happen. At that time, the temperature rises and the mechanical property is not the same as at 4 K anymore in both cases. The similar situation is considered in the CT specimen when the discontinuous crack growth happens. In this case, the crack extension would be affected by the temperature rise with the serration. Strictly saying, it is very difficult to obtain the actual toughness at 4 K.

When the crack initiated at the notch tip of JETT specimen, it gives the real toughness, in the engineering sense. However, since the notch tip blunting is accelerated by the serration, the crack initiation itself becomes harder to occur because of the loosen strain restraint. It results in giving higher fracture toughness and the Table 2 supports this consideration. When the cup & cone fracture occurs as in SUS316LN, the modification

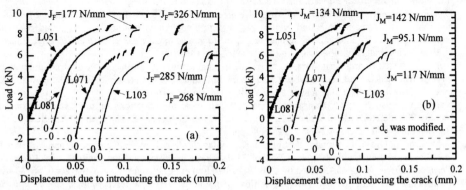

FIGURE 7. Results of 316LN. (a) Original data set. (b) Modified data by subtracting the deformation during serration.

121

gives small scatter as shown in Figure 7. In this case, it is not clear whether the J_M has a relation with J_{IC}. But it must be noticed that the relative comparison is useful for material screening.

CONCLUSION

The international round robin test on JETT has been carried out during last five years using 316LN and 9% Ni steel. This paper summarizes the activity performed by several institutes and laboratories. The main results are as follows:
(1) JETT is a very convenient test method, which is a tensile test with a circumferential notched bar, and gives relative fracture toughness in general.
(2) When the crack initiates at the notch tip, the real toughness is obtained. On the other hand, when the voids leads to a final fracture and cup & cone type fracture appears, the final J will loose the meaning of toughness and become a relative engineering parameter which is useful for material screening.
(3) The serration produces the additional blunting resulting in releasing the strain restraint at the notch tip and delaying the crack initiation. By subtracting the deformation with the serration, small scatter and real-like fracture toughness is obtained.

ACKNOWLEDGEMENTS

The authors greatly appreciate the collaboration of the participant institutes and their key persons who paid much effort for this program. This work has been performed within the framework of the VAMAS TWA 17 activity and supported by the Special Coordination of National Institute for Materials Science in Japan.

REFERENCES

1. ASTM E561-81. Standard practice for R-curve determination, *Annual book of ASTM standards*, ASTM, Philadelphia (1981).
2. ASTM E813-89. Standard test method for JIC, a measure of fracture toughness, *Annual book of ASTM standards*, ASTM, Philadelphia (1991).
3. Nyilas, A. and Obst, B., "A new test method for characterizing low temperature structural materials," in *Advances in Cryogenic Engineering*, 42A, edited by L. T. Summers, Plenum, New York, 1996, pp. 353-360.
4. Rice, J. R., Paris, P.C., and Merkle, J. G., "Some further results of J-integral analysis and estimates," in *ASTM Special Technical Publications* 536, (1973) pp. 231-245.
5. Ohtsuka, A., *Journal of the Society of Naval Architects of Japan*, 135, 307 (1974) (in Japanese).
6. Nishimura, A., Yamamoto, J. and Nyilas, A., "Fracture toughness evaluation of a round bar with a circumferential notch at cryogenic temperature", in *Advances in Cryogenic Engineering* 44A, edited by U. B. Balachandran et al., Plenum, New York, 1998, pp. 145-152.
7. Nyilas, A., Obst, B. and Nishimura, A., "Fracture mechanics investigations at 7 K of structural materials with EDM notched round and double edged - bars," in *Advances in Cryogenic Engineering* 44A, edited by U. B. Balachandran et al., Plenum, New York, 1998, pp. 153-160.
8. Nyilas, A., Obst, B. and D.R. Harries, D. R., "Mechanical investigations on aged stainless steel and Incoloy 908 materials at 4.2 K and 7 K," " in *Advances in Cryogenic Engineering* 44A, edited by U. B. Balachandran et al., Plenum, New York, 1998, pp. 17-24.
9. Nishimura, A., Ogata, T., Shindo, Y., Shibata, K., Nyilas, A., Walsh, R. P., Chan, J. W. and Mitterbacher, H., "Local fracture toughness evaluation of 316LN plate at cryogenic temperature," in *Advances in Cryogenic Engineering* 46A, edited by U. B. Balachandran et al., Plenum, New York, 2000, pp. 33-40.
10. Basinsky, Z. S., "The instability of plastic flow of metals at very low temperatures," *Proc. Roy. Soc.*, 240A:229 (1957)

EUROPEAN CRYOGENIC MATERIAL TESTING PROGRAM FOR ITER COILS AND INTERCOIL STRUCTURES

A. Nyilas,[1] A. Portone,[2] and H. Kiesel[1]

[1]Forschungszentrum Karlsruhe, Institut für
Technische Physik, Karlsruhe, Germany
[2] EFDA, Max Planck Institute for Plasma Physics, Garching,
Germany

ABSTRACT

The following materials were characterized for the use in the magnet structures of ITER: 1) Type 316LN cast materials having a modified chemistry used for a Model of the TF (Toroidal Field) outer intercoil structure were investigated with respect to tensile, fracture, fatigue crack growth rate (FCGR), and fatigue life behavior between 7 and 4 K. 2) For Type 316LN 80 mm thick plate used for the TFMC (Toroidal Field Model Coil) structure a complete cryogenic mechanical materials characterization was established. 3) For full size coil case mockups, repair weld properties of 240 mm thick narrow-gap welds were investigated to determine their tensile and fracture behavior. 4) For CSMC (Central Solenoid Model Coil) superconductor jackets, the fatigue lives of orbital butt welds made of Incoloy 908 and Type 316LN (aged and unaged) materials were determined up to one million cycles at 7 K. The results reveal to date that the FCGR of aged Type 316LN is inferior to Incoloy 908 material, whilst the fatigue life properties are comparable. However, for Type 316LN jacket structure considerable improvement of FCGR could be achieved by a solution heat treatment process. In addition, tensile and fatigue life tests performed with a new cryogenic mechanical test facility (630 kN capacity) are presented.

INTRODUCTION

This paper reports about the characterization of structural materials for the ITER (International Thermonuclear Experimental Reactor) superconducting coils. The main goal was to supply a database for the safe design of the superconducting Toroidal Field (TF) and

CP614, *Advances in Cryogenic Engineering:*
Proceedings of the International Cryogenic Materials Conference - ICMC, Vol. 48,
edited by B. Balachandran et al.
© 2002 American Institute of Physics 0-7354-0060-1/02/$19.00

Central Solenoid (CS) coils. For the TF coils two full scale mockups, called Model 1 and Model 2, representing a portion of the inner leg and part of the outer intercoil structure were produced. Model 1 is a forged [1] heavy part (Type 316LN) of 40 tons, which after forging was pierced and cut into two C-halves to insert the superconducting TF winding pack. The tensile and fracture data for the base metal were already published [1] and the present paper gives the FCGR results obtained at 7 K. The heavy weld closure (~240 mm deep) of the two halves could be successfully produced by narrow-gap welding and a set of repair welds were tested to determine the tensile and fracture properties. For Model 2 a 25 ton heavy piece of modified Type 316LN casting (higher manganese) was produced and representative samples in all three spatial orientation were tested. These data were compared with a one ton trial cast sample. Here, beside the static mechanical properties, also fatigue life and FCGR measurements were carried out to complete the database.

For the CS coil the fatigue properties of the superconducting cable jackets were extensively tested for both candidate materials, Incoloy 908 and Type 316LN. The results show so far [1] that the static properties of both alloys are comparable but the weld metal mechanical properties are worthwhile to be investigated further. Therefore, tensile measurements were performed using small 3 mm gauge length extensometers to establish weld metals database. Concerning cyclic behavior the Type 316LN (base metal) showed in the past a lower resistance against fatigue crack growth than Incoloy 908. Therefore, cyclic tests of the base and weld metals of the jackets were extensively carried out. Furthermore, to avoid weld metal cracking during the jacket compaction process, a low Nb content filler wire was used for the orbital butt welds of the extruded tubes made of Incoloy 908 material. Also a new series of weld samples were produced for the Type 316LN extruded jacket material. The FCGR data of the Type 316LN jacket were compared with several commercial 316LN base metals already used in cryogenic applications. These results show the possibility to improve the FCGR behavior of the extruded 316LN jacket in principal. Other than FCGR measurements, where the crack starts from a defined micro scale fatigue pre-crack, fatigue life tests show the material property in its original virgin state, where the total life time covers the crack nucleation, propagation, and failure process. Therefore, both materials were tested with usual smooth hour glass shaped flat and circular specimens carefully machined from the jacket materials weld section. The determination of the cryogenic materials properties of the TFMC intercoil structures, produced by thick plates of Type 316LN wrought material was an important part of the task.

In order to perform large scale loading investigations, the existing 200 kN screw driven machine [2] was upgraded with a four column servo hydraulic 630 kN load capacity testing machine. The first large scale tensile tests up to 20 mm Ø tensile specimens, made of Model 2 one ton cast material were performed in LHe. One of the goals of this installation was to supply cryogenic cyclic data of the full size CS jacket welded sections.

MATERIALS, SPECIMENS AND EXPERIMENTALS

The materials used for the cryogenic mechanical investigations were supplied as given in Table 1. The chemical compositions of all investigated base metals and so far available of the weld deposits are given in Table 2. The type of specimens used for tensile, fracture, FCGR, and fatigue life investigations were as follows: <u>Round smooth tensile</u>; gauge diameters of 4 mm Ø machined from plates, corner sections of extruded CS jackets, and transverse to orbital butt weld metals. Specimens of 4 mm Ø, 6 mm Ø, 12 mm Ø, and 20 mm Ø from the 1 ton trial cast steel, all in one spatial orientation. <u>Compact tension (CT)</u>;

TABLE 1. Origin of the investigated materials

- Representative blocks were provided from the 25 ton full scale Model 2 and one ton trial casting (modified Type 316LN material) manufactured at Usinor Industrie, France.
- For the full scale Model 1 a sample was procured from the final 40 ton square tube forging for FCGR tests.
- Two samples of 240 mm thickness made by narrow-gap welding for the Model 1 were provided by Belleli Energy, Italy. The weld sample No 1 was produced by GTAW hot wire process (gas tungsten arc welding with hot wire of type Siderfil 316LM) having a repair weld of ~ 50 mm deep welded by GMAW (gas metal arc welding with wire of type Siderfil 316LM), whereas weld sample No 2 was produced by SAW (submerged arc welding, wire type OE 20.16NC with flux OP76) process having a ~50 mm deep SMAW (shielded metal arc welding, type 20.16L) repair weld.
- 22 Incoloy 908 superconductor jacket material samples of CSMC with a length of 400 mm were provided by Ansaldo, Genoa. The jackets had an orbital butt weld in the mid section, welded with a new filler wire of low Nb content. Additional 22 samples produced from Type 316LN (Valinox) material were also provided with butt welds in the mid section
- 80 mm thick Type 316LN wrought plate material for the intercoil structure of the TFMC was provided by Noell-KRC, Germany.

ASTM proportional 40 x 43 x 4 mm CT specimens for the FCGR investigations. For weld metals the crack propagation test was conducted in weld length orientation. <u>EDM notched round bar;</u> 50 mm long with a shaft diameter of 6 mm and a circular notch around the girth. <u>Round smooth fatigue life tensile;</u> 80 mm long, smooth hour glass type specimen with 15 mm long threaded ends (M10 x 0.75). However, these specimens had a reduced section of 4 mm length (with 4 mm Ø) covering the entire weld and heat affected zone.

All investigations were carried out either at 7 K or in submerged liquid helium (LHe) according to the size of the specimens. Several tensile measurements with specimens having different gauge sections machined from provided samples were performed at 7 K and in LHe to ensure a size and test environment independence especially for the cast steels. Within this context, tensile measurements with 6, 12, and 20 mm gauge diameter specimens (for the one ton trail cast [1]) were performed in LHe with the newly installed 630 kN servo hydraulic tensile test machine (Type Schenck), whereas the 4 mm gauge diameter specimens were tested at 7 K using the servo hydraulic 25 kN capacity machine [3] (MTS).

TABLE 2. Chemical composition of the investigated cast stainless steel, forged Type 316LN, heavy weld metal repaired samples, base / weld metals of CS-jackets, and Type 316LN plate materials.

Materials	C	Si	Mn	P	S	Cr	Mo	Ni	Al	Cu	Nb	Ti	V	N
Cast steel Model 2	.016	.78	6.00	.022	.005	17.6	3.01	13.90	.005	.182	n. d.	n. d.	.057	.184
Forged tube Model 1	.018	.37	2.01	.028	.002	17.23	2.52	13.46	.009	n. d.	n. d.	n. d.	.06	.182
Repair weld No 1, GMAW	.010	.40	6.74	.012	.001	19.78	3.10	16.23	n .d.	.13	n. d.	n. d.	n. d.	.200
Repair weld No 2, SMAW	.025	.30	6.50	n. d.	n. d.	20.00	3.00	16.00	n. d.	n. d.	n. d.	n. d.	n. d.	.150
Incoloy 908 jacket base metal	.001	n. d.	.041	n. d.	n. d.	3.98	n. d.	49.00	.93	n. d.	2.92	1.74	n. d.	n. d.
Valinox jacket base metal	.012	.24	1.64	.025	n. d.	17.2	2.43	12.00	n. d.	.34	n. d.	n. d.	n. d.	.180
Incoloy jacket weld deposit	n. d.	n. d.	n. d.	n. d.	n. d.	4.00	n. d.	50.80	1.05	n. d.	.82	1.82	n. d.	n. d.
Valinox jacket weld deposit	.150	.23	3.98	.02	.03	18.5	3.44	15.1	n. d.	.13	n. d.	n. d.	n. d.	.150
TFMC Plate 80 mm thick	.020	.49	1.58	.027	.0008	17.18	2.53	11.89	n. d.	n. d.	n. d.	n. d.	n. d.	.160

TENSILE AND FRACTURE TOUGHNESS MEASUREMENTS

The results with tensile and fracture toughness tests are reported in Table 3. The first four rows of this Table give the data of 25 ton cast steel, which shows an anisotropic behavior. However, the obtained values for this 25 ton heavy casting is more homogenous than the one ton trial casting already reported [1], which showed a higher degree of anisotropy. The new values determined from the one ton trial casting with different size specimens, but from same orientation show the large scatter of the material (the rows 5–8). The data of the 20 mm Ø specimen (row 8) was determined using two independent extensometer systems (each of them with two extensometers) attached at different regions of the reduced section. It is obvious that even for the same specimen the results of yield strength and Young's modulus may vary depending on the measured zone. These results reveal the high inhomogeneous nature of the one ton trial cast steel (15% total elongation).

The tensile and fracture mechanical data obtained with the two repair welds of the narrow-gap weld joints show that weld sample No 1 welded with GMAW process has slightly lower properties than sample No 2 welded with SMAW concerning the tensile values, whereas the fracture toughness of weld sample No 2 is higher than No 1. However, both repair welds fulfill the 4 K design requirements ($K_{IC} > 130$ MPa√m).

For the CS jacket structures of Incoloy 908 and Type 316LN (Valinox) the weld metal tensile and fracture toughness data are given in the two rows (11 and 12) of Table 3. To ensure the measurement of weld metal data these tests were carried out using the double extensometer system with a gauge length of 3 mm, thus enabling the record of the strains only in the weld metal region. The low elongation (~5 %) of one of the tests for the Incoloy 908 weld metal can be attributed to weld defects, which were often observed for

TABLE 3. Tensile and fracture toughness results of modified 316LN casting (25 ton & one ton), repair weld samples, jacket materials, and Type 316LN 80 mm thick wrought plate sample at 7 K and 4 K.

Material / specimen orientation and positions	Young's Modulus GPa	Yield Strength MPa	Ultimate tensile strength MPa	Uniform Elongation %	Critical J JETT [4] N/mm	K_{IC} (converted) MPa√m
25 t cast steel Y, 7 K [a]	159 / 124	944 / 852	1086 / 1055	25 / 40	193 / 241	175 / 172
25 t cast steel Y, 4 K [a]	179 / 173	800 / 870	1231 / 1183	26 / 17	-	-
25 t cast steel X, 7 K [a]	162 / 176	911 / 847	1216 / 1048	33 / 27	242 / 202	198 / 188
25 t cast steel Z, 7 K [a]	214 / 168	877 / 869	1060 / 984	20 / 20	300 / 252	253 / 205
1 t cast steel, 4 Ø, 7 K	176 / 148	694 / 688	1144 / 858	21 / 32	-	-
1 t cast steel, 6 Ø, 4 K	126 / 190	695 / 766	964 / 1044	30 / 23	-	-
1 t cast steel, 12 Ø, 4 K	170	825	1255	25	-	-
1 t cast steel, 20 Ø, 4 K	113 / 158	676 / 738	1126 / 1126	34 / 9	-	-
Repair weld No 1	157 / 161	1064 / 1127	1471 / 1495	30 / 26	118 / 125	153 / 158
Repair weld No 2	192 / 186	1164 / 1190	1518 / 1527	22 / 25	131 / 84	161 / 130
316LN jacket weld	203 / 181	1143 / 1124	1476 / 1476	19 / 18	95 / 96	138 / 132
Incoloy 908 jacket weld	148 / 163	1038 / 1108	1377 / 1313	14 / 5	111 / 130	128 / 145
316LN / T-L surface*	209 / 193	1051 / 1002	1662 / 1631	40 / 43	251 / 306	226 / 250
316LN / T-L mid plane	204 / 206	1000 / 1015	1619 / 1641	42 / 40	280 / 255	239 / 228
316LN / S-T**	210 / 211	1009 / 1003	1551 / 1616	38 / 43	232 / 246	218 / 224
316LN / L-T surface***	210 / 192	991 / 1026	1620 / 1565	43 / 42	220 / 296	212 / 246
316LN / L-T mid plane	204 / 198	977 / 994	1614 / 1639	44 / 42	290 / 259	243 / 230

[a] X, Y, and Z are spatial orientation of the block as given in reference [1]

* T – L (Transverse long) orientation of specimen according to ASTM designation

** S - T (Short transverse)

*** L - T (Long transverse)

several specimens during fatigue life tests. These defects were in the range of 0.1 – 0.3 mm and result mainly from the welding process. Another fact is also the significantly low Young's modulus of the weld zone compared to the 316LN weld metal.

The last 5 rows of Table 3 present the data for the 80 mm thick Type 316LN wrought plate at 7 K at different orientations obtained by two specimens machined from the same position. The results show a well-balanced isotropic behavior of the measured tensile and fracture toughness values. In addition, the obtained data show a small scatter for the Young's modulus (mean = 204 GPa, standard deviation (SD) = 7 GPa), yield strength (mean = 1007 MPa, SD = 20 MPa), tensile strength (mean = 1616 MPa, SD = 34 MPa), and elongation (mean = 42 %, SD = 2 %). For fracture toughness the determined average value using the JETT [4] test method results to 232 MPa√m with a SD of 12 MPa√m.

FATIGUE LIFE AND FATIGUE CRACK GROWTH RATE TESTS

To characterize the cyclic properties of the structural materials mainly two type of investigations were carried out. These were the classical axial fatigue life experiments using small scale (4 mm Ø) specimens and FCGR tests under the load ratio (P_{min}/P_{max}) R = 0.1, all at 7 K. Major attention was focused on the CS jacket butt weld fatigue life characterization. In a similar way, the base metal of the cast 25 ton steel was investigated at 7 K with the same type of specimens in as received condition. The fracture surface obtained after the fatigue life tests of the CS jacket weld materials delivered valuable information about the weld performance and the determined defects were correlated to the fatigue life. Figure 1 shows the set of experimental results for Incoloy 908, Type 316LN (Valinox) weld metals, and cast steel. The jacket material Valinox was tested additionally with flat hour glass type specimens. Here the weld metal represented the thinnest wall area (4 - 5 mm) of the CS jacket section. All Incoloy 908 specimens were in aged condition (200h / 650°C at vacuum), whereas for Valinox, tests were conducted with both aged and unaged materials. The cast material 7 K fatigue life results show that the obtained line is significantly lower than that of common Type 316LN materials due to the lower ultimate strength. For the Valinox weld metals the values obtained with the flat specimens show a slightly lower fatigue life property compared to round specimens. Almost in all cases the failure started from the surface of the specimen. The lower cyclic performance is due to the rectangular cross section of the loaded area, where the sharp edge increases the possibility of an earlier crack nucleation. For Valinox there are no differences in the final results between aged and unaged conditions. The derived line gives the best estimate of the conducted test results. In fact, between Incoloy 908 and Valinox there is hardly any difference in fatigue life. For Incoloy 908 welds the fracture surfaces show clearly defects and the defect ·size dictates the achieved total cyclic number. However, even in practically defect free state both materials are comparable as these experiments confirm. The two dotted lines in Figure 1 represent the design approach based on cumulative damage rules (ASME Boiler and Pressure Vessel Code Section III), where the experimental curve is modified by applying a factor of two on stress or 20 in cycle number, whichever is more conservative [5]. The derived lines show that the allowable fatigue life is governed by stress rather than number of cycles. Therefore, for both materials (even with small weld defects) the allowable stress is within the range of 500 MPa for one million cycles at 4 K. Figure 1 shows also the failure of survived aged flat specimen with > 1 million cycles at second time loading (plot *) with elevated stress level of 1000 MPa, indicating the crack nucleation during the first cycling set at 900 MPa.

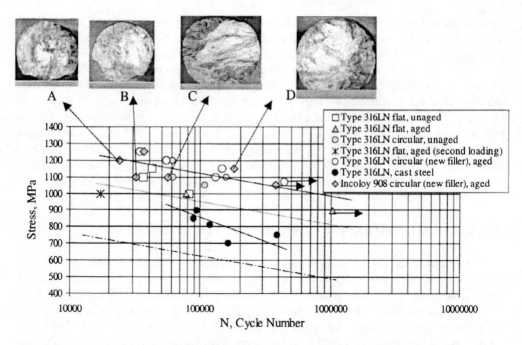

FIGURE 1. Fatigue life results with Type 316LN (aged / unaged), Incoloy 908 (aged) jacket material butt-welds, and cast steel at 7 K (load ratio, R = 0.1). The photographs A to D show the fracture surface of fatigue life specimens of Incoloy 908 weld sections. The specimen A failed by an internal small weld defect of ~0.3 mm. For specimens B and C failure started from the surface by small flaws of ~0.3 and ~0.2 mm lengths. For specimen D the failure started from the surface of the specimen without an initial weld defect.

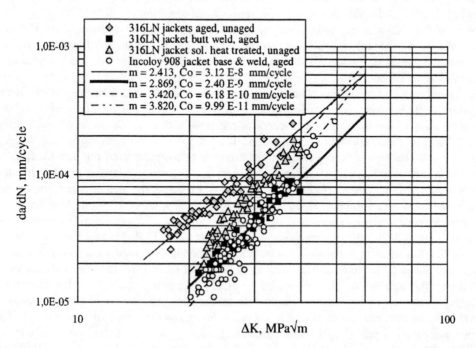

FIGURE 2. FCGR of Type 316LN (Mat.: Valinox, aged and unaged) and aged Incoloy 908 (base/weld) jacket materials at 7 K (R = 0.1). The lines for m and C_o show the mean of the determined Paris constants.

The FCGR characteristics of the Valinox and Incoloy 908 base and weld metals are shown in Figure 2. The determined average lines show the following facts: Between aged and unaged Type 316LN jacket materials there is no significant difference. For Incoloy 908 the weld and base metals show similar FCGR behavior. Type 316LN weld metal FCGR has a higher resistance against crack propagation above ΔK ~29 MPa$\sqrt{m}$, whilst below this stress intensity range the Incoloy 908 has a better performance. Solution heat treatment (1050 °C / ½ h, and water quench) of Type 316LN jacket material improves the FCGR behavior considerably if compared to jacket materials without any treatment. The reason of this finding seems to be the grain size, which in case of the extruded jacket material is very small (~70 μm) with a high amount of twins. By solution heat treatment the size of the grains are increased with fewer twins. An investigation program for the material 316LN showed that a variation of FCGR properties exist between different batches resulting from the microstructure. Figure 3 shows the obtained data for different groups of Type 316LN materials. In this diagram the old published data of 50 mm thick 316LN plate obtained with 24 mm thick CT specimens at 4 K are plotted [6]. In this work it was possible by using large specimens to determine the FCGR up to > 100 MPa$\sqrt{m}$, which in case of 4 mm thick specimens was not possible because of specimen collapse beyond the 40 MPa$\sqrt{m}$ regime. The test results with modified cast steel (Model 2), the forging (Model 1), and a commercial 20 mm thick plate (average grain size ~300 μm) are all in line with the old 4 K data. However, the 80 mm plate (TFMC intercoil structural material) is significantly inferior to Valinox (Figure 2). The investigated microstructure of the 80 mm plate indicated also a smaller average grain size (~120 μm) than usual with a high twin density.

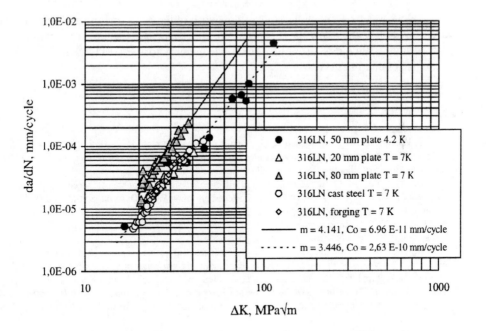

FIGURE 3. FCGR of Type 316LN base materials at 7 and 4 K (R = 0.1). The lines along with m and C_o show the mean of the determined Paris constants.

In addition to the above given jacket materials FCGR and fatigue life tests (small scale specimens) a fatigue life investigation program was started with full size jacket materials having a butt weld in mid section. The measurement took place inside the 400 mm diameter cryostat at LHe using the 630 kN servo hydraulic machine. A special equipment designed beneath the 4 K rig made a 4-point cyclic bend loading possible with peak loads at ~400 kN. The carried out preliminary tests at 4K show the applicability of such measurements at a frequency of circa 1 Hz with a LHe consumption of about 250 liters per day. The first fatigue life test results performed with full size jacket structures are expected in near future.

CONCLUSIONS

Type 316LN 80 mm thick plate, forging, repair welds of a deep narrow-gap welds, 25 ton modified cast steel, and candidate jacket materials (Incoloy 908 and 316LN) were characterized with respect to the cryogenic mechanical behavior using tensile, fracture toughness, FCGR, and fatigue life tests. The results indicate that the industrially produced structural materials and the welds fulfill the design requirements of ITER. In case of fatigue life both jacket materials show a similar behavior. For Type 316LN jacket materials the solution heat treatment process improved considerably the FCGR properties at 7 K.

ACKNOWLEDGEMENT

This work was performed within the framework of the Nuclear Fusion Project of the Karlsruhe Research Center and is supported by the European Communities within the European Fusion Technology Program. Acknowledgement to the EFDA team and to all private companies cited in the text.

REFERENCES

1. Nyilas, A., Harries, R. D., and Bevilacqua, G., "Status of European material testing program for ITER model coils and full size mockups", in *Advances in Cryogenic Engineering (Materials)* 46A, edited by U. B. Balachandran et al., Plenum, New York, 2000, pp. 443-450.
2. Krauth, H. and Nyilas, A., "Fracture toughness of nitrogen strengthened austenitic steels at 4 K", in *Fracture and Fatigue* in ECF 3, edited by J. C. Radon, Pergamon Press, Oxford (1980), pp. 119-128.
3. Nishimura, A., Yamamoto, J., and Nyilas, A., "Fatigue crack growth of SUS 316 and weld joint with natural crack at 7 K", in *Advances in Cryogenic Engineering (Materials)* 44A, edited by U. B. Balachandran et al., Plenum, New York, 1998, pp. 81-88.
4. Nyilas, A., Obst, B., and Nishimura, A., "Fracture mechanics investigations at 7 K of structural materials with EDM notched round and double edged-bars", in *Advances in Cryogenic Engineering (Materials)* 44A, edited by U. B. Balachandran et al., Plenum, New York, 1998, pp. 153-160.
5. Darleston, B. J. L., "The influence of fatigue on the energy scene", in *Fatigue 84,* Vol. III, edited by C. J. Beevers, Proc. of 2nd International Conference on Fatigue and Fatigue Thresholds, Birmingham (1984) pp. 1581-1595.
6. Nyilas, A., H. Krauth, Metzner, M., Munz, D., "Fatigue response of materials used in large superconducting magnets for fusion technology", in *Fatigue 84,* Vol. III, edited by C. J. Beevers, Proc. of 2nd International Conference on Fatigue and Fatigue Thresholds, Birmingham (1984) pp. 1637-1649.

FURTHER ASPECTS ON J-EVALUATION DEMONSTRATED WITH EDM NOTCHED ROUND BARS AND DOUBLE-EDGED PLATES BETWEEN 300 AND 7 K

A. Nyilas,[1] A. Nishimura,[2] and B. Obst[1]

[1] Forschungszentrum Karlsruhe, ITP
D-76021 Karlsruhe, Germany
[2] National Institute for Fusion Science
Toki, Gifu 509 - 5292 Japan

ABSTRACT

Based on recent experiences with respect to JETT (J - $\underline{E}$valuation on $\underline{T}$ensile $\underline{T}$est) several materials were measured to investigate the potential of this novel technique with the aim to determine the fracture toughness of materials. The investigated materials covered a wide field of toughness levels ranging from ~2 MPa√m to ~300 MPa√m. These materials were as follows; OFHC annealed copper, CuNiSi, sintered NdFeB, FeCo, 9Ni steel, Al 6061-T6, wrought 316LN, and modified Type 316LN cast steels. The tests comprised measurements of EDM notched round bars as well as double edged flat tensile specimens between 300 K and 7 K. Wherever possible the tests were cross checked with ASTM standard compact tension or single edged notched specimens using the standard procedures. Main attention was paid to seek a solution towards a realistic integration path focusing the materials fracture point, comprising tests with high ductility and medium toughness materials represented by copper and copper alloys. Scanning electron microscopic investigations at the notch tip of loaded and subsequently unloaded specimens at the maximum load position could reveal the crack initiation at the notch tip immediately after the start of necking. A concept was driven for reliable evaluation and estimation of the materials critical J and the related fracture toughness.

INTRODUCTION

Since the first publications [1-5] started mid of 90s' much work has been spent to

CP614, *Advances in Cryogenic Engineering:*
Proceedings of the International Cryogenic Materials Conference - ICMC, Vol. 48,
edited by B. Balachandran et al.
© 2002 American Institute of Physics 0-7354-0060-1/02/$19.00

FIGURE 1. The equipment represents the multi-specimen rig fixed inside the helium flow cryostat allowing to shift the four specimens on a track using a small remote controlled electrical driven motor. All four EDM (Electro Discharge Machining) notched 6 mm Ø specimens are separately clamped with two high resolution extensometers necessary for the average displacement record during the loading.

scrutinize the test methodology and the evaluation of the developed JETT method with respect to its weakness. During the several stages of the development it could be recognized that depending on failure type (ductile, brittle, or mixed mode), specific attention had to be paid concerning the test and evaluation route. Large series of tests accompanied with different type of materials were necessary to elucidate the potential of the technique. For this purpose a new multi-specimen 4 K rig allowing to test four specimens one after the other was designed and successfully used inside the existing cryogenic test facility [3]. Figure 1 shows the built-in arrangement inside the helium flow cryostat, which could test four specimens between 300 and 7 K on a daily rate basis.

Equipped with this test device several hundreds of specimens were measured during the years to gain substantial experience, necessary to improve the fundamental technique behind it. To cover the entire range of different materials, measurements were performed varying from very brittle, to intrinsically medium toughness with high ductility, and to high toughness materials. At the present stage it is possible to use the JETT method for materials screening tests as well as to obtain local fracture toughness of the materials necessary for material development and design purposes. In addition, the analytical-based approach to determine the fracture toughness from the EDM notched round bars also shows an excellent agreement between the experimental data and the predicted load displacement curves [6].

The load displacement records of materials can be divided into four generic types. Figure 2 shows the original raw data obtained at cryogenic temperatures between 77 K and 7 K of these four classes. Here GMAW represents the weld metal (Gas Metal Arc Weld) of a heavy weld structure consisting of Type 316LN material. As shown, the load displacement curve exhibits discontinuous serrated type yielding prior to failure. The load drops accompanied with large sudden displacements have to be accounted during the integration using the equ. 1 derived by Rice at al.[7]. Here r is the net section radius of the circumferential-notched specimen and P is the corresponding load. The displacement of the localized small zone dc is calculated for each loading step by subtracting the shaft displacements from the experimentally determined extensometer displacements [2]. The integration of the entire curve will thus overestimate the critical J. It could be verified by

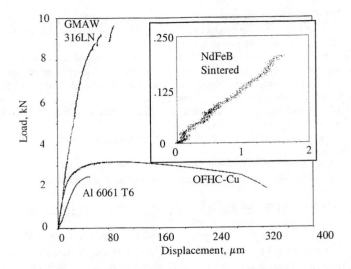

FIGURE 2. Different type of energy absorption responses during measurement of load versus displacement depending on materials ductility and failure mode at cryogenic temperatures. All records are test results obtained at 7 K except the 77 K measurement of sintered NdFeB material (shown as an insert having the same axis dimensions in kN and μm).

several experiments using a modified procedure that a sound matching of the results with the obtained standard J_{IC} values for different high toughness materials is given.

$$J = \frac{1}{2 \cdot r^2 \cdot \pi} \left[3 \cdot \int_0^{dc} P \cdot d(dc) - P \cdot dc \right] \tag{1}$$

The oxygen free high conductivity (OFHC) annealed copper material shows contrary to the latter one a large necking after reaching the maximum load owing to the intrinsic high ductility even at very low temperatures. An integration of the entire curve using the portion beyond the maximum load will naturally result to a large overestimation of the critical J obtained by the equ. 1. Therefore attempts have been made to investigate the notch/crack tip of the high ductile/medium toughness materials and derive a concept about the integration path. The tempered high strength aluminum material Al6061 T6 is a typical material failing at the maximum load, thus enabling a clear integration path. The results of these series are also compared with J_{IC} values obtained by E399 standard and found to show a relatively good matching between the two methods.

The last class of tested materials comprised extreme brittle materials (FeCo and sintered NdFeB). The insert of Figure 2 shows the raw data of the obtained load-displacement curve of this material at 77 K. The general problem of such brittle materials is the mismatch of the axial loading line. Here, although the two extensometers average out the displacement information, any bending of the specimen by a slight mismatch of the axial line will result to an early failure of the specimen.

MATERIALS, SPECIMENS AND EXPERIMENTALS

The supplied materials for the critical J-test investigations are given in Table 1.

TABLE 1. Origin of the investigated materials

-1000 mm long 6 mm Ø oxygen free high conductivity copper rod purchased from the vendor Goodfellow, England. Before machining, the rod was annealed at 300 °C for one hour at vacuum.

-Flat rolled 3 mm thick CuNiSi sheet (with 2.5 % Ni and %0. 6 % Si) manufactured by MSX-Call, France and provided by Imperial College, London. Prior to the machining of the specimens the material was optimized with a precipitation hardening treatment. This optimization process consisted of an age hardening at 475 °C following the solution heat treatment process [8] at 820 °C.

-Small bars (50 x 14 x 10 mm and 80 x 15 x 15 mm) of sintered NdFeB material (Type N35H Zn coated unmagnetized) were provided by Siemens AG, Germany.

-A plate consisting of 9 Ni steel material with 20 mm thickness, cut from a round block of ~300 mm Ø was provided by Siemens, Germany.

-7 mm Ø rods from FeCo (50/50) material were provided by Siemens, Germany.

-A block of 200 x 200 x 150 mm consisting of aluminum alloy 6061 in tempered state T6 was provided by Space Cryo, Oxford, England.

-Modified Type 316LN material provided from a 25 ton casting [9] by Creusot Loire Industries, France.

The type of specimens were ASTM proportional CT specimens with 60 x 63 x 15 mm for the Al 6061 T6 and 45 x 43 x 3 mm CT for the 3 mm thick CuNiSi plate material, EDM notched round bars of 50 mm length with a shaft diameter of 6 mm and a net diameter after notching around the girth of about one third of the shaft diameter, double edged bars with EDM notch with 78 mm length and 2 mm thick specimens for NdFeB material with a width of 10 mm and a net width of 2 mm after notching, and single edged notched (SEN) with 80 x 15 x 7.5 mm for NdFeB material with an EDM notch of ratio a/W= 0.5 (a is the crack length and W the width).

Aluminum Alloy 6061-T6

From the provided block two round JETT specimens from each spatial orientation were machined and designated according to the ASTM as L-T, S-T, and T-L. The machined side grooved CT specimens were measured in L–T and S-T orientation. All results measured at 7 K are given in Table 2 and for the reason of clarification the last column of this Table consists of an illustration of the sample positioning related to the designated orientations. The results show, considering the same crack plane orientation of JETT and CT specimens, a good matching of the obtained fracture toughness values using different type of methods. The JETT specimens with T-L orientation have the same crack plane as measured by the CT specimens oriented in T-S direction. Similar to the latter case the JETT specimens oriented as T-S have the same crack plane of the CT specimen measured with L-T. In fact, both obtained E399 values slightly overestimate (about 15 %) the JETT values regarding the results. The lowest fracture toughness value according to the JETT technique is measured in L-T orientation (vertical to the rolling direction). The overall lower values of the JETT method can be attributed to the general low fracture toughness of this alloy, which in case of a slight machine/specimen mismatch give rise to a bending strain at the EDM notch tip, consequently resulting to an early failure of the specimen. This phenomenon is more pronounced if the materials toughness further decreases, which was the case with the investigated materials of high brittleness.

Alloys NdFeB and FeCo

Three flat EDM notched double-edged specimens were machined from the NdFeB sintered material bars. Two measurements from these series were carried out at 295 K. The obtained results with respect to critical J - values at 295 K were in line with low scatter,

TABLE 2. Fracture mechanical test results with different methods applied to the alloy Al 6061 T 6 at 7 K.

Specimen orientation	Type of test	Critical J N/mm	Fracture Toughness MPa√m
Long-Transverse, L-T	JETT	10.1	27
Long-Transverse, L-T	JETT	5.4	20
Transverse-Long, T-L	JETT	12.3	30
Transverse-Long, T-L	JETT	13.5	31
Transverse-Short, T-S	E399	-	36
Transverse-Short, T-S	JETT	14.5	33
Transverse-Short, T-S	JETT	17.9	36
Long-Transverse, L-T	E399	-	42

whereas the test conducted at 77 K resulted a low value compared to 295 K. Young's modulus values were also determined, which was necessary for the conversion of the obtained critical J values into K_{IC}. These were 136 GPa at 77 K and 145 GPa at 295 K. Finally, an ASTM standard SEN specimen was 3 point loaded at 295 K according to E399 standard. The results of these measurements are given in Table 3. The standard SEN measurement yielded almost a factor of three higher K_{IC} value, indicating that the JETT results show a large discrepancy. The reason for this large variation is the brittleness of the material, which under loading shows a high sensitivity to any notch tip bending. Before reaching the pure axial loading the bending of the notch tip results to a mode I failure according to the fracture mechanics. In the last column of the Table 3 the individual responses of each extensometers attached to the double-edged flat specimen (for 295 K test) are shown in a diagram. In fact, the records in this diagram confirm the superimposing of the bending and the axial displacement during specimen loading. As a consequence, the determination of the exact value can not be achieved using this type of a procedure with such a brittle material unless the specimen / rig mismatch is minimized. The determined K_q value by the E399 test is also not valid considering the fact that the standard E399 bears also problems with very brittle materials. In addition, a pre-cracking of the SEN specimen was not possible and the crack started for the present test from the machined EDM notch rather than a crack during the loading. Regarding these uncertainties an overestimation of the determined K_q value is much more likely. Other than the material NdFeB the material FeCo has a relatively high fracture toughness and this material was investigated at different temperatures using round bar JETT specimens. Prior to the JETT tests the materials Young's modulus was determined by tensile tests, too. The measured Young's modulus between 77 K and 7 K varied between 207 GPa and 212 GPa. The obtained critical J results indicate here also a large scatter of the values. The analysis of the raw data showed, however, that the obtained lower critical J values exhibit higher bending of the specimens during the loading compared to the high critical J results. Therefore, it is reasonable to conclude that the lower values underestimate in general the materials fracture behavior. To investigate the fracture toughness by ASTM E399 standard procedure a small CT sample provided from a former batch was machined. After pre-cracking the specimen at 295 K the material was measured in liquid helium. The determined result with K_{IC} = 12 MPa√m was valid according to the ASTM E399 criteria and confirms the value of the measured 7 K test using the JETT test procedure.

CuNiSi and OFHC Ductile Materials

The fracture toughness measurements for the 3 mm thick CuNiSi plate comprised standard ASTM E 813-89 unloading compliance tests and JETT tests with specially

135

machined specimens from the plate between 300 K and 7 K, having a 2 mm Ø circular EDM cut in the mid of the flat specimen. The carried out J-test measurements at 295 K and at 7 K resulted in non valid values because of achieved small physical crack extensions below 0.5 mm. The small travel length of the clip-on-gauge could not record the needed large load line displacements of > 3 mm. Finally, a developed new clip-on-gauge with a travel length of 10 mm could perform a J-test with an achieved average physical crack length of 1.4 mm at 77 K. The evaluation of the resistance curve yielded a J_{IC} of 200 N/mm. Recent J_Q measurements performed at 77 K given by the reference [8] show a much higher value (400 N/mm) than the above obtained result. However, as stated in this work [8] the obtained J-values (multi-specimen test method) do not fully satisfy the requirements of the standard procedure and the true J_{IC} was expected to be much lower than this value.

Measurements performed with the JETT test using the flat 3 mm thick specimens with circular net sections are shown in Figure 3. The record of the four tests show that by decreasing the temperature the total displacement increases. Within this context smooth specimen tensile tests confirmed this phenomenon of ductility increase of the material by temperature decrease, too. This behavior affects also the materials fracture toughness, where the material shows a higher performance against cracking at low temperatures. The increase of ductility, however, contributes for the current JETT test a large necking of the specimens net section. An integration path up to the specimen failure, therefore, will overestimate the critical J-value considerable. To investigate this condition the specimens were unloaded short after the ultimate load. The notch tips analyzed by scanning electron microscope (SEM) show clearly the existence of large cracks initiated by the loading. Therefore, it was decided to follow the integration path according to the equ. 1 up to the maximum load position. Figure 4 shows the SEM pictures obtained from these test series at 300 K and 200 K. A similar observation was also obtained for the test performed at 77 K. Similar to those CuNiSi material measurements four annealed OFHC copper JETT specimens were tested at 295 K and at 7 K. Prior to these investigations benchmark tests with this material confirmed that before the maximum load position there was no evidence of a crack initiation. Cracks opened in each case shortly after reaching the maximum load. Therefore, for ductile materials with tendency to neck, the integration path should be taken up to the maximum load position to avoid an overestimation. In Table 4 the obtained results with various materials are compiled.

Structural Materials

Among the measured structural materials at the present work 9 Ni steel represents

TABLE 3. Fracture mechanical properties measured with different specimens between 295 K and 4.2 K for NdFeB and FeCo materials. Load-displacement diagram of the NdFeB flat specimen is given as an insert in fifth column (295 K). The picture shows initial distortions during loading because of sample bending.

Specimen type & material	Temperature K	Critical J N/mm	K_{IC} MPa√m	
JETT flat, NdFeB	295	0.023	1.8	
JETT flat, NdFeB	295	0.024	1.8	
JETT flat, NdFeB	77	0.010	1.2	
ASTM SEN, NdFeB	295	-	4.3	
JETT, FeCo	77	0.6 / 0.4	11/ 6	
JETT, FeCo	20	0.3 / 0.9	8 / 14	
JETT, FeCo	30	0.4 / 0.9	6 / 14	
JETT, FeCo	7	0.4 / 0.7	9 / 12	
ASTM CT, FeCo	4.2	-	12	

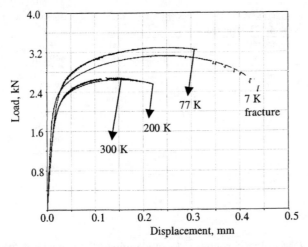

FIGURE 3. Load-Displacement records of CuNiSi material at four different temperatures. The arrows indicate the unloading points of the tests. The specimen tested at 7 K was loaded until fracture.

a joint activity within the framework of the VAMAS TWA 17. However, the measured JETT data belong here to a similar batch provided by the cited industry, whilst the two CT bound J_Q data obtained with 15 mm thick side grooved CT specimens result from the distributed 9 Ni steel of VAMAS batch. These fracture mechanical results are given in Table 4. The major improvement for the evaluation procedure using the equ. 1 is the consideration of the observed serrations during the integration. By closing the serration related displacement portions to a continuous curve according to the number of the load drops and the achieved each individual displacement jumps during the loading the total energy of the load versus displacement curve is decreased. These modifications served a better matching of the results obtained at low temperatures especially for structural materials such as Type 316LN, owing to serrated type of load displacement behavior [9]. Several former curves belonging to the different type of 316LN materials were once again re-evaluated with the discarded serration related jumps. The results showed that for several measured data sets where an explanation for the obvious overestimation lacked to date the modification could give a reasonable final answer.

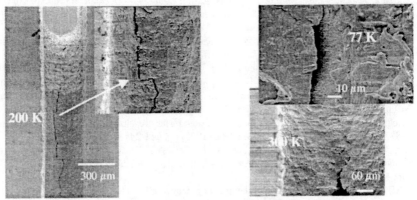

FIGURE 4. SEM images taken from the three unloaded round bar JETT specimens which were tested at 300 K, 200 K, and 77 K. The notch tips of these specimens show large cracks initiated during the loading.

TABLE 4. Fracture mechanic properties measured with different specimens between 300 K and 7 K for CuNiSi, OFHC copper, and 9 Ni steel materials.

Material	Temperature K	Critical J JETT, N/mm	Critical J_Q CT, N/mm	K_{IC} MPa√m
CuNiSi	300	93	-	111
CuNiSi	200	115	-	124
CuNiSi	77	215	200	169
CuNiSi	7	240	-	185
OFHC Cu	300	25 / 26	-	56 / 56
OFHC Cu	7	76 /74	-	98 / 97
9 Ni steel	7	26 / 20	-	72 / 66
9 Ni steel	4.2	-	17 / 17	61 /61

CONCLUSIONS

Using the EDM notched round and double edged flat specimens four different type of materials were investigated between 300 K and 7 K with failure occurrences at the elastic line, at the elastic-plastic maximum load position, ductile failure after necking, and failure after serrated yielding. These investigations showed that for each class of failure by applying different approaches for the critical J equation integration, developed by J. R. Rice [7] sound results can be established compared with to date existing standard fracture toughness test methods.

REFERENCES

1. Nyilas, A. and Obst, B., "A new test method for characterizing low temperature structural materials", in *Advances in Cryogenic Engineering (Materials)* 42A, edited by L. T. Summers, Plenum, New York, 1996, pp. 353-360.
2. Nyilas, A., Obst, B., and Nishimura, A., "Fracture mechanics investigations at 7 K of structural materials with EDM notched round and double edged-bars", in *Advances in Cryogenic Engineering (Materials)* 44A, edited by U. B. Balachandran et al., Plenum, New York, 1998, pp. 153-160.
3. Nishimura, A., Yamamoto, J., and Nyilas, A., "Fatigue crack growth of SUS 316 and weld joint with natural crack at 7 K", in *Advances in Cryogenic Engineering (Materials)* 44A, edited by U. B. Balachandran et al., Plenum, New York, 1998, pp. 81-88.
4. Shibata, K., Kadota, T., Kohno, Y., Nyilas. A, and Ogata, T., "Mechanical properties of a boron added superalloy at 4 K and magnetic effect", in *Advances in Cryogenic Engineering (Materials)* 46A, edited by U. B. Balachandran et al., Plenum, New York, 2000, pp. 73-80.
5. Nyilas, A., Shibata, K., Specking, W., and Kiesel, H., "Fracture and tensile properties of boron added Ni-base superalloy at 7 and 4.2 K, and the effect of 13 Tesla field", in *Advances in Cryogenic Engineering (Materials)* 46A, edited by U. B. Balachandran et al., Plenum, New York, 2000, pp. 81-88.
6. Shindo, Y., Mano, Y., Horiguchi, K., and Sugo, T., "Cryogenic fracture toughness determination of a structural alloy weldment by notch tensile measurement and finite element analysis", in *Journal of engineering Materials and Technology*, Vol. 123, (2001) pp. 45-50.
7. Rice, J. R., Paris, P. C., and Merkle, J. G., "Some further results of J-integral analysis and estimates", in *ASTM Special Technical Publications 536*, (1973) pp. 231-245.
8. Ageladarakis, P. A., O'Dowd, N. P., and Webster, G. A., "Tensile and fracture toughness tests of CuNiSi at room and cryogenic temperatures", in *JET-R(99)01*, Joint European Torus JET Report (1999), pp. 1-15
9. Nyilas, A., Harries, R. D., and Bevilacqua, G., "Status of European material testing program for ITER model coils and full size mockups", in *Advances in Cryogenic Engineering (Materials)* 46A, edited by U. B. Balachandran et al., Plenum, New York, 2000, pp. 443-450.

EFFECT OF A MAGNETIC FIELD ON CRACK LENGTH MEASUREMENT OF 9NI STEEL BY UNLOADING COMPLIANCE METHOD

C. Nagasaki, K. Matsui and K. Shibata

Department of Metallurgy, The University of Tokyo, Tokyo 113-8656, Japan

ABSTRACT

In the fracture toughness measurement of ferro-magnetic materials in high magnetic fields, it has not been clarified whether we can use the same formulas as ones used in a non-magnetic field. Therefore, it is necessary to understand the magnetic effect in the fracture toughness measurement of ferro-magnetic materials. As the first step, crack length was measured at 4 K by unloading compliance method in the testing of CT specimen of 9% nickel steel. The same formula was used in the magnetic field of 0 and 8 Tesla. They were compared with the length measured by optical fractography. The magnetic field had little effect on the crack length measurement by the unloading compliance method. The small amount of retained austenite in Q T heat-treated specimen did not exhibit a magnetic effect on the crack length measurement by the unloading compliance method.

INTRODUCTION

Fracture toughness J_Q and tearing modulus of Incoloy 908 alloy was increased by an application of 8 Tesla in compact tension testing at liquid helium temperature [1]. As the matrix phase (austenite) and precipitates of Incoloy 908 are stable, no phase transformation occurs even under deformation at very low temperatures. The cause of the increase in J_Q is not structure or phase transformation.

On the other hand, tearing modulus of SUS 304L steel is decreased by an application of 8 Tesla [2]. In the case of SUS 304L steel, martensitic transformation occurs during deformation at very low temperatures. The decrease in the fracture toughness can be attributed to martensitic transformation induced by a magnetic field. When the test is carried out under such a condition as the martensitic transformation is suppressed, the magnetic effect on fracture toughness disappears [2,3]. Under suitable conditions, an increase in fracture toughness was observed by an application of high magnetic fields [3].

In the previous report [4], it was clear that magnetic field of 8 Tesla had no effect on Young's modulus, yield stress, tensile stress and total elongation of Incoloy 908. However, the magnetic effect on fracture toughness of the alloy could not been discussed in compact

CP614, *Advances in Cryogenic Engineering:*
Proceedings of the International Cryogenic Materials Conference - ICMC, Vol. 48,
edited by B. Balachandran et al.
© 2002 American Institute of Physics 0-7354-0060-1/02/$19.00

tension test because a large crack propagated suddenly before the number of plots was sufficient in the J-a curve.

Therefore, the reason for the increase of the fracture toughness of Incoly 908 in a magnetic field has not been clarified. The following reasons are worthy to be considered.

(1) ΔE effect

The stress applied on ferromagnetic steels causes additional elongation accompanied with spontaneous magnetization. The application of magnetic field causes elongation of the material by magnetostriction [5]. Hence, Young's modulus is increased in a high magnetic field when the magnetic domain boundary movement is saturated before loading. If Young's modulus is increased in a high magnetic field, an increase in fracture toughness is expected.

(2) Effect of magnetic force due to gradient of a magnetic field

In a solenoid-type magnet, the top and bottom pull-rods are pulled in the direction of the magnet center. As a result, the specimen is compressed even when the load cell does not detect any load. Therefore, additional tensile load is necessary for the nucleation and growth of the crack.

(3) Effect of magnetic force due to interference among magnetic field, deformation, and stress field [6].

Shindo [7] has developed the theory by proposing the model applied in high magnetic fields. Experimental certification, however, seems not to be sufficient.

It has recently reported that K_{IC} and J-R curve of Incoloy 908 is not significantly affected by magnetic field as high as 14 T [8].

In the present paper, the effect of a high magnetic field on the crack length estimated by unloading compliance method was investigated in testing at 4 K. In this experiment, 9% nickel steel was used as a representative of ferro-magnetic materials because the crack propagates straightly and its valid length could be measured, while the crack in Incoloy 908 tended to show a curved propagation front.

EXPERIMENTAL PROCEDURE

Material and Specimens

Chemical composition of the 9% nickel steel used in this experiment is shown in Table 1. The plates forged from ingots were received after quenching and tempering (QT) heat treatment from NKK Corporation. The plates were water-quenched after holding at 1083 K for 15 min in the Q treatment and air-cooled after holding at 848 K for 15 min in the T treatment. The plates re-heated at 1083 K and water-quenched were also prepared: this heat treatment will be called Q.

The half-size specimens for compact tension (CT) test were machined from the plates. The size and geometry of the specimens are shown in Fig. 1. The specimens were pre-cracked at room temperature under controlled cyclic loading at stress ratio of 0.1 and at frequency of 50 Hz. The total pre-cracking length a_0 was aimed to be 15 mm, as nominal width ratio a_0 / w is 0.6. In addition, side grooving of the CT specimens was performed by 1 mm.

TABLE 1. Chemical composition of the 9% nickel steel used (mass%).

C	Si	Mn	P	S	Ni	Al
0.05	0.2	0.61	0.001	0.001	9.18	0.82

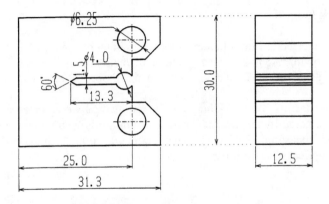

FIGURE 1. Size and geometry of specimen for compact tension test (mm).

Cryostat with Magnet, Load Cell and Extensometers

Schematic illustration of the cryostat equipped with the magnet is shown in Fig. 2. The magnet was produced by Japan Magnet Technology. The height is 270 mm and the bore diameter is 74 mm. The specimen was placed at the center of the magnetic field (110 mm from the bottom) and the load was applied through the pull-rod of 30 mm diameter, type 310S stainless steel.

Three extensometers (Shepic Corp., USA) were attached to the specimen at around 120-degree intervals. Extensometer for CT test had been already calibrated in a previous investigation [1,2]. Magnetic effect on the out-put of the load cell was checked in the previous report [4].

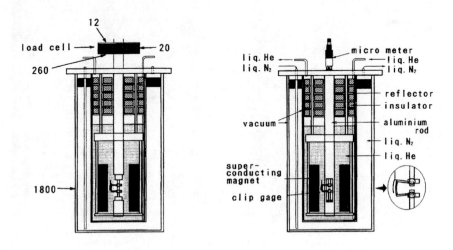

FIGURE 2. Schematic illustration of the cryostat equipped with the magnet.

Measurement of Crack Length

The specimens were tested in liquid helium. Crack length was estimated by unloading compliance method following ASTM E813-87. The specimen was loaded and held at a certain load for 30 s. Then the specimen was unloaded by 30% of the load at which unloading started.

The specimens after testing by unloading compliance method were heated at 673 K for an hour in air and fractured by tensile deformation at liquid nitrogen temperature. The final crack length was measured following usual standard through thickness in the fractured specimens by optical observation.

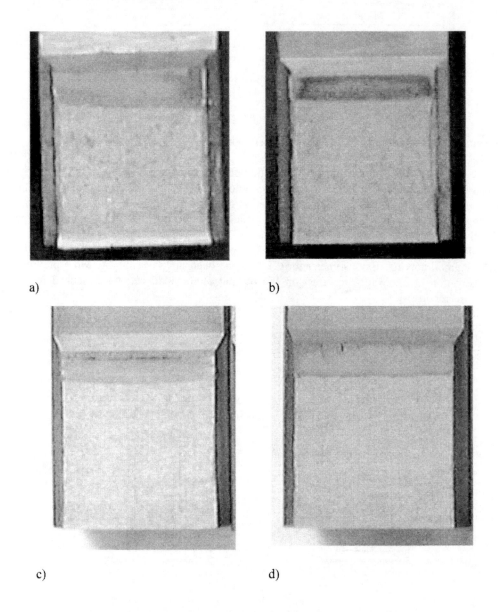

a)

b)

c)

d)

FIGURE 3. Fracture surfaces of the CT specimens; a) Q T heat treatment, 0 T, b) Q T heat treatment, 8 T and c) Q heat treatment, 0 T and d) Q heat treatment, 8 T.

RESULTS AND DISCUSSION

Crack Length Measurement in CT Specimen

In the QT and Q heat-treated specimens, the fractured specimens after testing at 4 K in the magnetic fields of 0 and 8 T are shown in Fig. 3. The width direction of the specimens is vertical and the thickness direction is horizontal. Fatigued pre-cracks of about 1.5 mm length are exhibited. In the lower side of the pre-cracks, the black zones of narrow strip exist. These are crack propagation area by the crack length measurement by unloading compliance method.

In QT heat-treated specimen, the final crack length at each position through thickness is measured by optical observation and compared with the crack length estimated by unloading compliance method, as shown in Fig. 4. Every value measured by optical observation is placed inside of the range between 93% and 107% of the average value. The value estimated by unloading compliance method is lower than that measured optically.

In the Q heat treatment specimen, the crack length at each position through thickness after testing is compared with the length by unloading compliance method, as shown in Fig. 5. Every value measured optically is placed inside of the range between 93% and 107% of the average value.

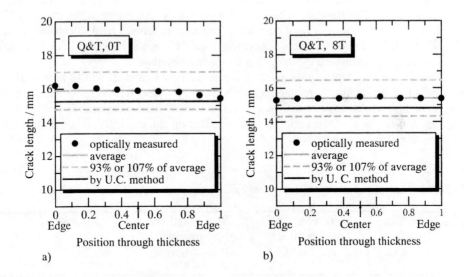

FIGURE 4. Distribution of crack length through thickness tested at 4 K in the magnetic fields of a) 0 T and b) 8 T (QT heat treatment).

143

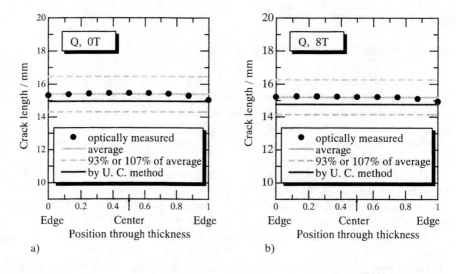

FIGURE 5. Distribution of crack length through thickness tested at 4 K in the magnetic fields of a) 0 T and b) 8 T (Q heat treatment).

In the QT and Q heat treatment specimens, the results of crack length estimated by unloading compliance method a_{UC} are compared with optically measured crack length a_{Op}, as shown in Tables 2 and 3. In both heat treatment conditions or in the magnetic field of 0 and 8 T, a_{UC} is smaller than a_{Op}. The ratio of a_{UC} to a_{Op} is compared in the magnetic field of 0 and 8 T. The ratio is a little smaller in the high magnetic field compared with no magnetic field, but the difference is extremely small. High magnetic field has little effect on the crack length estimated by unloading compliance method.

There is no difference in the ratio between QT heat treatment and Q heat treatment. The amount of paramagnetic retained austenite in the QT heat-treated specimen is thought to be about 5%, while austenite does not retained in the Q heat treatment. Therefore, such a small amount of the retained austenite has no effect on the crack length measurement by the unloading compliance method.

TABLE 2. The crack length by UC method and optical observation in the Q T heat-treated specimen.

Magnetic field (Tesla)	Crack length by UC method a_{UC} (mm)	Crack length by optical observation a_{Op} (mm)	a_{UC}/a_{Op}	
0	14.63	14.93	0.980	
0	15.26	15.90	0.962	mean=0.971
8	14.57	15.42	0.948	
8	14.61	15.32	0.954	
8	14.81	15.40	0.961	mean=0.954

144

TABLE 3. The crack length by UC method and optical observation in the Q heat-treated specimen.

Magnetic field (Tesla)	Crack length by UC method a_{UC} (mm)	Crack length by optical observation a_{Op} (mm)	a_{UC}/a_{Op}	
0	14.72	15.26	0.965	
0	14.95	15.39	0.979	mean=0.972
8	14.77	15.20	0.972	
8	14.30	15.26	0.937	
8	15.06	15.39	0.979	mean=0.963

CONCLUSIONS

Magnetic effect on the measurement of crack length at liquid helium temperatures were examined using the 9% nickel steel receiving QT and Q heat treatment. The high magnetic field of 8 T has no effect on the crack length measured by unloading compliance method in the test at 4 K.

ACKNOWLEDGEMENTS

The authors are grateful to NKK Corporation for melting the 9% Ni steel and to the Cryogenic Center of the University of Tokyo for providing with much convenience to perform the experiments at 4 K. This research was financially supported by the Special Coordination of Japan Ministry of Education, Culture, Sports, Science and Technology.

REFERENCES

1. Kadota, T., Tanaka, T., Kohno, Y., Shibata K., Nakajima, H. and Tsuji, H., "Effect of an 8 Tesla Magnetic Field on Mechanical Properties of Incoloy 908 Alloy at 4 K," in *Advances in Cryogenic Engineering (Materials)* 44, Plenum, New York, 1998, pp. 1-8.
2. Tanaka, T., Kadota, T., Kohno, Y. and Shibata, K., "Effects of Stress Level in Pre-Cracking on Fracture Toughness of SUS304L Steel in an 8 Tesla Magnetic Field at 4 K," in *Advances in Cryogenic Engineering (Materials)* 44, Plenum, New York, 1998, pp. 137-144.
3. Chan, J. W. D., Chu, D., Tseng, C. and Morris, J. W., "Cryogenic Fracture Behavior of 316LN in Magnetic Fields up to 14.6 T," in *Advances in Cryogenic Engineering (Materials)* 44, Plenum, New York 1994, pp.1215-1221.
4. Shibata, K., Kadota, T., Kohno, Y., Nyilas, A. and Ogata, T., "Mechanical Properties of a Boron Added Superalloy at 4 K and Magnetic Effect," in *Advances in Cryogenic Engineering (Materials)* 46, Plenum, New York, 2000, pp.73-80.
5. Bozorth, R. M., in *Ferromagnetism*, Van Nostrand, New York, 1951.
6. Shindo, S. Y., Shoji, T. and Saka, M., "A Magneto-Fracture Mechanics Approach to the Structural Integrity Assessment System of a Super-conducting Magnet for a Fusion Reactor," in *Electromagnetomechanical Interactions in Deformable Solids and Structures*, 1987, p.131.
7. Shindo, Y., *J. Appl. Mech.*, **44**, pp. 47-50 (1977).
8. Clatterbuck, D. M., Chan, J. W. and Morris, J. W. Jr., *Mater. Trans. JIM*, **41**, pp. 888-892 (2000).

LOCAL FRACTURE TOUGHNESS OF STAINLESS STEEL WELDS AT 4 K BY J-EVALUATION ON TENSILE TEST

T. Ogata[*] T. Yuri[*], M. Saito[**], and Y. Hirayama[**]

[*]National Institute for Materials Science
Tsukuba, Ibaraki 305-0047, Japan
[**]Mitsubishi Heavy Industry, Ltd.
Nagasaki, Nagasaki, 851-0301, Japan

ABSTRACT

Local fracture toughness of top, middle, bottom, and heat-affected zones of weld joints of SUS304L and SUS316L have been evaluated by a new testing procedure of J-evaluation on tensile test (JETT) with circumferentially notched round bar specimen. Delta (δ)-ferrite contents of those welds were changed between 0% and 10%. The tests were carried out at 293, 77, and 4 K. In the case of austenitic stainless steels, being too ductile at room temperature, this test method is effective at low temperatures. Fracture toughness at 4 K obtained by JETT was a little bit higher than that obtained by conventional CT specimens and was lower in weld metals than in base metals and decreased as the content of delta ferrite increased in both 304L and 316L. Scatter in the values of fracture toughness among the locations in weld metals were less than 50 % and were considered to be caused by effects of δ-ferrite grains.

INTRODUCTION

Austenitic stainless steels are widely use in the cryogenic applications because of its ductility even at the cryogenic temperatures and hydrogen environment. However, about 5% δ-ferrite is introduced to prevent hot cracking during welding, but δ-ferrite is brittle at low temperatures. So, it is very important to evaluate the mechanical properties of structural materials including weld metals at low temperatures in the applications of cryogenic technology. Fracture toughness of cryogenic structural materials is one of indispensable properties which is required for the design of large scale facilities, such as superconducting magnets or for clean energy to transport and store liquid hydrogen, and is usually evaluated at cryogenic temperatures according to ASTM E 813 or JIS Z 2284. But the standard specimen size of these standards is one-inch compact tension and it is not

CP614, *Advances in Cryogenic Engineering:*
Proceedings of the International Cryogenic Materials Conference - ICMC, Vol. 48,
edited by B. Balachandran et al.
© 2002 American Institute of Physics 0-7354-0060-1/02/$19.00

good for the test in liquid helium. Further it is hard to obtain valid fracture toughness in the case of weld joints and thin plate. A new fracture toughness test method, proposed by Rice et al. [1], was developed by Nyilas [2,3] and examined by Nishimura [4,5] so far. In this study, we are going to evaluate the fracture toughness of weld joints of austenitic stainless steels and the effect of δ-ferrite on the fracture toughness using this newly developed testing method.

EXPERIMENTAL PROCEDURE

Material and specimens

The materials used in this study were the commercial austenitic stainless steel SUS304L, and SUS316L. The chemical compositions of base materials are listed in TABLE 1. The weld joints were made by joining plates (28t mm x 200w mm x 1000L mm) with TIG welding. The contents of Cr and Ni of filler metals were regulated using the Delong diagram and prototype welding electrodes were newly designed and fabricated so that the strength level of weld metals would be almost the same and the amount of δ-ferrite would be approximately 0%, 5% and 10%. The chemical compositions of weld metals and amount of ferrite (measured by ferrite meter and not only δ-ferrite) are also listed in TABLE 1. The weld metals also satisfy the JIS 1 grade by non-destructive inspections (screening tests and radiant ray inspections).

Sampling location of the JETT specimens in the weld joint is shown in FIGURE 1. In order to evaluate the difference of fracture toughness in the weld joints, a set of specimens were machined from weld metal, heat-affected zone, and base metal. The specimen is a round bar with circumferential notch and the dimensions of specimen are shown in FIGURE 1. The direction of the specimen is along the welding direction. The shaft diameter was 6 mm and the notch radius of 0.06 mm was induced by an electro-discharge machining.

TABLE 1. Chemical compositions of base and weld metals of SUS304L and SUS316L (wt%).

a) base metal

	C	Si	Mn	P	S	Ni	Cr	Mo	N	O
SUS304L	0.017	0.56	0.87	0.031	0.0021	9.05	18.33	0.17	0.0412	0.0028
SUS316L	0.022	0.52	0.85	0.026	0.0005	12.09	17.59	2.13	0.0380	0.0015

b) SUS304L weld metal

	C	Si	Mn	P	S	Ni	Cr	Mo	N	O	δ-ferrite
0% δ	0.012	0.41	1.53	0.021	0.002	11.65	17.26	0.02	0.033	0.0035	0-0.2
5% δ	0.011	0.40	1.52	0.022	0.003	10.50	18.46	0.01	0.035	0.0050	5.6-6.6
10% δ	0.014	0.40	1.51	0.021	0.003	9.74	19.08	0.01	0.035	0.0040	9.5-12

c) SUS316L weld metal

	C	Si	Mn	P	S	Ni	Cr	Mo	N	O	δ-ferrite
0% δ	0.011	0.41	1.46	0.020	0.002	13.90	17.01	2.03	0.034	0.0045	0-0.3
5% δ	0.017	0.42	1.45	0.021	0.002	12.82	18.15	2.04	0.034	0.0065	4.5-5.6
10% δ	0.013	0.41	1.43	0.019	0.002	12.08	19.03	2.03	0.030	0.0025	8.9-11

Test Procedure

The conventional fracture toughness according to ASTM E 813-89 was evaluated with one-inch thickness compact tension (CT) specimen. The notch and fatigue crack were induced along the welding direction at the center of weld metal.

The JETT and the conventional fracture toughness tests were carried out 293, 77, and 4 K. In the JETT, a crosshead speed of testing machine was 0.1 mm/min and two extensometers were used to cancel the bending component. A conventional equation (1) of Rice et al. [1] was applied.

$$J = \frac{1}{2\pi r^2}\left[3\int_0^{dc} Pd(dc) - P \cdot dc\right] \qquad (1)$$

where r is a radius of crack tip, P is a tension force and dc is a displacement between the load points due to introducing the crack and dc was calculated by the equation (2) below:

$$dc = d_{ex} - (G.L.)P / (E \pi R^2) \qquad (2)$$

where d_{ex} is a measured displacement by two extensometers, G.L. is a gage length, E is a Young's modulus, and R is a radius of the specimen. The Young's modulus of weld joints is usually different from that of base metal but it also varies with solidification structure. In this study, a typical value of 205 GPa for SUS304L base metal at 4 K was used also as a typical value of weld joints for the calculation.

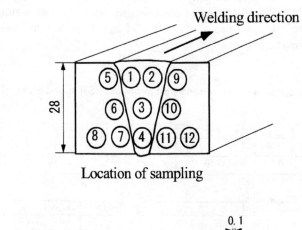

Location of sampling

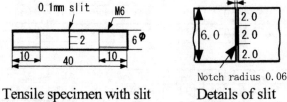

Tensile specimen with slit   Details of slit

FIGURE 1. Sampling location and dimension of JETT specimen

RESULTS AND DISCUSSIONS

Load-displacement curves

Load-displacement curves of base and weld metals of SUS304L and SUS316L at 4 K are shown in FIGURE 2. The curves of base metal and heat-affected zone were almost same and showed a larger deformation for a narrow deforming region. At 293 K deformations progressed after the necking at the notch occurred, which means that the materials tested were too ductile for even a notch tensile test and that this test method is effective only at low temperatures in the case of materials that have low strength and high toughness at room temperature. In weld metals, the amount of deformation decreased with an increase of δ-ferrite, which also resulted in a decrease of the calculated fracture toughness J-Rice.

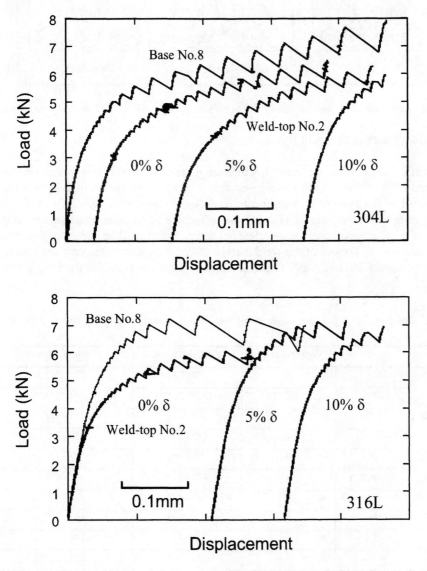

FIGURE 2. Load-displacement curves of base and weld metals of SUS304L and SUS316L at 4 K.

149

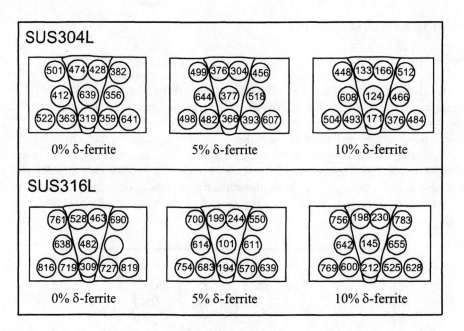

FIGURE 3. Distribution of fracture toughness J-Rice (N/mm) in the weld joints of 304L and 316L at 4 K.

Fracture toughness in the weld joints

FIGURE 3 shows distribution of fracture toughness J-Rice (N/mm) in the weld joints of 304L and 316L at 4 K. Fracture toughness was lower in weld metals than in base metals and decreased as the amount of delta ferrite increased in both 304L and 316L. One of the objectives of this research was to examine the difference of fracture toughness among the locations of weld joints, however, no clear tendency was found so far among them; the fracture toughness of bottom of the weld metal showed a smaller value than the others but it was not systematic and also those of heat-affected zone were smaller than those of base metals.

TABLE 2. Comparison of fracture toughness of weld joints obtained by JETT and CT specimen.

Material	δ -ferrite (%)	Temperature(K)	K_{Ic}(J) (MPa√m)	K_Q(J-Rice) (MPa√m)
SUS304L	Base*	4	252 - 283	270 - 362
	0%	77	334 - 381	333 - 376
	0%	4	302 - 323	255 - 362
	5%	4	135 - 153	250 - 278
	10%	4	97 - 108	159 - 187
SUS316L	Base*	4	496 - 549	342 - 410
	0%	77	419 - 442	466 - 475
	0%	4	282 - 342	251 - 329
	5%	4	173 - 177	144 - 224
	10%	4	125 - 138	172 - 217

*Base for CT was machined from hot-rolled plates and for JETT includes heat affected zone (No.5-12).

TABLE 2 lists the fracture toughness of weld joints obtained by the conventional tests using CT specimen (K_{Ic}(J)) and JETT (K_Q(J-Rice)) with their minimum and maximum value; the number of CT specimen was 2. The fracture toughness (K_{Ic}(J)) of SUS304L weld metal containing 10% δ-ferrite at 293 K was high at 400 MPa√m and more, but this fell rapidly with decrease of the test temperature and dropped to around 100 MPa√m at 4 K. The fracture toughness of SUS316L weld metal containing 10% δ-ferrite at 293 K was extremely high, but this again fell rapidly with decrease of the test temperature to around 120 MPa√m at 4 K. Furthermore, concerning SUS316L weld metals, unstable fracturing was recognized under certain test conditions. It was found with both SUS304L and SUS316L weld metals that, by reducing the amount of δ-ferrite, the fall of weld metal fracture toughness caused by decrease of the test temperature can be reduced. When the δ-ferrite content is 0 %, the fracture toughness was found to be approximately 300 MPa√m at 4 K and still high even at 4 K.

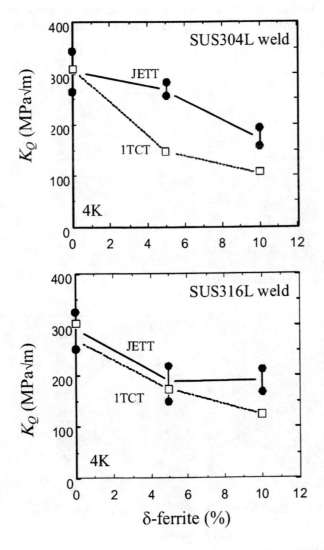

FIGURE 4. Influence of δ-ferrite on the fracture toughness of weld metal of SUS304L and 316L at 4 K.

Fracture toughness obtained by JETT (K_Q(J-Rice)) in TABLE 2 was converted from the J-Rice value (N/m) with the Young's modulus (205 GPa) and was a little bit higher than the K_{Ic}(J) for both base and weld metals, which means that the K_{Ic}(J) of weld joints often results in invalid but represents the lower fracture toughness. Fracture toughness was lower in weld metals than in base metals and decreased as the content of δ-ferrite increased in both 304L and 316L, and scatter in the values of fracture toughness among the locations in weld metals were less than 50 %. FIGURE 4 shows an influence of δ-ferrite on the fracture toughness of weld metal of SUS304L and 316L at 4 K. The effect of 5% δ-ferrite on K_Q(J-Rice) was not clear in SUS304L but significant in SUS316L.

Fracture surface

FIGURE 5 presents SEM images of fracture surface of JETT at 4 K for SUS304L. In 0% δ-ferrite, a ductile fracture surface was observed and covered with almost dimple. In 10% δ-ferrite, the surface reveled a dendritic solidification structure of welding and δ-ferrite was observed in final solidified region, which indicates the effect of δ-ferrite on the decrease of fracture toughness.

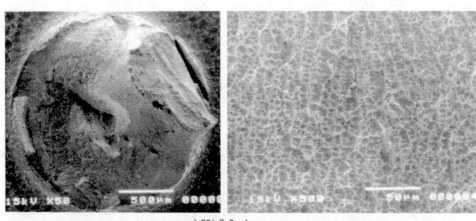

a) 0% δ-ferrite

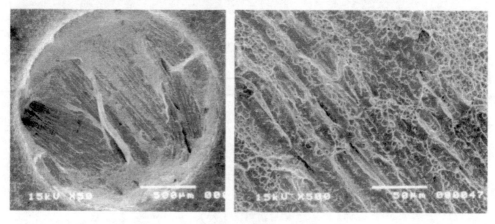

b) 10% δ-ferrite

FIGURE 5. SEM images of fracture surface of JETT at 4 K for SUS304L.

In 5% δ-ferrite, an intermediate fracture surface covered with dimple partly was observed and is corresponding to the change of fracture toughness.

Fracture surfaces in SUS316L showed a similar tendency and the effect of δ-ferrite were observed also in the surface.

Shindo [6] reported that the fracture toughness prediction accuracy is roughly equal to ±25% in JN1 weld metal, the fracture toughness K_Q(J-Rice) in these weld metals scattered 50 % at most. This scatter was considered to be caused by effects of solidification structure and/or δ-ferrite grains in the section of specimen with the diameter of 2 mm at the bottom of the notch.

CONCLUSIONS

- Fracture toughness obtained by JETT was a little bit higher than that obtained by conventional CT specimens.
- In the case of materials that have low strength and high toughness at room temperature, such as austenitic stainless steels, this test method would be effective only at low temperatures.
- Fracture toughness was lower in weld metals than in base metals and decreased as the content of δ-ferrite increased in both 304L and 316L.
- Scatter in the values of fracture toughness among the locations in weld metals were less than 50 %. This scatter was considered to be caused by effects of solidification structure, or δ-ferrite grains, in the section of specimen with the diameter of 2 mm at the bottom of the notch.

REFERENCES

1. Rice, J.R., Paris, P.C., and Merkle, J.G., "Some further results of J-integral analysis and estimates," in *ASTM Special Technical Publications* 536, (1973), pp.231
2. Nyilas, A. and Obst, B., "A new test method for characterizing low-temperature structural materials, " in *Advances in Cryogenic engineering (materials)* 42A, 1996, pp.353-360
3. Nyilas, A., Obst, B., and Nishimura, A., "Fracture Mechanics Investigations at 7 K of Structural Materials with EDM Notched Round and Double Edged-Bars, " in *Advances in Cryogenic Engineering* 44, 1998, pp153-160.
4. Nishimura, A., Yamamoto, J., and Nyilas, A., "Fracture toughness evaluation at cryogenic temperature by round bar with circumferential notch," in *Advances in Cryogenic engineering (materials)* 44A, 1998, pp.145-152
5. Nishimura, A., Ogata, T., Shindo, Y., Shibata, K., Nyilas, A., Walsh, R. P., Chan, J. W., and Mitterbacher; H., "Local Fracture Toughness Evaluation of SUS316LN Plate at Cryogenic Temperature," in *Advances in Cryogenic Engineering* 46, 2000, pp. 33-40.
6. Shindo, Y., Mano, Y., Horiguchi, K., and Sugo, T., "Cryogenic Fracture Toughness Determination of a Structural Alloy Weldment by Notch Tensile Measurement and Finite Element Analysis," in *Journal of Engineering Materials and Technology* 123, 2001, pp. 231-245.

MAGNETIC EFFECT ON YOUNG'S MODULUS
MEASUREMENT OF CP-TI AT 4K
(RESULT OF ROUND ROBIN TEST)

K. SHIBATA,[a] A. NYILAS,[b] Y.SHINDO[c] and T.OGATA[d]

[a] Dept. of Metallurgy, The Univ. of Tokyo, Tokyo 113-8656, Japan
[b] Forschungszentrum Karlsruhe, Institut für Technische Physik,
   Karlsruhe, Germany
[c] Dept. of Materials Processing, Tohoku Univ., Sendai 980-8579, Japan
[d] National Institute for Materials Science, Tsukuba 305-0047, Japan

ABSTRACT

From a practical viewpoint, it is convenient if Young's modulus can be determined without any special technique or device. There are standards to determine Young's modulus. However, the research about magnetic effects on the measurement of Young's modulus is not enough and the standard in a magnetic field has not been established especially at cryogenic temperatures. In the present research, four institutes measured Young's modulus of specimens machined from a unique commercial purity titanium plate with and without an application of magnetic fields up to 13 Tesla at liquid helium temperature, and the obtained values are compared. All participants used two or three extensometers and the slopes of the stress-strain curves obtained with them were averaged. The magnetic effect on the value of Young's modulus was not observed. But some deviation in the value was observed among participants. To use the averaging method and longer gage length is recommended.

INTRODUCTION

Young's modulus is one of important mechanical properties of structural materials used in cryogenic engineering. It is convenient from the viewpoint of practice if Young's modulus can be determined by usual tensile test, without any special method or device. Test method for determining Young's modulus is specified in some standards like ASTM E111. However, no standard exists in high magnetic fields. Therefore, the present round robin test was planned. Commercially pure titanium (CP-Ti) was chosen as co-used specimen in this round robin test. CP-Ti is paramagnetic around 4K[1]. Therefore, any magnetic effect like ΔE effect cannot be expected on Young's modulus. If any magnetic effect in Young's modulus measurement of CP-Ti is observed, it can attribute to some

CP614, *Advances in Cryogenic Engineering:*
Proceedings of the International Cryogenic Materials Conference - ICMC, Vol. 48,
edited by B. Balachandran et al.
© 2002 American Institute of Physics 0-7354-0060-1/02/$19.00

problems in measuring the modulus, for instance, strain or stress measurements. K. Shibata [2] and A. Nyilas[3] have already reported some results using the same CP-Ti. In the present paper, two new participants joined and results obtained by these four participants are analyzed.

EXPERIMENTAL

Specimen

A 30mm thick plate of CP-Ti (JIS, Class3) was granted by Nippon Steel Corporation. Figure 1 shows the procedure of producing the CP-Ti. Table 1 is the chemical composition in mass %. Table 2 is its tensile properties at 4K and 0 Tesla. Blanks were cut from the 30mm thick plate in the T-L direction and machined to specimens. Table 3 shows the geometry of specimens used at each university and institute.

Magnet

All participants used superconducting magnets. The characteristics of the magnets are

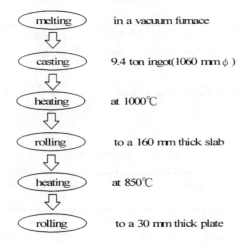

melting	in a vacuum furnace
casting	9.4 ton ingot(1060 mm ϕ)
heating	at 1000℃
rolling	to a 160 mm thick slab
heating	at 850℃
rolling	to a 30 mm thick plate

FIGURE 1. Procedure of producing CP-Ti material.

TABLE 1. Chemical composition of CT-Ti （mass%）.

Fe	C	O	N	H	Ti
0.05	0.006	0.22	0.008	0.0013	Bal.

TABLE 2. Tensile properties at 4K and 0 Tesla.

0.2% proof stress	Tensile strength	Total elongation
998MPa	1190MPa	24%

(after the University of Tokyo)

shown in Table 4. The magnet used at NIMS is a liquid He free-type, while others are operated in liquid He. The magnet used at FZK is a split-type and stress is applied in the perpendicular direction to a magnetic field. On the other hand, others are solenoid-type and stress is applied in the parallel direction to a magnetic field.

Table 5 exhibits the distance between the centers of the magnet and the load cell. In the case of the University of Tokyo, the magnetic fields are 12, 20 and 260 Gauss near the top, the center and the bottom of the load cell, respectively.

Extensometer

Strain was measured by averaging method. In this method, two or three extensometers are used and the outputs are averaging. Characteristics of the extensometers used by participants are shown in Table 6. These extensometers were calibrated in zero and magnetic fields following the continuous (at FZK) or incremental (at others) method before measuring Young's modulus.

TABLE 3. Geometry of specimens.

Participants	Diameter	Length of reduced region	Gage length	Threaded end
Univ. of Tokyo	6mm	40mm	25.4mm	M12
FZK	2.6mm	24mm	14.5mm	M4
Tohoku Univ.	7mm	45mm	25.4mm	M14
NIMS	6mm	33mm	8.0mm	M12

TABLE 4. Superconducting magnets used by each participant.

Participants	Solenoid/split	Cooling	max. field	Direction of a field to tensile load
Univ. of Tokyo	solenoid	dipped in liq. He	8T	parallel
FZK	split	dipped in liq. He	13T	perpendicular
Tohoku Univ.	soplenoid	dipped in liq. He	8.7T	parallel
NIMS	solenoid	refrigerator	6T	parallel

TABLE 5. Distance from the magnet to the load cell.

Participants	Distance between the center of the magnet and the load cell
Univ. of Tokyo	710mm
FZK	2300mm
Tohoku Univ.	1100mm
NIMS	670mm

Figure 2 shows the calibration curves obtained at Tohoku University and NIMS, respectively. The effect of a magnetic field on the calibration curve is very small and the linearity is good for the extensometers used at Tohoku University. The extensometers used at the University of Tokyo showed similar characteristics[2]. As for the extensometers used at FZK also showed a good linearity but a little larger magnetic effect on the slope (Table 7); the output voltage increased by about 1.4% whereas about 0.06% for the extensometer used at Tohoku University. In the case of the extensometers used at the University of Tokyo, an increase of about 0.4% was observed in an 8 Tesla magnetic field.

On the other hand, larger noise and lower linearity were observed in the extensometer used at NIMS, while an effect of a magnetic field on the output voltage was not almost observed. The noise may be attributed to the mechanical vibration induced by the compressor of the refrigerator. Figure 3 showed the calibration curves written with a pen recorder.   The curve for #1 extensometer shows some deflection in a 6 Tesla field.

TABLE 6.   Extensometer used at each institute.

Participants	Number of extensometer	Material	Manufacturer
Univ. of Tokyo	3	Ti alloy	John A. Shepic
FZK	2	Ti alloy	Arman Nyilas
Tohoku Univ.	3	Ti alloy	John A. Shepic
NIMS	2	Ti alloy	Arman Nyilas

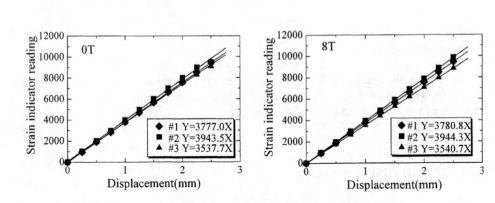

FIGURE 2.   Calibration curves of the extensometer used at Tohoku University.

TABLE 7.   Effect of a magnetic field of 13 Tesla on the output voltage per 1mm displacement (V/mm) of the extensometers used at FZK.

Field	Extensometer	run 1	run 2	run 3	run 4	run 5	mean
0 Tesla	#1	1.6157	1.6219	1.6295	1.6047	—	1.6180
	#2	1.7062	1.6823	1.6853	1.6887	—	1.6906
13 Tesla	#1	1.6391	1.6511	1.6427	1.6343	1.6299	1.6394
	#2	1.7082	1.7082	1.7077	1.7100	1.7083	1.7085

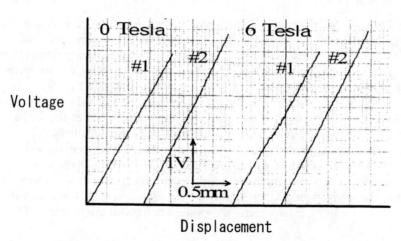

FIGURE 3. Calibration curves of the extensometers use at NIMS.

TABLE 8. Tensile machine and deformation rate.

Participants	Max. load	Stroke drive	Deformation rate
Univ. of Tokyo	100kN	hydraulic servo	0.04mm/s
FZK	100kN	step motor	0.023mm/min
Tohoku Univ.	100kN	hydraulic servo	0.02mm/s
NIMS	50kN	hydraulic servo	0.02mm/s

Measuring condition

All participants measured Young's modulus during continuous loading in liquid helium. Table 8 shows characteristics of the tensile machine and deformation rate. The elongation and the load were measured using analog and /or digital recorders, and converted to stress and strain.

RESULTS AND DISCUSSION

Values of Young's modulus obtained by all participants are shown in Fig.4 and Table 9. The mean values at 0 and a magnetic field are 120.3 and 121.0GPa, respectively. That is to say, the effect of magnetic fields on the value of Young's modulus seems to be negligibly small. However, the standard deviation seems to be fairly large comparing with the value shown in Table 10, the results of the round robin tests under VAMAS project[4]. This large standard deviation is due to small Young's modulus obtained at NIMS.

Figure 5 shows load-elongation curves obtained at NIMS. The ultimate strengths obtained are 1166MPa at 0 Tesla and 1204MPa at 6 Tesla. These values are statistically same with the value 1190MPa obtained at the University of Tokyo (Table 2). Therefore, the small value of Young's modulus cannot be attributed to the load cell. The gage length

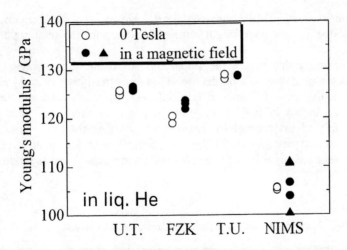

FIGURE 4. Comparison of the values of Young's modulus measured by each participant in zero and magnetic fields. U.T.: the University] of Tokyo, FZK:Forschungszentrum Karlsruhe, T.U.: Tohoku University, NIMS : National Institute for Materials Science

TABLE 9. Young's modulus measured by each participant at zero and magnetic fields.

Participants	Young's modulus at zero Tesla (GPa)		Young's modulus in a magnetic field (GPa)		
Univ. of Tokyo	125.6		126.6	8T	
	125.6		126.5	8T	
	124.9		125.8	8T	
	125.8	mean=125.5	126.3	8T	mean=126.3
FZK	120.5		122.0	13T	
	119.0		123.1	13T	
			123.7	13T	
		mean=119.8	123.7	13T	mean=123.1
Tohoku Univ.	128.1		128.7	8T	
	129.0		128.8	8T	
	129.0	mean=128.7			mean=128.8
NIMS	105.7		103.9	4T	
	105.1		106.7	4T	
	105.6		108.4	4T	
			100.3	6T	
			110.9	6T	
		mean=105.5	110.6	6T	mean=106.5
mean	120.3		118.5		
S.D.	9.4		9.8		
mean*	125.3		125.5		
S.D.*	3.5		2.3		

* others than the results at NIMS

S.D. standard deviation

of the specimen used at NIMS seems too short. However, it has remained not clear whether this shortness of the gage length is the main reason for the small value of Young's modulus or not.

Before concluding this paper, the importance of using the averaging method in measuring Young's modulus will be discussed. Figure 6 shows stress-strain diagrams obtained at the University of Tokyo. It is usual case that each extensometer exhibits a different slope as shown in the left diagram in Fig.6. In this case, Young's modulus obtained using only #1 extensometer is smaller by 10% than the mean Young's modulus obtained using three extensometers. By the way, Young's modulus obtained using only #2 and #3 extensometer are larger by 0.5% and 0.6% than the mean Young's modulus.

TABLE 10. Summary of tensile tests performed under VAMAS project. [4]

Materials		Yield strength (MPa)	Ultimate strength (MPa)	Young's modulus (GPa)
Al 2219-T87	mean	474	668	78.9
	S.D.	9	8	5.7
Ti-6SI-4V	mean	1311	1443	123
(ELI)	S.D.	12	9	3
316LN	mean	1065	1714	199
(2nd)	S.D.	15	28	9.2
316LN	mean	1039	1687	207
(1st)	S.D.	46	48	13

S.D: standard deviation

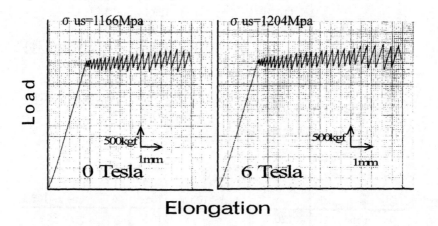

FIGURE 5. Load-elongation curves obtained at NIMS.

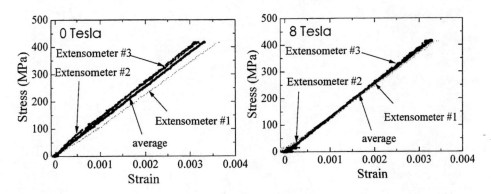

FIGURE 6.   Example of stress-strain curves obtained at the University of Tokyo.

CONCLUSIONS

The results and discussion can be concluded as follows:
(1) The magnetic effect on the value of the Young's modulus was not observed.
(2) The value obtained at NIMS was much smaller than the value obtained by other participants. It is not clear whether the small gage length is the reason for the lower Young's modulus. But such a small gage length is not recommended.
(3) Measuring Young's modulus using the averaging method is much superior to the method using only one extensometer.

ACKNOWLEDGEMENT

The authors are grateful to Nippon Steel Corporation for providing the CP-Ti, and the universities and institutes to which the authors are belonging for giving much convenience to perform the experiments. This research was financially supported by the Special Coordination of Japan Ministry of Education, Culture, Sports, Science and Technology.

REFERNCES

1. Umezawa, O. and Ishikawa, K., Electrical and Thermal Conductivities and Magnetization of Some Austenitic Steels, Titanium and Titanium Alloy at Cryogenic Temperatures, *"Cryogenics"*, **32**(1992), p.873
2. Shibata, K., Kadota, T., Kohno, Y., Nyilas, A., and Ogata, T., Mechanical properties of a boron added superalloy at 4K and magnetic effect, *Advances in Cryogenic Engineering(Materials)*, **46**(2000), pp. 73-80
3. Nyilas, A., Shibata, K., Specking, W., and Kiesel, H., Fracture and Tensile Properties of Boron Added Ni-base Superalloy at 7 and 4.2K, and the Effect of 13 Tesla Field, *Advances in Cryogenic Engineering(Materials)*, **46**(2000), pp.65-72
4. Ogata, T., and Evans, D., VAMAS Tests of Structural Material on Aluminum Alloy and Composite Material at Cryogenic Temperatures, *Advances in Cryogenic Engineering (Materials)*, **42**(1996), pp.277-284

CRYOGENIC DEFORMATION
AND FRACTURE OF
METALS AND ALLOYS

STRENGTHENING MECHANISM OF 316LN STAINLESS STEEL AT CRYOGENIC TEMPERATURES

L.F. Li[1], K. Yang[2], and L.J. Rong[2]

[1]Cryogenic Laboratory, Technical Inst. of Phys & Chem.
Chinese Academy of Sciences, Beijing 100080, China
P. O. Box 2711, Beijing 100080, P. R. China

[2]Institute of Metal Research, Chinese Academy of Sciences
Shenyang 110015, China

ABSTRACT

Using vacuum metallurgical technique, we have fabricated the metastable austenitic stainless steel, 316LN which is aimed to be used as structural materials for superconducting magnets in Tokamak equipment in China. We have studied the basic properties of due materials at cryogenic temperatures, such as tensile strength, yield strength and Young's modulus. The effects of various elements on mechanical behaviors are discussed.

INTRODUCTION

The candidate materials for structural application in superconducting magnet such as the magnet cases, cable tube, and supports are stainless steel[1] and advanced ceramics[2]. Recently, in China, a project named HT7-U was started in the city of Hefei. In HT7-U, NbTi coils use a rectangular cable-in-conduit conductor with a 316LN jacket. The schematic diagram of the jacketing and forming process for a 10kA CICC was shown in figure 1. The size of a single 316LN jacket tube shall be 25mm in out-diameter, 23mm in inner-diameter, and over 10m in length; the mechanical properties shall meet following requirements: for yield strength, shall be over 350 MPa, ultimate strength, over 650MPa, elongation, over 30% at room temperature, and for yield strength shall be over 1200MPa, ultimate strength, over 1600MPa, elongation, over 30% at liquid helium temperature.

According to these requirements, we selected 316LN as the suitable candidate CICC jacketing tube material. 316LN was widely investigated in the world involving cryogenic application, due to its good mechanical properties at cryogenic temperatures [4-6].

CP614, *Advances in Cryogenic Engineering:*
Proceedings of the International Cryogenic Materials Conference - ICMC, Vol. 48,
edited by B. Balachandran et al.
© 2002 American Institute of Physics 0-7354-0060-1/02/$19.00

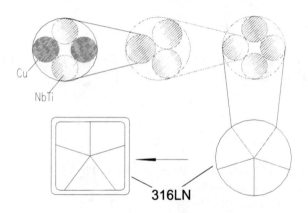

FIGURE 1. Schematic diagram of the jacketing and forming for a 10kA test CICC.

316LN is a kind of metastable austenitic stainless steel. It will undergo a stress-induced transformation from austenite to martensite at cryogenic temperature, and the resulting transformation will enhance strength and toughness. In general, transformation austenitic steels can be classified into athermal and isothermal types, which means different amounts of martensite phase and microstructure. In athermal transformation, the amount of martensite after the initial occurrence of the transformation increases little during a constant holding temperature, but, in isothermal transformation, the amount increases substanitially during isothermal holding. In the view of microstructure, one is α' martensite, and the other is ε martensite. α' phase will cause problems at high magnetic field due to its ferromagnetism. For example, the 304 steel, the amount of α' phase increases with the increase of magnetic filed, however, in 316LN steel, the amount of α' induced by magnetic fields was very small[6], and thus, it is suitable for fusion reactor application.

Since the requirement in strength of jacketing tube in HT7-U project is very high, and the commercial available 316LN cannot meet the requirement, therefore, we try to improve its low temperature tensile properties by changing the chemical composition, especially in the balance of C and N.

MATERIAL AND EXPERIMENTS

The material used in this study was obtained by vacuum melting method. The chemical composition is listed in table 1. Tensile specimens were 3mm in diameter with a gage length of 25mm. All specimens were solution-treated at 1338K for 1 hour and then quenched. Tensile tests were carried on at 298K, 77K and 4.2K. For each temperature, five specimens were tested. A 10 ton capacity screw-driven machine was used for all tensile tests, with a crosshead velocity of 0.5mm/min. Yield strength was evaluated at 0.2% strain offset. The microstructure of the steel was single-phase austenite, with an average grain size of 20 μm. The fractured face analysis was performed on a S-360 microscope. TEM observation was performed on a Philips EM 420 electron microscope.

TABLE 1 Chemical composition of materials used in this test (in mass %, Fe balance)

C	Si	Mn	S	P	Cr	Ni	Mo	N
0.010	0.21	1.54	0.006	0.007	19.33	10.09	2.13	0.20

RESULTS AND DISCUSSIONS

Figure 2 shows a typical tensile curve of 316LN specimen at 4.2K. From the load – time curve, it is found that the load which corresponds stress in the sample changes in a wave-like way. This behavior reveals the phase transformation under stress, i.e., stress induced transformation, and in return, the transformation induced strengthening and plasticity. Table 2 shows the experimental results of as-studied 316LN steel at 298K, 77K and 4.2K.

Table 2 Tensile properties of 316LN stainless steel

Temperature K	Spec. #	Yield strength MPa	Ultimate strength MPa	Elongation %
4.2	1	1359	1795	59
4.2	2	1334	1768	54
4.2	3	1260	1680	45
4.2	4	1266	1706	44
4.2	5	1296	1703	45
77	6	890	1340	54
77	7	910	1362	49
77	8	905	1355	50
77	9	932	1390	49
77	10	921	1381	51
298	11	410	700	56
298	12	413	720	55
298	13	415	718	53
298	14	401	680	46
298	15	406	678	46

The results show great increases in both yield strength and ultimate strength compared with the results at room temperature. This is due to the stress induced transformation contributions. Moreover, these results are also higher than those reported on this material by cryogenic structural materials subgroup in the Versailles Project on Advanced Materials and Standards(VAMAS) in 1994[4]. Here, we believe that the reason for those divergent results is due to the slight differences of nitrogen and carbon contents. The function of nitrogen addition is to increase strength because nitrogen is a strengthening element in stainless steel. Nitrogen content is about 0.18% in VAMAS's materials, the nitrogen content in our material, however, is 0.2%.

In the view of stress-induced martensitic transformation, the contributions on strength and toughness are highly dependent on composition and grain size because they affect the martensitic start temperature(Ms). If the Ms temperature is slightly lower than testing temperature, the mechanical gain from stress-induced transformation will be higher. Although we haven't measured the Ms temperature for the studied material, we believe that

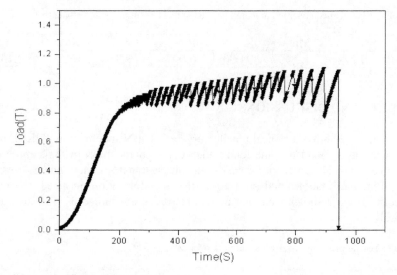

FIGURE 2. Typical tensile curve for 316LN at 4.2K

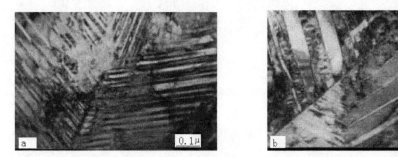

FIGURE 3 TEM observations on fractured samples at 4.2K (a) and 77K (b).

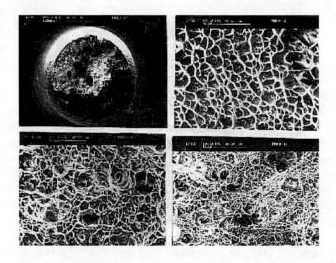

FIGURE 4 SEM pictures of fracture surface at 4.2K which show well toughening fractograph

it is much lower because the carbon content in our material is only 0.01%, and it is well known that with the decrease of carbon content in stainless steel, the Ms temperature will decrease. In order to know the stress induced phase transformation and the contribution to strength, TEM observations were carried out on the samples fractured at 4.2K (figure 3a) and 77K (figure 3b). From TEM pictures, it is indicated that the stress-induced martensitic phase is in plate-like form, however, the width of martensitic plate in the sample fractured at 4.2K (see figure 3a) becomes fine and narrow . This phenomenon might be related to the further transformation from large size martensitic plate appeared at 77K(see figure 3b) when the stress is applied and the temperature is down. The fine martensitic plate like structure will benefit the increase in strength.

We haven't measured the fracture toughness, however, from figure 4 in which it shows the SEM observation results on the fractured surface of 316LN at 4.2K, it indicates the fracture mechanism is typical tough style.

CONCLUSIONS

By controlling carbon and nitrogen contents, the 316LN steel was fabricated with good mechanical properties at 4.2K, thus satisfying the requirements for fusion reaction application used as CICC jacketing tube material. The low carbon content and high nitrogen content are main agents to improve the mechanical properties for metastable austenitic style steel.

ACKNOWLEDGEMENTS

The authors greatly appreciate the help in mechanical measurements by L.Z. Zhao, W.H. Wang and Y.F. Xiong, and the financial supports from the Institute of Plasma, CAS and National Natural Science Foundation of China, the contract number is 19732001.

REFERENCES

1. Chan, J.W., Chu, D., Tseng, C., and Morris, J.W., "Cryogenic Fracture Behavior of 316LN in Magnetic Fields up to 14.6T," in *Advances in Cryogenic Engineering* 40B, edited by R.P. Reed et al, Plenum Press, New York, 1994, pp. 1215-1221.
2. Li, L.F., Li, Y.Y. and Meriani, S. , "ZrO2-CeO2 Alloys as Candidate Structural Materials for Cryogenic Application," J.Am. Ceram. Soc. **80** , pp. 1005-1008(1997).
3. Drexler, E.S., Simon,N.J. and Reed, R.P., "Strength and Toughness at 4K of Forged, Heavy-Section 316LN, " in *Advances in Cryogenic Engineering* 40B, edited by R.P. Reed et al, Plenum Press, New York, 1994, pp. 1199-1206.
4. Ogata,T, Nagai, K. and Ishikawa, K., "VAMAS Tests of Structural Materials at Liquid Helium Temperature," in *Advances in Cryogenic Engineering* 40B, edited by R.P. Reed et al, Plenum Press, New York, 1994, pp. 1191-1198.
5. Nishimura, A., Ogata,T., Shindo, Y., Shibata, K., Nyilas, A., Walsh, R.P., Chan, J.W., and Mitterbacher, H., "Local Fracture Toughness Evaluation of 316LN Plate at Cryogenic Temperature," in *Advances in Cryogenic Engineering* 46A, edited by Balachandran et al., Kluwer Academic/ Plenum Publishers, 2000, pp. 33-40.
6. Shibata,K., Kurita, Y., Shimonosono, T., Murakami, Y., Awaji, S. And Watanabe K., "Effects of High Magnetic Fields on Martensitic Transformation at Cryogenic Temperatures for Variously Heat Treated Stainless Steels," in *Advances in Cryogenic Engineering* 40B, edited by R.P. Reed et al, Plenum Press, New York, 1994, pp. 1207-1213.

PLASTIC STRAIN INDUCED DAMAGE EVOLUTION AND MARTENSITIC TRANSFORMATION IN DUCTILE MATERIALS AT CRYOGENIC TEMPERATURES

C. Garion, B.T. Skoczen

CERN, LHC Division
CH-1211, Geneva 23, Switzerland

ABSTRACT

The Fe-Cr-Ni stainless steels are well known for their ductile behaviour at cryogenic temperatures. This implies development and evolution of plastic strain fields in the stainless steel components subjected to thermo-mechanical loads at low temperatures. The evolution of plastic strain fields is usually associated with two phenomena: ductile damage and strain induced martensitic transformation. Ductile damage is described by the kinetic law of damage evolution (cf. [1]). Here, the assumption of isotropic distribution of damage (microcracks and microvoids) in the Representative Volume Element (RVE) is made. Formation of the plastic strain induced martensite (irreversible process) leads to the presence of quasi-rigid inclusions of martensite in the austenitic matrix. The amount of martensite platelets in the RVE depends on the intensity of the plastic strain fields and on the temperature. The evolution of the volume fraction of martensite is governed by a kinetic law based on the accumulated plastic strain (cf. [2]). Both of these irreversible phenomena, associated with the dissipation of plastic power, are included into the constitutive model of stainless steels at cryogenic temperatures. The model is tested on the thin-walled corrugated shells (known as bellows expansion joints) used in the interconnections of the Large Hadron Collider, the new proton storage ring being constructed at present at CERN.

INTRODUCTION

The Fe-Cr-Ni stainless steels are commonly used to manufacture components of superconducting magnets and cryogenic transfer lines since they retain their ductility at low temperatures and are paramagnetic. The nitrogen-strengthened stainless steels of series 300 belong to the group of metastable austenitic alloys. Under certain conditions the steels undergo martensitic transformation at cryogenic temperatures that leads to a considerable evolution of material properties and to a ferromagnetic behaviour. The martensitic

CP614, *Advances in Cryogenic Engineering:*
Proceedings of the International Cryogenic Materials Conference - ICMC, Vol. 48,
edited by B. Balachandran et al.
© 2002 American Institute of Physics 0-7354-0060-1/02/$19.00

transformations are induced mainly by the plastic strain fields and amplified by the high magnetic fields. The stainless steels of series 300 show at room temperature a classical γ-phase of face centred cubic austenite (FCC). This phase may transform either to α' phase of body centred tetragonal ferrite (BCT) or to a hexagonal ε phase. The most often occurring γ-α' transformation leads to formation of the martensite dispersed in the surrounding austenite matrix. In the course of the strain-induced transformation the martensite platelets modify the FCC lattice leading to local distortions. The volume fraction of the martensite (ξ) depends on the chemical composition, temperature, stress state, plastic strains and exposure to a magnetic field. It is well known that solutes like Ni, Mn and N considerably stabilise the γ-phase. For instance the strain-induced martensitic content in the grades 304LN, 316LN at low temperatures is much lower than in the grades 304L, 316L for the same level of plastic strain [3]. Simultaneously, a plastic strain driven by evolution of ductile damage occurs in the stainless steel under thermo-mechanical loads at cryogenic temperatures. This dissipative and irreversible process leads to creation of microcracks and microvoids and to a "softening" of the material. The intensity of isotropic damage fields is described by a scalar parameter D, related directly to the accumulated plastic strain p. Damage evolution starts above the so-called damage threshold p_D corresponding to the minimum accumulated plastic strain necessary to initiate the process. When the damage parameter reaches its critical value D_{cr} an onset and propagation of a macrocrack is observed. Further development of the macrocrack is governed by the fracture mechanics.

Both the martensite inclusions, microcracks and microvoids coexist in the same volume of material (RVE, FIG 1). The volume fraction of martensite is expressed by the following ratio:

$$\xi = \frac{dV_\xi}{dV} \ ; \ 0 \le \xi \le 1 \tag{1}$$

where V_ξ denotes the volume of the martensitic phase and V stands for the total volume.

Damage is represented by a scalar damage variable, that is defined as a surface density of microvoids and microcracks. If dS_D denotes the surface of intersection of the micro-voids and micro-cracks with a given plane within the RVE and dS stands for the total intersection surface then the damage parameter is defined as:

$$D = \frac{dS_D}{dS} \ ; \ 0 \le D \le 1 \tag{2}$$

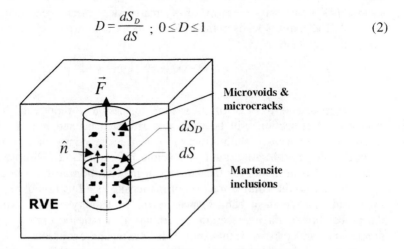

FIGURE 1. The Representative Volume Element with martensite inclusions and damage

171

The increase in martensite fraction promoted by plastic deformation can be detected by measuring the magnetic permeability μ. The evolution of μ at low temperature corresponding to monotonic straining as well as to the cyclic loads for 304L and 316L stainless steels was investigated by Suzuki et al. [4]. Tensile properties of stainless steels at low temperatures are strongly influenced by the plastic strain induced martensitic transformation. As a result of the transformation the initially homogenous γ-phase loses its homogeneity because of the inclusions of the harder martensite phase. The martensite platelets embedded in the soft austenite matrix provoke local stress concentration and block the movement of dislocations. Therefore the onset of the martensitic transformation leads to an increase of the strain hardening. On the other hand, the intensity of ductile damage can be detected by measurements of the evolution of modulus of elasticity. It is assumed that the elastic properties of the dual-phase material are not modified by the martensitic transformation (the elastic properties of the martensite and of the austenite are quite similar, cf. [5], [6]). Since the effective modulus of elasticity is defined as (cf. [1]):

$$\tilde{E} = E(1 - D) \tag{3}$$

the evolution of D can be identified via measurements of $\tilde{E}$ on the unloading paths. A simple monotonic tensile test is sufficient to determine the evolution of damage associated with the accumulated plastic strain. Thus, both the intensity of martensitic transformation and the evolution of damage fields can be measured by using the simple and well established methods.

In the present paper the relevant kinetic evolution laws will be formulated and the complete set of constitutive equations allowing analysis of both phenomena at cryogenic temperatures will be developed (cf. [2]). The constitutive model is tested on the thin-walled corrugated shells (bellows expansion joints), used in the interconnections of the LHC at CERN, subjected to thermo-mechanical loads at cryogenic temperatures.

KINETICS OF MARTENSITIC TRANSFORMATION

Transformation kinetics has been developed by Olson and Cohen [7]. The authors attribute the strain-induced martensite nucleation sites to the shear-band intersections (the shear-bands being in the form of ε' martensite, mechanical twins or stacking-fault bundles). The analysis leads to the following equation for the volume fraction of martensite versus plastic strain:

$$\xi_{\alpha'} = 1 - \exp\left\{-\beta\left[1 - \exp\left(-\alpha\varepsilon^p\right)\right]^n\right\} \tag{4}$$

where α represents the rate of shear-band formation, β represents the probability that a shear-band intersection will become a martensite site and n is a fixed exponent. The transformation curves (volume fraction of martensite versus plastic strain) show a typical sigmoidal shape with saturation levels below 100 % (FIG. 2). Olson and Cohen developed a one-dimensional model for the kinetics of martensitic transformation, called OC model. The evolution of the volume fraction of martensite as a function of plastic strain is derived by considering the shear band formation, the probability of shear-band intersections and the probability of an intersection generating a martensitic embryo. In this model, only temperature and plastic strain control the martensite evolution. Different improvements have been brought to this model, covering the influence of the stress state [5] and the strain

172

rate [6]. However, a considerable number of parameters have to be identified for these models. In the present paper, a simplified model - developed by Garion and Skoczen [2] for cryogenic applications – is applied. The volume fraction of martensite ξ can be presented in the following form:

$$\xi = \xi\left(p, T, \underline{\dot{\varepsilon}}^P, \underline{\sigma}\right) \tag{5}$$

where p is the accumulated plastic strain defined by:

$$p = \int_0^t \sqrt{\frac{2}{3} \underline{\dot{\varepsilon}}^P : \underline{\dot{\varepsilon}}^P} \, d\tau \tag{6}$$

$\underline{\dot{\varepsilon}}^P$ denotes the plastic strain rate and $\underline{\sigma}$ is the stress tensor. Under isothermal conditions and for a given strain rate, the classical sigmoidal curve has the following form (FIG. 2):

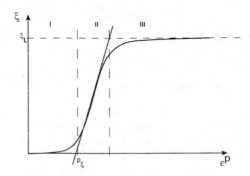

FIGURE 2. Volume fraction of martensite versus plastic strain at cryogenic temperatures (4K, 77K)

The curve is decomposed into 3 phases: *phase I* corresponds to a nonlinear increase of the martensitic content with the plastic strain, *phase II* - the α' volume fraction (ξ) is linearly related to plastic deformation (ε^P) and *phase III* corresponds to a saturation effect (ξ_L). A simplified law of evolution of the martensite content has been proposed (cf. [2]) for the phase II under the following form:

$$\dot{\xi} = A\left(T, \underline{\sigma}, \underline{\dot{\varepsilon}}^P\right) \dot{p} H\left(\left(p - p_\xi\right)\left(\xi_L - \xi\right)\right) \tag{7}$$

where A is a function of the temperature, the stress and the plastic strain rate, p_ξ is the accumulated plastic strain threshold (to trigger the formation of martensite), ξ_L is the martensite content limit, over which the martensitic transformation rate is considered equal to 0. Finally, H represents the Heavyside function ($H(x)=1$ if $x \geq 0$; $H(x)=0$ if $x < 0$).

KINETICS OF EVOLUTION OF DUCTILE DAMAGE

The damage variable, as introduced by Kachanov [8], has been defined on the basis of the irreversible thermodynamic processes leading to nucleation and growth of the micro-voids and micro-cracks in the entire volume of a sample. Here all types of voids and cracks (inter- and trans-granular) that spoil integrity of the material are accounted for. The damage parameter, as defined by Eq. 2, is a non-negative state variable satisfying the following inequality:

$$0 \le D \le D_{cr},\tag{8}$$

where $D = 0$ for a non-damaged material and $D = D_{cr}$ for a total decohesion of the sample (usually $D_{cr} \le 1$, FIG. 3).

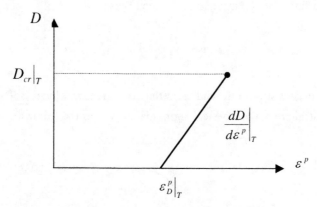

FIGURE 3. Linearized evolution of damage at a given temperature T: $\varepsilon_D^p\big|_T$ denotes damage threshold

Since the damage evolution is a "dissipative" phenomenon similar to the evolution of the plastic strain fields, a similar formulation for the potential of dissipation is applied (a continuous, convex scalar function of the dual variables):

$$\dot{D} = \frac{\partial F}{\partial Y}\lambda \ ; \ F = \frac{S}{(s+1)(1-D)}\left(\frac{Y}{S}\right)^{s+1}\tag{9}$$

where S represents the strength energy of damage (material parameter) and s has to be identified by using the experimental data. Y denotes the strain energy density release rate expressed by:

$$Y = \frac{1}{2}K\underline{\underline{\varepsilon}}^e\underline{\underline{\varepsilon}}^e\tag{10}$$

Such a formulation yields the following kinetic law of damage evolution (cf. Lemaitre [1]):

$$\dot{D} = \left(\frac{Y}{S}\right)^s \dot{p}H(p - p_D)\tag{11}$$

where p_D is the so-called damage threshold.

CONSTITUTIVE EQUATIONS OF ELASTO-PLASTIC MATERIAL WITH EVOLUTION OF DAMAGE AND MARTENSITE CONTENT

The variables associated with the state functions can be derived from a state potential. It is assumed that the state potential is assimilated with the Helmholtz free energy and that for an elasto-plastic material and an isothermal process it can be expressed as:

$$\Psi = \Psi(\underline{\underline{\varepsilon}}^e, r, \underline{\alpha}, D) = \Psi(\underline{\underline{\varepsilon}} - \underline{\underline{\varepsilon}}^p - \underline{\underline{\varepsilon}}^{th} - \underline{\underline{\varepsilon}}^{PT}, r, \underline{\alpha}, D)\tag{12}$$

or, in the developed form:

$$\Psi = \frac{1}{\rho}\left(\frac{1}{2}(1-D)\underline{\underline{K}}\,\underline{\varepsilon}^e\,\underline{\varepsilon}^e + \frac{1}{3}H(\xi)\underline{\underline{\alpha\alpha}}\right) + \Psi_r \tag{13}$$

where $\underline{\underline{\varepsilon}}^{PT}$ is the distortional strain field associated with the phase transformation, H is the kinematic hardening modulus, $\underline{\alpha}$ is the back strain and Ψ_r stands for the part of the Helmholtz potential associated with the isotropic hardening. Here, the assumption is made that both the density ρ and the elastic stiffness tensor $\underline{\underline{K}}$ depend weakly on the volume fraction of martensite ξ. Based on the Helmholtz state potential the following set of constitutive equations can be derived:

- kinetics of the martensitic transformation:

$$\dot{\xi} = A\!\left(T,\underline{\underline{\sigma}},\dot{\underline{\varepsilon}}^p\right)\dot{p}H\big((p-p_\xi)(\xi_L-\xi)\big) \tag{14}$$

- kinetics of damage evolution:

$$\dot{D} = \left(\frac{Y}{S}\right)^s \dot{p}H(p-p_D) \tag{15}$$

- the constitutive law:

$$\underline{\underline{\sigma}} = \underline{\underline{K}} : \left(\underline{\underline{\varepsilon}} - \underline{\underline{\varepsilon}}^p - \underline{\underline{\varepsilon}}^{th} - \underline{\underline{\varepsilon}}^{PT}\right) \tag{16}$$

- the yield surface:

$$f_r\!\left(\underline{\tilde{\sigma}},X,R\right) = \sqrt{\frac{3}{2}\big(\underline{\tilde{s}}-\underline{X}\big):\big(\underline{\tilde{s}}-\underline{X}\big)} - \sigma_y - R = 0 \tag{17}$$

- the normality rule:

$$d\underline{\underline{\varepsilon}}^p = \frac{3}{2}\frac{\underline{\tilde{s}}-\underline{X}}{J_2\!\left(\underline{\tilde{\sigma}}-\underline{X}\right)}\frac{d\lambda}{1-D} \tag{18}$$

- the kinematic hardening:

$$\underline{\dot{X}} = \frac{2}{3}H(\xi)\underline{\dot{\varepsilon}}^p(1-D) = \frac{2}{3}H(\xi)\underline{\dot{\alpha}} \tag{19}$$

where:

$$H(\xi) = C(\xi) + 3\beta(1-\xi)\big[\mu_{MT}(\xi) - \mu_{t0}\big] \tag{20}$$

- the isotropic hardening:

$$\dot{R} = b(\xi)[R_\infty(\xi) - R]\dot{p}(1-D) = b(\xi)[R_\infty(\xi) - R]\dot{r} \tag{21}$$

where:

$$b(\xi) = (1-\xi)(1-\beta) \tag{22}$$

$$R_\infty(\xi) = 3\big[\mu_{MT}(\xi) - \mu_{t0}\big] \tag{23}$$

Here, $C(\xi)$ is the hardening modulus of biphase material, β stands for the Bauschinger parameter and μ_{MT} denotes the tangent modulus resulting from the Mori-Tanaka homogenisation. Also, in the above given set of equations the concept of effective stress and effective deviatoric stress has been used:

$$\underline{\underline{\tilde{\sigma}}} = \frac{\underline{\underline{\sigma}}}{1-D} \quad ; \quad \underline{\tilde{s}} = \frac{\underline{s}}{1-D} \tag{24}$$

It is worth pointing out that no coupling between the evolution of damage and volume fraction of martensite is assumed at the level of the kinetic evolution laws. This assumption may turn out to be too strong and therefore needs to be verified via further experimental research under cryogenic conditions.

NUMERICAL IMPLEMENTATION AND RESULTS

The model presented in the previous chapter has been numerically implemented in the FE code CASTEM 2000. The method of the type "return mapping" is used to integrate the constitutive equations for an active plastic process. The return mapping algorithm is based on the elastic-plastic split, by first integrating the elastic equations to obtain an elastic predictor, which is used as initial condition for the plastic return. The constitutive model is tested on the thin-walled stainless steel corrugated shells, known as bellows expansion joints, used in the interconnections of the LHC. The bellows are subjected to particularly severe loads (thermomechanical cycles between ambient and 1.9K, internal pressure up to 2 MPa) therefore the analysis of accumulation of plastic strains, damage and volume fraction of martensite is of particular interest for the reliability of the collider. Half convolution of a typical bellows is shown in FIG. 4 (axisymmetric FE model). The bellows is subjected to an axial stretch of 100% of its initial length at 77K.

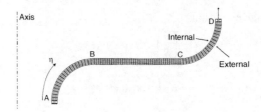

FIGURE 4. Model of half-convolution of a cryogenic bellows

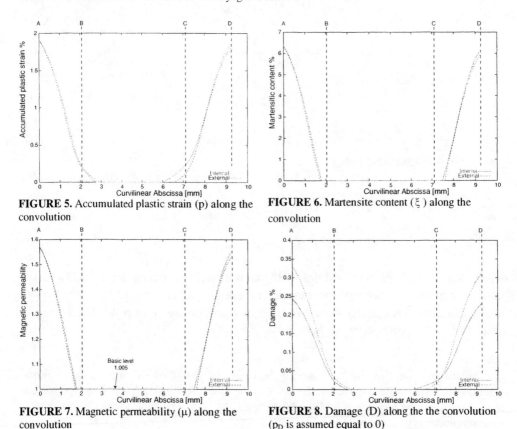

FIGURE 5. Accumulated plastic strain (p) along the convolution

FIGURE 6. Martensite content (ξ) along the convolution

FIGURE 7. Magnetic permeability (μ) along the convolution

FIGURE 8. Damage (D) along the the convolution (p_D is assumed equal to 0)

The accumulation of plastic strains and the corresponding evolution of the volume fraction of martensite is shown in FIG.5 and FIG.6, respectively. The magnetic permeability (corresponding to the field of 1000 Oe) is presented in FIG.7. Finally, the evolution of the damage parameter along the convolution is shown in FIG.8.

CONCLUSIONS

In the present paper a constitutive model of plastic flow, accompanied by the strain induced martensitic trasformation and the evolution of isotropic ductile damage at cryogenic temperatures, has been presented. The description of the plastic strain induced martensitic transformation is based on the kinetics developed by Olson and Cohen. However, a new kinetic law of evolution of the volume fraction of martensite, corresponding to phase II of the sigmoidal transformation curve, has been proposed. The evolution of isotropic ductile damage is based on the kinetic law developed by Lemaitre. The accumulation of ductile damage is also driven by the evolution of plastic strains. Both the martensitic transformation and the evolution of ductile damage are irreversible and dissipative phenomena that occur at cryogenic temperatures and lead to the coexistence of martensite inclusions, microcracks and microvoids. The mutual influence of phase transformation and ductile damage is not yet well determined. Therefore, both kinetic laws are not coupled. This particular problem needs further extensive study. Also, a further research shall be directed towards the identification of the parameters of both kinetic laws at different cryogenic temperatures.

Application of the numerical model to the analysis of evolution of plastic strains, volume fraction of martensite and ductile damage in the cryogenic bellows, designed for the interconnections of the LHC, shows a strong concentration of damage and martensite at root and at crest of the convolutions. This yields a risk of propagation of a macrocrack once the damage parameter will reach its critical level (D_{cr}). Also, an enhanced level of the magnetic permeability at root and at crest can be expected. Both effects have to be confirmed by the appropriate measurements and should be taken into account when designing a reliable cryogenic mechanical compensation system.

REFERENCES

1. Lemaitre, J., *A Course on Damage Mechanics*, Springer Verlag, Berlin, 1992.
2. Garion, C., Skoczen, B., *ASME Journ. Appl. Mech.*, under review, 2001.
3. Morris, J. W., Chan, J. W. and Mei, Z., *Cryogenics* **32** *ICMC supplement*, pp. 78-85 (1992).
4. Suzuki, K., Fukakura, J. and Kashiwaya, H., *Journ. of Test. and Eval.* **16**, pp. 190-197 (1988).
5. Stringfellow, R. G., Parks, D. M. and Olson, G. B., *Acta. Metall. Mater.* **40**, pp. 1703-1716 (1992).
6. Levitas, V.I., Idesman, A.V. and Olson, G.B., *Acta. Mater.* **47**, pp. 219-233 (1999).
7. Olson, G. B. and Cohen, M., *Metallurgical Transactions* **6A**, pp. 791-795 (1975).
8. Kachanov, L.M., *Introduction to Continuum Damage Mechanics*, Martinus Nijhoff Dordrecht, 1986.

VALIDATION OF A SMALL PUNCH TESTING TECHNIQUE TO CHARACTERIZE THE CRYOGENIC FRACTURE PROPERTIES OF AUSTENITIC STAINLESS STEEL WELDS

K. Horiguchi and Y. Shindo

Department of Materials Processing,
Graduate School of Engineering, Tohoku University,
Aoba-yama 02, Sendai 980-8579, Japan

ABSTRACT

This study was performed to demonstrate the feasibility of performing liquid helium temperature (4 K) small punch (SP) tests on austenitic stainless steel weld for Large Helical Device (LHD) superconducting magnets. The SP specimens ($10 \times 10 \times 0.5$ mm) were prepared from the different locations of electron-beam weld in type 316 stainless steel to examine the variation of the fracture properties in the weld and its heat affected zone. Previously proposed correlations for SP and elastic-plastic fracture toughness test methods were applied to predict a SP test-based fracture toughness from equivalent fracture strain. The correlation was also found between the room temperature Vickers hardness of the heat affected zone and its fracture toughness at 4 K. A finite element analysis was performed to convert the experimentally measured load-displacement data into useful engineering information. The maximum strain energy density was calculated and correlated with equivalent fracture strain.

INTRODUCTION

Type 316 stainless steel welds are used for the main structural supports of the superconducting magnet systems needed for Large Helical Device (LHD) [1]. These weldments must be capable of withstanding high stresses at liquid helium temperature (4 K). Structural failures most commonly occur at welds, originating from fabrication defects or service-induced cracking. Hence, there has been immense interest in evaluating the cryogenic fracture properties of welds to ensure safety during all anticipated operating and accident conditions.

CP614, *Advances in Cryogenic Engineering:*
Proceedings of the International Cryogenic Materials Conference - ICMC, Vol. 48,
edited by B. Balachandran et al.
© 2002 American Institute of Physics 0-7354-0060-1/02/$19.00

At present, elastic-plastic fracture toughness J_{IC} is usually determined using compact tension (CT) specimens, which are rather expensive to manufacture and require relatively sophisticated laboratory equipment to test. The first cryogenic standard for J_{IC} test appeared in 1998 (JIS Z 2284) [2] as a direct result of the U.S. - Japan Cooperative Program for Development of Test Standards. The standard uses CT specimens with fatigue precracking to provide a sharp crack front. In some specific cases, however, the use of this standardized test specimen may be impractical, if not impossible. For instance, it is well known that weldments contain regions of different fracture properties. To evaluate the fracture toughness of weldments, a test specimen much smaller than the standard CT specimen would be extremely desirable. In an effort to overcome specimen size problems, recent studies have been conducted in the development of small size specimen tests which yield fracture property data. Shindo et al. [3] have examined the use of small punch techniques to estimate fracture toughness of austenitic stainless steels and weld metals at 4 K, and have assessed correlations between equivalent fracture strain, SP energy, and J_{IC}. In order to explain the experimental results, a finite element analysis is also performed to convert the experimentally measured load-displacement data into useful engineering information. The maximum strain energy density is calculated and correlated with J_{IC}. Using circumferentially notched bar specimens, Shindo et al. [4] have also investigated the cryogenic fracture toughness of austenitic stainless steels and weld metals through experimental and analytical characterizations.

The principal objective of this research is to establish the suitability of a novel small specimen testing technique, the small punch (SP) tests, for cryogenic fracture characterization of austenitic stainless steel weld and its heat affected zone. SP tests were performed with thin plate specimens at 4 K. A method outlined by Shindo et al. [3] is applied to predict a SP test-based fracture toughness from equivalent fracture strain. Experimental relationships between cryogenic fracture toughness and room temperature Vickers hardness are also discussed. A finite element analysis was performed to compute the strain energy density. Correlations between maximum strain energy density and equivalent fracture strain are assessed.

EXPERIMENTAL PROCEDURE

Materials and Specimens

Electron-beam weld in type 316 stainless steel was supplied from the National Institute for Fusion Science (NIFS) in the form of a 75-mm-thick plate. This weld was measured at 3 to 5 % delta ferrite. The yield strength σ_{YS}, ultimate tensile strength σ_{TS}, effective yield strength $\sigma_Y = (\sigma_{YS} + \sigma_{TS})/2$, and Young's modulus E obtained for the test materials at 4 K are shown in Table 1. Hardness measurements were performed using a Vickers indenter with a test load of 2 N at room temperature.

TABLE 1. Mechanical properties of test materials at 4 K.

	σ_{YS} (MPa)	σ_{TS} (MPa)	σ_Y (MPa)	E (GPa)
Base metal	621	1642	1132	202
Weld metal	671	1626	1149	195

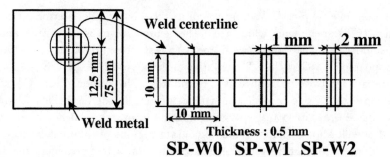

FIGURE 1. Specimen locations.

Small, thin plate specimens of $10 \times 10 \times 0.5$ mm used for SP tests were sliced. The SP specimen locations are shown in Fig. 1. The SP specimens extracted from three different locations (SP-W0, SP-W1 and SP-W2) were oriented with the thickness direction parallel to the welding direction, and located at 12.5 mm from the top surface of the weld to the center of the specimen.

SP Testing Method

Using a 10 kN screw-driven test machine, all SP tests were conducted in a cryostat with the sample immersed in liquid helium at 4 K. The punch and the specimen holder, designed for SP tests, are shown in Fig. 2. The SP specimen holder consisted of an upper and lower die and four clamping screws. Using this specimen holder clamped, it was possible to prevent specimens from cupping upward during punching, and therefore the plastic deformation was concentrated in the region below the punch. The bore diameter of the lower die (3.4 mm) was designed such that the deformed SP specimens were not subjected to friction forces arising from the contact of the SP specimens with the inner wall of the lower die hole. The punch deformation was performed perpendicularly to the plate with a 2.4 mm steel ball. The crosshead speed was 0.2 mm/min. Displacement was measured by measurement of the motion of the punch relative to the lower die using a clip gage. After testing, the fracture surfaces were examined by scanning electron microscopy (SEM).

FINITE ELEMENT ANALYSIS

The criterion used for fracture is the strain energy density (strain energy absorbed per unit volume) required to produce crack initiation in a solid, uncracked specimen [5]. The measurement of the maximum strain energy density is made essentially via a finite element analysis of the SP test using the commercially available ANSYS computer code. The incremental strain theory and von Mises yield criterion were used to analyze the continuum stress-strain deformation properties of the material. The constitutive stress-strain model used is a Ramberg-Osgood model. The model and model parameters are defined as follows :

$$\sigma = E\varepsilon \quad \text{for} \ \varepsilon < \varepsilon_{YS} \tag{1}$$

$$\sigma = \sigma_{YS} \quad \text{for} \ \varepsilon = \varepsilon_{YS} \tag{2}$$

$$\sigma = D(\varepsilon - \varepsilon_{el})^n \quad \text{for} \ \varepsilon \geq \varepsilon_{YS} + \varepsilon_{pl} \tag{3}$$

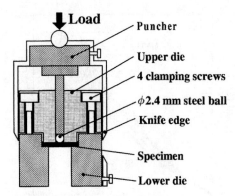

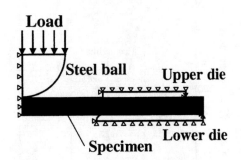

FIGURE 2. Schematic of SP test device.

FIGURE 3. Finite element model of the SP test.

where σ is the stress, ε is the total strain consisting of the elastic portion ($\varepsilon_{el} = \sigma/E$) + the plastic portion (ε_{pl}), ε_{YS} is the strain at yield, given by σ_{YS}/E, D is a constant stress, and n is a strain-hardening exponent.

We denote the rectangular Cartesian coordinates of a point by $x_i(i=1,2,3)$. The strain energy density W at a given instant during deformation is [6] :

$$W = \int_0^{\varepsilon_{ij}} \sigma_{ij} d\varepsilon_{ij} \qquad (4)$$

where σ_{ij} is the stress tensor, ε_{ij} is the strain tensor, and the usual summation convention over repeated indices is applied. The analytical model is an axisymmetric model as shown in Fig. 3. The idealization consisted of blocks of eight-node isoparametric quadrangular structural solid elements. The total number of nodes and elements in the finite element analysis are 7232 and 11027, respectively. The absorbed strain energy density W_{SP} is computed over a single element volume up to the peak load in the load-displacement curve.

RESULTS AND DISCUSSION

Using weld transverse sections, hardness measurements were performed across the fusion zone. A typical hardness profile is shown in Fig. 4. The main point to note here is that the hardness within the fusion zone was considerably less than that of the heat affected zone. The load-displacement curves are shown in Fig. 5. Serrations are formed by repeated bursts of unstable plastic flow (discontinuous yielding) and arrests. Note that at 4 K the material's specific heat is very low so that plastic deformation may cause significant heat evolution. The unstable plastic flow is a free-running process occurring in localized regions of the specimen at higher than nominal rates of strain with internal specimen heating.

Fracture surfaces of the SP test specimens were examined using SEM. Figure 6 shows the fracture appearance and fractograph of SP-W1 specimen. Failure occurred along the circumferential edge of the bulge caused by the punch extrusion. A ductile tearing region can be observed on the fracture surface with void formations. All test specimens failed in a ductile manner with fracture surfaces exhibiting typical ductile dimple morphologies.

181

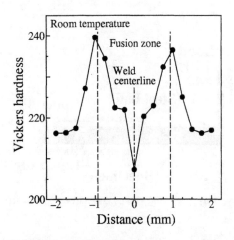

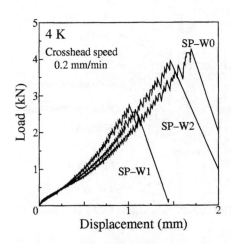

FIGURE 4. Plot of hardness versus distance across the transverse section.

FIGURE 5. Load - displacement curves.

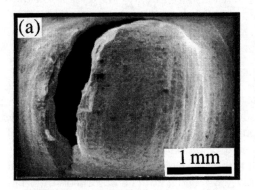

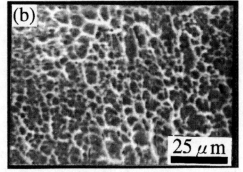

FIGURE 6. Fracture appearance and fractograph of SP-W1 specimen. (a) fracture morphology and (b) fracture surface.

In the 4 K SP tests of austenitic stainless steels and weld metals, an approximate definition of equivalent fracture strain $\bar{\varepsilon}_{qf}$ was adopted [3] :

$$\bar{\varepsilon}_{qf} = 0.0756\left(\frac{\delta_{max}}{t_0}\right)^{1.83} \tag{5}$$

where δ_{max} is the displacement at peak load and t_0 is the initial thickness. The SP test-based fracture toughness $J_{IC}(SP)$ can be estimated from the equivalent fracture strain as [3]

$$J_{IC}(SP) = 850.9\bar{\varepsilon}_{qf} - 65.4 \tag{6}$$

where J_{IC} has units of kJ/m^2. The values of the SP test-based fracture toughness $J_{IC}(SP)$ were estimated as 374.5 (SP-W0), 129.5 (SP-W1) and 257.1 (SP-W2) kJ/m^2, respectively. The J_{IC} value of weld metal according to the JIS Z 2284 procedure was 356.3 kJ/m^2. Using 20 % side-grooved 1TCT specimens, the J_Q (provisional value of J_{IC}) and J_{IC} values were obtained from conventional JIS Z 2284 standard tests as shown in Appendix. Excellent agreement between SP test-based and measured fracture toughness values was obtained. The SP-W1 specimen having a relatively high Vickers

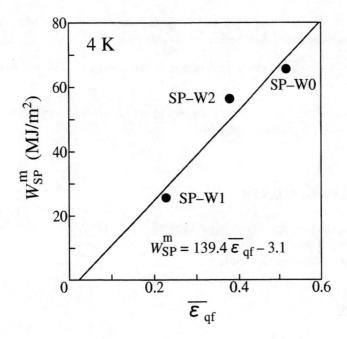

FIGURE 7. Maximum Strain energy density W_{SP}^m versus equivalent fracture strain $\overline{\varepsilon}_{qf}$.

hardness level at room temperature failed in the heat affected zone and showed a lower cryogenic fracture toughness.

The analytical fracture behavior of the SP specimen was evaluated using the strain energy density. The load - displacement curve and strain energy density were obtained from the finite element analysis. Figure 7 shows the maximum strain energy density W_{SP}^m versus equivalent fracture strain $\overline{\varepsilon}_{qf}$. The strain energy density as a function of applied load is computed via finite element analysis. The W_{SP}^m occurs at the observed crack initiation location. We found excellent correlation between the W_{SP}^m and $\overline{\varepsilon}_{qf}$. The best-fit equation of data in Fig. 7 is

$$W_{SP}^m = 139.4\overline{\varepsilon}_{qf} - 3.1 \tag{7}$$

where W_{SP}^m has units in GJ/m^2. Measurement of W_{SP}^m provides quantitative fracture toughness information.

CONCLUSIONS

The cryogenic fracture toughness has been characterized for type 316 austenitic stainless steel weld and its heat affected zone by small punch test. Analytical-based approaches to determining the maximum strain energy density have been also developed. The following conclusions can be drawn from this study :

1. For weld metal, the fracture toughness estimation from the equivalent fracture strain agrees very well with the measured fracture toughness value.

2. The heat affected zone with high Vickers hardness at room temperature shows lower fracture toughness, compared to other regions.

3. A linear correlation between maximum strain energy density and equivalent fracture strain is found.

4. The small punch test is a promising tool by which to directly evaluate weld fracture properties at cryogenic temperatures.

ACKNOWLEDGEMENTS

This research was carried out in collaboration with the National Institute for Fusion Science (NIFS). The authors thank Professor A. Nishimura at NIFS for supplying us with the material used in the study.

APPENDIX

The fracture specimens were 25-mm-thick (B) compact tension (CT) specimens of a geometry described in JIS Z 2284. The specimen width, W, and width-to-thickness ratio, W/B, were 50 mm and 2.0, respectively. Other dimensions are shown in Fig. A1. The weld metal specimens obtained from electron-beam weld in 75-mm-thick plate were oriented such that loading was perpendicular to the weld and crack extension was along the weld centerline, and located at 12.5 mm from the top surface of the weld to the center of the specimen (CT-S), at 37.5 mm (CT-C), at 62.5 mm(CT-B). These specimens were precracked at room temperature with final stress intensity factor range $\Delta K = 30$MPa$\sqrt{\text{m}}$. The precracking was continued to an original crack length to width ratio $a_0/W = 0.6$. After precracking, side-grooves were machined on specimens to a net thickness (B_N) reduction of 20 %.

A 100 kN servo-hydraulic test machine adaptable for cryogenic service was used in all 4 K single-specimen J-integral tests. J_{IC} tests were performed with the specimen and clip gage completely submerged in liquid helium environment at a constant crosshead rate of 0.2 mm/min. The load reduction during unloading was 10 % of the maximum load. Using a crack length versus compliance correlation, the crack length at each unloading was inferred.

The physical crack size a_p and physical crack extension Δa_p were measured with an optical microscipe. The dimension a_p was taken as the average of nine measurements a_i ($i = 1 \sim 9$) taken at equally spaced points across the thickness direction between the load line and the final crack front after testing. Similarly, the final value of physical crack extension Δa was taken as the average of nine similar measurements between the original crack front and the end of stable crack growth. Predicted crack extension Δa was determined from the last unloading compliance. Crack extension data are listed in Table A1. As defined in JIS Z 2284, the difference between the crack extension Δa predicted by elastic compliance at the last unloading and the average physical crack extension a_p does not exceed $0.15\Delta a_p$ and none of the nine physical measurements a_i differ by more 7 % from the average physical crack size a_p. The weld metal specimen

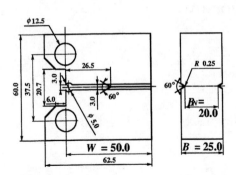

FIGURE A1. Compact tension specimen used for J_{IC} test.

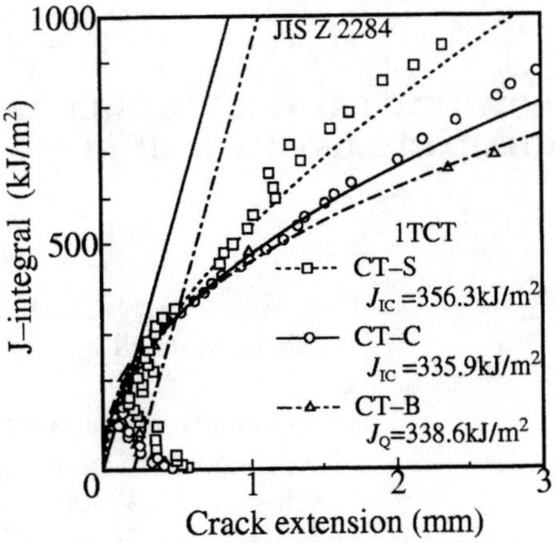

FIGURE A2. J-resistance curves for compact tension specimens.

TABLE A1. Measured and predicted values of crack extension.

	Δa (mm)	Δa_p (mm)	$[(\Delta a - \Delta a_p)/\Delta a_p] \times 100$ (%)	$[(a_i - a_p)/a_p] \times 100$ (%)
CT-S	2.33	2.56	-8.85	$-1.29 \sim 3.30$
CT-C	2.35	3.84	-6.60	$-2.32 \sim 2.78$
CT-B	2.74	4.01	-31.6	$-1.21 \sim 1.56$

satisfied the crack extension and crack size requirements. Figure A2 shows the J-integral versus crack extension curves (J-resistance curves).

REFERENCES

1. Nishimura, A., Tobler, R. L., Tamura, H., Imagawa, S. and Yamamoto, J., *Fusion Engineering and Design* **42**, pp.425-430 (1998).
2. JIS Z 2284, *Method of Elastic-Plastic Fracture Toughness J_{IC} Testing for Metallic Materials in Liquid Helium*.
3. Shindo, Y., Horiguchi, K., Sugo, T. and Mano, Y., *ASTM Journal of Testing and Evaluation* **28**, pp.431-437(2000).
4. Shindo, Y., Mano, Y., Horiguchi, K. and Sugo, T., *ASME Journal of Engineering Materials and Technology* **123**, pp.45-50(2001).
5. Foulds, J. R., Woytowitz, P. J., Parnell, T. K. and Jewett, C. W., *ASTM Journal of Testing and Evaluation* **23**, pp.3-10(1995).
6. Gensheimer, V., Kirby, G. C. and Jolles, M. I., *Engineering Fracture Mechanics* **33**, pp773-785(1989).

LOW TEMPERATURE TENSILE AND FRACTURE CHARACTERISTICS OF HIGH PURITY NIOBIUM

R. P. Walsh[1], Ke Han[1], V. J. Toplosky[1]
and R. R. Mitchell[2]

1. National High Magnetic Field Laboratory
Florida State University
Tallahassee, FL 32306

2. Los Alamos National Laboratory
Los Alamos, NM

ABSTRACT

A materials test program has been performed on high purity niobium to support the structural design and analysis of the Accelerator Production of Tritium (APT) cavity assembly. Tensile tests at 295, 77, and 4 K and fracture toughness tests at 4 K were performed to characterize the mechanical properties of the materials as a function of temperature. The tensile test results are used to evaluate the materials and as input for general design analysis. The fracture toughness test results are used to develop allowable flaw size criteria related to design and fabrication issues. The results of the materials characterization program are reported here to provide mechanical properties data in areas where the previous lack of low temperature data prohibited a confident design.

INTRODUCTION

A materials test program conducted by the National High Magnetic Field Laboratory (NHMFL) in collaboration with the Engineering Sciences and Applications Division of Los Alamos National Laboratory (LANL) is reported here. The research was conducted to generate mechanical property data for niobium that can be confidently used for the design and construction of Superconducting-Radio Frequency (SCRF) cavities. The strength and toughness are of concern at temperatures near absolute zero (the SCRF operating temperature). Tensile tests at 295, 77, and 4 K and fracture toughness tests at 4 K are performed to characterize the mechanical properties of the materials as a function of temperature. Microstructural and fractographic analyses were conducted to further characterize the materials.

CP614, *Advances in Cryogenic Engineering:*
Proceedings of the International Cryogenic Materials Conference - ICMC, Vol. 48,
edited by B. Balachandran et al.
© 2002 American Institute of Physics 0-7354-0060-1/02/$19.00

TABLE 1. List of Test Materials

ID	Material Information	Plate Thickness	Tests	Rockwell Hardness 15T	Grain Size μm
L -1	Nb RRR 250	3.17 mm and 4 mm	T, CT	63	58
L -2	Nb RRR 40	3.17 mm (0.125 in.)	T	64	80
L -11	Nb RRR 40	19.05mm (0.75 in.)	T, CT	75	46
C -1	Nb RRR 250	3.17mm (0.125 in.)	T	50	72
C -2	Nb RRR 40	3.17mm (0.125 in.)	T, CT	66	63

BACKGROUND

Materials having a body-centered-cubic (BCC) crystal structure, such as niobium, are known to undergo ductile-to-brittle transitions making them undesirable for cryogenic applications. The occurrence of ductile or brittle fracture at a given temperature depends on whether the material's yield strength or fracture strength is reached first. As the temperature is lowered, the lattice resistance to slip progressively increases until at some point the fracture strength may be reached before slip can occur, resulting in brittle fracture [1,2]. The tensile properties and the ductile-to-brittle transition temperature (DBTT) are dependent on metallurgical variables such as grain size, processing history and chemistry. Previous studies of pure niobium have shown that it exhibits a DBTT above 100 K [3, 4].

The DBBT of a material is a qualitative number that is usually determined by performing Charpy impact tests over a temperature range. For engineering design purposes, the fracture toughness of the material is the quantitative property needed. There are two commonly used ASTM test methods (E399 and E813) for measuring the fracture toughness of metals. ASTM E399 is used for plane strain conditions where linear-elastic fracture behaviour is observed. ASTM E813 (J–integral (J_{IC}) test method,) requires elastic-plastic fracture behaviour. The J_{IC} determined from E813 can be converted to obtain an estimate of $K_{IC,}$ but care must be taken when comparing the two values due to assumptions necessary for the conversion [10]. The E399 method is usually the preferred test method because it can provide a direct measurement of K_{IC} but is more difficult to perform successfully because of the need for large test samples. For ASTM E399, the required thickness must be $> 2.5*(K_{IC}/\sigma_{ys})^2$. The ASTM E813 minimum required sample thickness must be $> 25*(J_Q/\sigma_{ys})$. For a low toughness (30 $Mpa*m^{0.5}$ to 50 $Mpa*m^{0.5}$) material with yield strength (450 MPa to 760 MPa), like annealed Nb at 4 K, the required sample thickness for E399 ranges from 4 mm to 19 mm, and for E813 is about 2 mm.

MATERIALS

Tables 1 and 2 list the test materials with a brief description and their compositions. There are two plates of Nb RRR250 grade material and three plates of Nb RRR40 grade material. The Residual Resistivity Ratio (RRR) of the Nb material is related to purity such that; the greater the RRR of the material is, the greater the purity of the material is. The quantity RRR is typically defined as the ratio of the resistivity at 273 K to the resistivity at 4.2 K. The RRR values of these materials were determined elsewhere and since niobium is superconducting above 4.2 K special techniques are employed that are not described here.

TABLE 2. Material Compositional Analyses in wt. %

ID	Material	C	H	N	O	Ta	Zr	Ti	Nb
L-1	Nb RRR 250	0.0025	<0.0003	<0.002	<0.004	<0.01	<0.005	<0.004	>99.953
L-2	Nb RRR 40	0.0028	<0.0003	<0.002	0.006	0.043	<0.005	<0.004	>99.917
L-11	Nb RRR 40	0.0038	<0.0003	0.003	0.006	0.033	0.0052	<0.004	>99.925
C-1	Nb RRR 250	<0.002	<0.0003	<0.002	<0.004	0.094	<0.005	<0.004	>99.869
C-2	Nb RRR 40	0.014	<0.0003	0.0036	0.008	0.2	<0.005	<0.004	>99.746

Fe < 0.0035, for all materials

Mo and W < 0.003, for all materials

Si and Hf < 0.005, for all materials

The materials are in the annealed condition but the only certain documentation of annealing is for the material L-1 which was annealed at 1013 K for 2 hrs. In general the materials have an equiaxed grain structure with grain sizes ranging from 46 µm to 80 µm and have low hardness as expected for the fully annealed condition.

TEST PROCEDURES

Tensile tests are conducted according to ASTM E1450. The specimens are machined as shown in Figure 1B with the tensile axis parallel to the rolling direction. Strain is monitored using clip-on extensometers or bonded resistance strain gages. Ambient room temperature accounts for the 295 K test temperature while the 77 K and 4 K temperatures are obtained with boiling liquid nitrogen or liquid helium respectively. Tensile tests are performed in displacement control at a rate of 0.5 mm/min. The elongation is determined from 25 mm gage length scribe marks.

Fracture toughness tests are conducted according guidelines provided in ASTM E399 and ASTM E813. Both methods can use a Compact-Tension (CT) sample (Figure 1A) which are machined in the TL orientation. All precracking was performed at 77 K at a

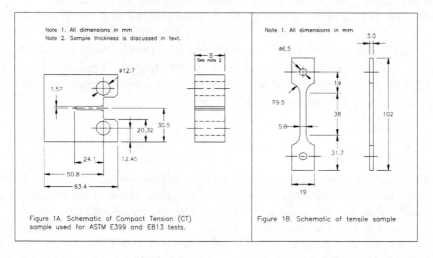

Figure 1A. Schematic of Compact Tension (CT) sample used for ASTM E399 and E813 tests.

Figure 1B. Schematic of tensile sample

FIGURE 1. Test Specimen Schematics

maximum stress intensity of 15 to 17 Mpa*m$^{0.5}$. Four of the five test materials (L-1, L-2, C-1 and C-2) have typical thickness (< 4 mm) used for the construction of the APT-SCRF cavities. Sample thickness less than 4 mm requires the use of test method ASTM E813. Material L-11 was included in the test program to allow direct K_{IC} measurements using ASTM E399.

A detailed description of specimen size requirements is given in Reference 8 and is briefly summarized here. The accuracy with which a K value that is measured in a test, predicts the fracture toughness of the material, is dependent on how well the stress intensity factor describes the region around the crack tip. The relative accuracy of K increases with an increase in pertinent dimensions since the relative size of the plastic zone decreases. In order for a K_{IC} test to be valid these characteristic dimensions must exceed a size, proportional to; Constant*$(K_{IC}/\sigma_{ys})^{\wedge 2}$. The constant of 2.5 has been determined empirically from tests on more conventional engineering materials tested at room temperature and is recommended by the E399 Test Standard. The applicability of the K value that is measured is ultimately dependent on material properties that influence the plastic zone as well as the specimen geometry. The specimen dimension requirements are reviewed here to assess their relativity to governing 4 K tests of niobium. The multiple of 2.5 appears to be large for such a brittle material as pure Nb, and can probably be reduced as the test results show.

RESULTS

The average tensile properties are shown in Table 3 along with data found in the literature for pure Nb [4 and 7]. The Nb tensile properties that are strongly temperature dependent. The dramatic increase in strength that occurs upon cooling from room temperature to cryogenic temperatures is evident in the stress-strain curves (Figure 2) for Materials L-1 (Nb RRR250) and C-2 (Nb RRR40). These characteristics are also observed on the tensile property plots shown in Figures 3 and 4. The 4 K strengths are about 10 times the 295 K yield and tensile strengths. Over the same temperature range the ductility decreases at least 50 % as shown in Figure 4. The decrease in ductility is an indication of a ductile to brittle transition and an associated decrease in toughness. The typical overall stress vs strain curves shown in Figure 2 are for relative comparison, the strain data are obtained from the test machine's displacement transducer.

The tensile properties as a function of temperature of the two Nb RRR250 (L-1 and C-1) and the two Nb RRR40 metals (L-2 and C-2) are very similar as is observed in the tables and figures discussed above. A notable difference in test results is that the RRR40 materials exhibit low elongation to failure. Another, important deviation in behavior is the dramatic decrease in performance exhibited by the L–2 RRR40 material. The ductility of this RRR40 material is severely compromised at decreased temperature as can be seen in all of the tensile properties vs temperature plots. In 4 K tensile tests the material is nearly linear-elastic, fractures before reaching the 0.2% offset yield strength, and has negligible elongation and reduction of area. The comparable C-2 RRR40 material behaves much differently, in that it retains strength and ductility at 4 K.

The decreased low temperature strength and ductility of the L-2 material must be related to metallurgical variables. It is known that higher interstitial content contributes to solid solution strengthening, however, the segregation of the interstitial atoms to grain may offset the effect. There is not a trend relating the tensile or yield strength to the interstitial content. If the heat treatment conditions (time and temperature) are the same, the relative

TABLE 3: Nb Tensile Test Results

Material	Temp K	# of Tests	Young's Modulus GPa	Yield Strength MPa	YS Std Dev. MPa	Tensile Strength MPa	TS Std Dev. MPa	Red. in Area %	Elongation %
L-1	295	2	100	67	4	172	1	88	57
RRR 250	77	2	115	618	12	642	3	72	30
Base	4	4	118	658	3	929	6	29	16
C-1	295	2		57	3	162	2	95	50
RRR 250	4	3		455	4	889	4	26	26
L-2	295	2	113	76	0	171	1	81	41
RRR 40	77	2	120	443	4	502	7	7	3
Base	4	3	122			468	57	1	2
C-2	295	2		98	11	187	8	96	26
RRR 40	4	3		767	9	964	50	6	7
L-11	4	2		1055		1068	14		2
Ref. 7	295			44		155			59
RRR 200	77			444		591			25
Base	4			500		875			15
Ref. 4	295					270		96	40
~99.5 % Nb	77					951		73	18
Base	4					655		0	0

distribution of the interstitial elements would be about the same. The lack of correlation of the tensile properties to the composition indicates that the heat treatment conditions of the different materials are probably different resulting in different amounts of interstitial element segregation to the grain boundaries. There is correlation between the relative 4 K ductility of the materials and the tensile fracture characteristics. The RRR250 base materials L-1 and C-1 exhibit a clean fracture at 45 degrees to the tensile axis, which is a fracture process attributed to shear stress (Figure5A). The RRR40 base materials L-2 and C-2 tensile test specimens have features of brittle fracture at 4 K with transgranular cleavage fracture that is perpendicular to the tensile axis (Figure 5B).

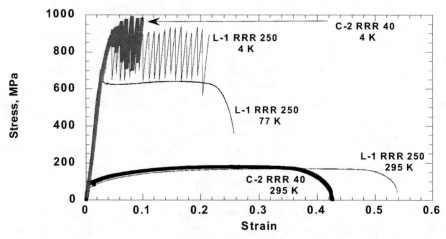

FIGURE 2. Typical Stress vs Strain curves of Nb RRR250 and RRR40 at the three test temperatures.

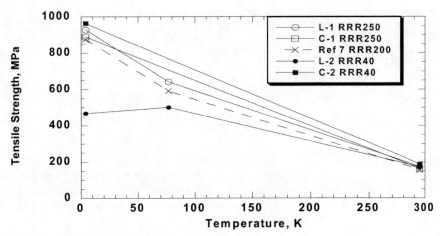

FIGURE 3. Nb Tensile Strength (average) vs. Temperature.

Higher magnification SEM images of the fracture-surfaces of samples L-1 and L-2 in Figures 6A and 6B show the contrast between the ductile dimple rupture of material L-1 and the brittle cleavage fracture of material L-2.

ASTM E399 fracture toughness tests results (only three of the five materials were tested) are shown in Table 4 along with data from Reference 7. The data reported is the conditional stress intensity factor K_Q, because of problems satisfying ASTM validity requirements (for all tests $P_{Max}/P_Q > 1.1$). The materials behaved in too brittle a manner for ASTM E813 but had just enough ductility to invalidate the plane-strain requirements of ASTM E399. Fracture toughness has been found to vary inversely with yield strength

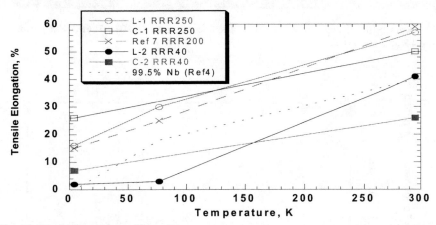

FIGURE 4. Nb Tensile Elongation vs Temperature

191

FIGURE 5A. At 4 K RRR250 (Mat'l L-1) has a shear fracture at 45° to tensile axis. **FIGURE 5B.** At 4 K RRR40 (Mat'l L-2) has brittle cleavage fracture perpendicular to the tensile axis.

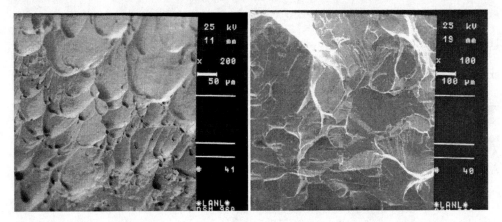

FIGURE 6A. SEM photograph of L-1 tensile 4 K fracture surface. **FIGURE 6B.** SEM photograph of L-2 tensile 4 K fracture surface.

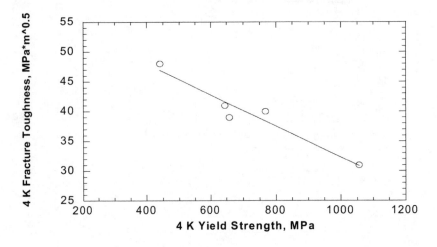

FIGURE 7. The 4 K yield strength vs. Fracture Toughness Relationship for pure Nb.

192

TABLE 4. Nb Yield Strength and Fracture Toughness data plotted in Figure 7.

Material	CT Sample thickness	Average Yield Strength	Average Fracture Toughness	Fracture Toughness Scatter		
	mm	Mpa	Mpa*m^.5	No of test	High	Low
L-1	4	658	39	1	39	39
C-2	9.5	767	40	2	39	40
L-11	19.05	1055	31	4	29	32
KEK	3	440	48	2	46	50

within groups of closely related metals [9]. The data are plotted in Figure 7 show that an inverse relationship between strength and toughness appears to exist. The fracture surfaces of all of the fracture toughness specimens exhibit brittle features of transgranular cleavage.

CONCLUSIONS

The materials characterization program was successful in providing low temperature mechanical properties data for SCRF cavity design and analysis. The data presented here contributes greatly to the extremely small low temperature database available for pure Nb. The test results show that Nb is a complex material whose low temperature properties are dependent on the material processing history and metallurgical state. Based on the tensile properties, one of the five Nb materials (L-2) appears to exhibit a DBTT that is above 77 K. Unfortunately, material L-2 was not available for fracture toughness tests. The identification of the material by its RRR value is not adequate for specifying or predicting the low temperature mechanical properties since "identical" RRR materials exhibit a range of properties.

REFERENCES

1. G. E. Dieter, "Mechanical Metallurgy", 3rd ed, McGraw-Hill Publishing Co., 1986, pp 268- 269
2. E. T. Wessel, "Some Basic and Engineering Considerations Regarding the Fracture of Metals at Cryogenic Temperatures", Behaviour of Materials at Cryogenic Temperatures, ASTM STP 387, 1966, p. 32
3. E. T. Wessel and D. D. Lawthers, "The Ductile-to-Brittle Transition in Niobium" in The Technology of Columbium, Wiley, NY 1958, p66.
4. E. T. Wessel, L. L. France, and R. T. Begley, "The Flow and Fracture Characteristics of Electron-Beam-Melted Columbium"
5. Rao,M.G. and Kneisel, P., Mechanical Properties of High RRR Niobium at Cryogenic Temperatures, Adv. In Cryo. Engr., Vol. 40, pp.1383-1390.
6. R. P. Walsh, R. R. Mitchell, V. J. Toplosky and R. C. Gentzlinger, "Low Temperature Tensile and Fracture Toughness Properties of SCRF Cavity Structural Materials" in Proceedings of 9th Workshop on RF Superconductivity, Santa Fe, NM 1999
7. K. Ishio, K. et. al., "Fracture Toughness and Mechanical Properties of Pure Niobium and Welded Joints for Superconducting Cavities at 4K " in Proceedings of 9th Workshop on RF Superconductivity, Santa Fe, NM 1999
8. W. F. Brown, Jr. and J. E. Srawley, "Plane Strain Crack Toughness Testing of High Strength Metallic Materials" ASTM Special Technical Publication No. 410, published by ASTM 1966.
9. R. P. Reed and A. F. Clark, "Materials at Low Temperatures" ASM 1983, p. 254

NON-METALLIC MATERIALS

THERMAL CONDUCTIVITY MEASUREMENTS OF EPOXY SYSTEMS AT LOW TEMPERATURE

F. Rondeaux, Ph. Bredy and J.M. Rey

CEA Saclay DSM/DAPNIA/STCM
91191 Gif sur Yvette Cedex – France

ABSTRACT

We have developed a specific thermal conductivity measurement facility for solid materials at low temperature (LHe and LN2). At present, the Measurement of Thermal Conductivity of Insulators (MECTI) facility performs measurements on epoxy resin, as well as on bulk materials such as aluminium alloy and on insulators developed at Saclay. Thermal conductivity measurements on pre-impregnated fiber-glass epoxy composite are presented in the temperature range of 4.2 K to 14 K for different thicknesses in order to extract the thermal boundary resistance. We also present results obtained on four different bonding glues (Stycast 2850 FT, Poxycomet F, DP190, Eccobond 285) in the temperature range of 4.2 K to 10K.

INTRODUCTION

Epoxy resin and fiberglass composites are used in the manufacturing of superconducting magnets of large dimensions, in particular for the electrical insulation and the mechanical bonding of conductors in the winding of well-known "fully epoxy-impregnated coils". The anisotropic constitution and the various elaboration modes of these resin-fiberglass end in a large disparity in their physical properties, in particular in their thermal properties at low temperatures. For large magnet projects where inadequate design may lead to dramatic consequences in terms of performances, cost and time, each "non-traditional" material used at low temperatures must be carefully tested. For this reason, our laboratory, involved in the construction of several magnets for LHC, has developed a thermal conductivity experiment (MECTI) to characterize different materials such as pre-impregnated fiberglass-epoxy composite and epoxy glues around liquid helium temperature.

Nowadays, commercial epoxy glues are used in the construction of superconducting coils, as thermal linkage and structural bonding for the cooling circuits. For some of them, the thermal conductivities have been measured [1,2]. Others products could be potentially interesting, however their thermal properties are not systematically given in the range of 4

CP614, *Advances in Cryogenic Engineering:*
Proceedings of the International Cryogenic Materials Conference - ICMC, Vol. 48,
edited by B. Balachandran et al.

to 15 K. To examine their possibilities, we have retained four glues to be tested with MECTI.

PRINCIPLE OF THE MEASUREMENT

Our thermal conductivity measurements are based on a steady-state method with a longitudinal heat flow. Heat flow is then considered as one-dimensional along the sample. Pure conduction Fourier's law is given by equation (1) :

$$Q = \frac{S}{l} \cdot \int_{Tc}^{Th} \lambda(T).dT \tag{1}$$

where Q is the heat flux, S the transverse cross section (assumed constant), l the sample length, T_c and T_h the cold and the hot temperature of the sample, respectively, and $\lambda(T)$ its thermal conductivity. Average thermal conductivity λ_{av} between T_c and T_h is

$$\lambda_{av} = \frac{1}{(T_h - T_c)} \cdot \int_{Tc}^{Th} \lambda(T).dT = \frac{Q \cdot l}{S \cdot (T_h - T_c)} \tag{2}$$

If λ is a linear function of T, equation (2) is simplified to :

$$\lambda_{av} = \lambda[\frac{T_h + T_c}{2}] = \frac{Q \cdot l}{S \cdot (T_h - T_c)} \tag{3}$$

As experimental results will show, the thermal conductivity of our resin is not a linear function of T. Nevertheless, for small temperature differences and small variations of λ (T) (as measured), the equation (3) can be applied in our measurements and the induced error is less than 1 to 2 % [3].

DESCRIPTION OF THE SET-UP

As already carried out in other laboratories, our apparatus is developed to measure thermal conductivity in steady-state conditions[4]. The sample is located between a cold heat sink cooled by a cryogenic bath (here, liquid helium used) on one side and a heat source on the other as described in Figure 1. Heat flow is assumed to be transverse and one-dimensional between the two faces of the sample. High vacuum around the sample stops heat leaks via convection and the use of small diameter thermalized manganin wires for instrumentation further limits thermal losses by conduction. Conduction losses are also minimized on the sample holder by using nylon and G10 pieces behind the heater and stainless steel tie rods. Radiation heat transfer is negligible around liquid helium temperature because vacuum tank is immersed in the liquid He bath. Thermal differential dilatations are compensated by using Belleville washes. Calculated heat flow losses from heater outside sample are less than 0.3 % in the 0-55 mW range. Accuracy on difference temperature measurement is about +/- 8 mK in the 4-15 K range.

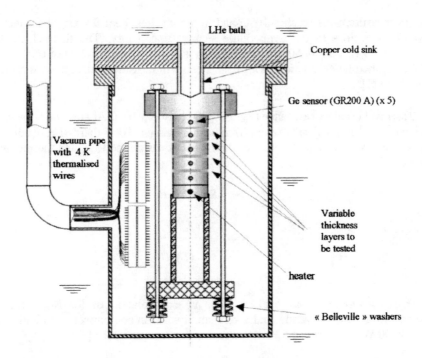

FIGURE 1. Experimental apparatus.

Samples are made with several layers of the product to be tested, separated by blocks of 1050 Aluminium alloy 10 mm thick. Temperature sensors are located in the middle of these blocks. Sensitive elements used are Germanium resistance sensor XGR-200A-3000-2 from Lake Shore Cryotronics. Temperature difference inside aluminium blocks is small enough to be negligible in comparison with the one in the layer. Material thickness is measured with a precision of 0.02 mm.

MEASUREMENTS ON THE PRE-IMPREGNATED FIBER-GLASS EPOXY COMPOSITE

Each composite sample is made with 5 blocks of aluminium separated with several layers of pre-impregnated tape IVA® (97 33/2)[5] (0.32 mm thick per layer). Thickness of composite was varied to determine thermal boundary resistance. Aluminium blocks are first sand-blasted and degreased by an ultrasonic method in alcohol, dried, then assembled with the pre-preg. Composite is afterwards polymerized in a vacuum oven (thermally cycled at 80°C for 4 h and at 120°C for 5 h). Insulation thicknesses between 5 blocks are about 0.26, 0.95, 1.9 and 2.8 mm. Three samples have been fabricated. Figure 2 shows the typical cross-section and picture of these samples.

The tests have been carried out on three multi-layer samples (#1,2,3) with 6 different heat fluxes (up to 50 mW), the smallest composite thickness being located on the cold sink side, except for the sample 1 which has been also tested in the opposite position (results 1a and 1b respectively).

Figure 3 presents the results, which are very consistent and show a low noise. The largest error bars observed correspond both to the thinner layer (for which the error on the thickness measurements reaches 10%) and to the smallest heat fluxes, where temperature differences are close to the temperature sensors uncertainty. The thermal conductivity measured at 4.5 K is about 50 mW/m.K, which is much lower than that of G10 [6] initially used for our thermal calculations. Our results confirm measurements obtained by LASA of INFN-Milan [7].

Thermal boundary resistance ($R_h = \Delta T.S/Q$) between 1050 Al alloy and the composite is estimated from our results with different thicknesses. Total thermal resistance R_{tot} for each layer is assumed to be the sum of thermal bulk resistance R_{bulk} and thermal boundary resistances R_h.

$$R_{tot} = R_{bulk} + 2 \cdot R_h \tag{4}$$

with
$$R_{bulk} = \frac{l \cdot (T_h - T_c)}{\int_{Tc}^{Th} \lambda(T).dT} \tag{5}$$

For a given temperature ($T \approx 5.5$ K), an extrapolation of the $R_{tot}(l)$ law at zero thickness (see figure 4) leads to find a very small value in comparison with bulk properties ($1/R_h > 2000$ W/m^2.K around 5 K, consistent with previous works [8]).

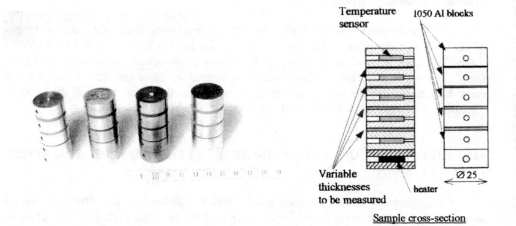

Figure 2. Multi layers composite samples and cross-section scheme

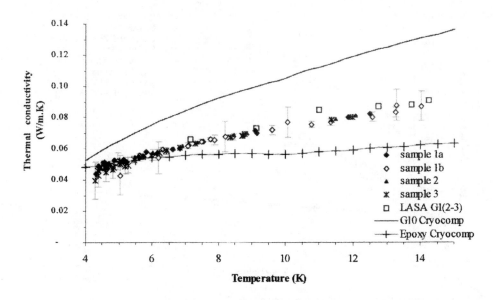

FIGURE 3. Thermal conductivity of pre-impregnated multi-layers composite

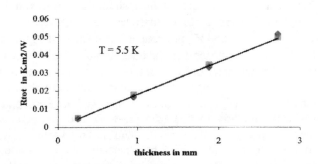

FIGURE 4. Estimation of the Al/composite thermal boundary resistance.

MEASUREMENTS ON VARIOUS BASED-EPOXY GLUES

Four epoxy glues have been measured at cryogenic temperatures using MECTI; i.e Eccobond 285 [9], Stycast 2850FT [10], Poxycomet F [11], DP 190 [12]. Eccobond and Stycast are traditionally used in many cryogenic assemblies where good mechanical behaviour and electrical insulation are needed. Their thermal conductivities at low temperature are known to be relevant to cryogenics contacts. To increase its thermal conductivity, metal particles addition is employed as a solution. Unfortunately, this is unfavorable from the electrical insulation point of view. Poxycomet series is one of such glues (up to 80% Al) with a good mechanical behavior in aluminium assembly at low temperatures but with an unknown thermal conductivity. DP190 is often used in low temperature assembly [13] and it combines electrical insulation (up to 30 kV/mm) with good mechanical properties at low temperatures (remains sufficiently "soft" at low temperatures). Its thermal conductivity is also unknown in this range of temperature.

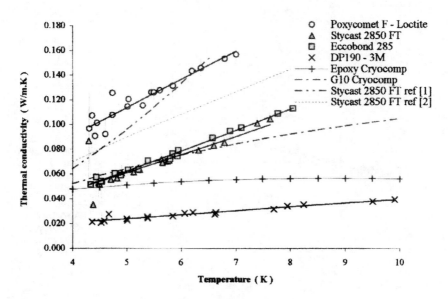

FIGURE 5. Thermal conductivity of epoxy-based glues.

Using 2 samples for each glue with 4 aluminium blocks and 3 layers of equal thickness (e ≈ 0.7 mm), thermal conductivity has been measured. Results are shown in figure 5. Near 4 K, Stycast 2850 FT preparation (with Catalyst 9) follows a $\lambda(T)=a\,T^n$ law with a ≈ 0.0095 and n ≈ 1.2, closer than the quasi similar linear behavior already observed with this catalyst [2] but in contrast to a ≈ 0.0053 and n ≈1.8 found in another work (with catalyst 11) [1]. All over the range 4-8 K, experimental thermal conductivity values are 30 % lower than values in [1] and [2]. In this range, $\lambda(T)$ of Eccobond 285 (with catalyst LV24) behaves very similarly to that of Stycast. DP190 has poor and rather constant thermal conductivity, between 0.02 and 0.03 W/m.K. This value is lower than pure epoxy resin, probably because it contains 50% hardener. As presumed with its high charge in aluminium, (≈80% in mass), Poxycomet F has a thermal conductivity twice higher than Stycast 2850 FT, which follows also a linear law in this range as 0.023T. The increase in thermal performance is finally poor for this so-called "pasty aluminium".

CONCLUSIONS

Pre-impregnated epoxy-fiberglass composite measurements show a significant difference in its thermal conductivity from the values of G10 found in literature [4], i.e from 16 to 30 % less between 4 and 10 K. When values of G10 are used in simulation calculations instead of those of pre-impregnated composite, this difference must be taken into account. Thermal boundary resistance between the composite and Al has been evaluated at less than 5.10^{-4} K.m^2/W which can be considered negligible in our analysis.

Bonding glues measurements give additional data to be used in low temperature assemblies. A thermal conductivity range from 0.02 W/m.K up to 0.1 W/m.K at 4 K has been measured for the 4 commercial epoxy- based glues.

ACKNOWLEDGEMENTS

The authors are very grateful to J-C. Paladji, D.Thomas and S.Raude for their technical contribution during the experimental program.

REFERENCES

1- C.L Tsai, H. Weinstock, W.C. Overton Jr, "Low temperature thermal conductivity of Stycast 2850 FT", in *Cryogenics* , **18**, September 1978 pp. 562-563.
2- A. Siri, G. Sissa, "Low temperature measurements of thermal diffusivity in composite Epoxies", Proceedings ICEC 7, (1978), pp 499-504 .
3- J.G. Hust, A.B. Lankford, *Int. Journal. Thermophysic.*, 3, part I, **67**,1982
4- F.Broggi, L.Rossi, "Test of an apparatus for thermal conductivity measurements of superconducting coil blocks and materials at cryogenic temperatures", in *Rev. Sci. Instrum.* **67** (9), September 1996, pp 3193-3200.
5- I.V.A - GEC ALSTHOM F-69883 Meyzieu
6- Cryocomp v 3.06 Cryodata Inc , Eckels Engineering, Florence SC, USA 29501. *G10 "normal to cloth layer"*
7- M.Damasceni, L.Rossi, S.Visona, "Measurements at cryogenic temperature of the thermal conductivity of the ground and turn insulation of a model of the ATLAS BT coils"., *internal report* LASA/ATLAS/50 (1998).
8- E.Gmelin, M.Asen-Palmer, M.Reuther, R.Villar, "Thermal boundary resistance of mechanical contacts between solids at sub ambient temperatures", J.Phys. D: Appl.Phys. vol **32**, 1999, pp.R19-R43
9- Eccobond® 285 + catalyst 24LV, National Starch & Chemical SA, F-69400 Villefranche sur Saône (Emerson & Cuming, Inc, Canton, Mass 02021, USA).
10- Stycast 2850FT + catalyst 9, National Starch & Chemical SA, F-69400 Villefranche sur Saône (Emerson & Cuming, Inc, Canton, Mass 02021, USA).
11- Poxycomet F, epoxy charged with aluminum, Loctite. North. America, 1001 Trout Brook Crossing, Rocky Hill, Connecticut 06067 , USA.
12- Scotch-Weld DP190, 3M Center, St. Paul MN,55144-1000, USA.
13- B. Hervieu, CEA, "Cible Daphne", private communication.

SOME ENGINEERING PROPERTIES OF COTTON-PHENOLIC LAMINATES

R. P. Walsh and V. J. Toplosky

National High Magnetic Field Laboratory
Florida State University
Tallahassee, FL 32306

ABSTRACT

Although cotton/phenolic laminates are commonly used at cryogenic temperatures as structural and insulating materials, the available low temperature materials properties data is limited. We have reviewed the existing low temperature database for cotton/phenolic and have identified areas of need. We have conducted a materials test program on the two common types (linen and canvas) of cotton/phenolic laminates to add to the existing database and to generate new data in areas where needed. Also included is a comparison of cotton/phenolic engineering properties to the properties of NEMA G-10 CR glass-cloth reinforced laminate. The properties studied here are tensile and compressive strength, elastic modulus, shear properties and thermal expansion characteristics over the temperature range from 295 K to 4 K.

INTRODUCTION

High pressure cotton/phenolic laminates are versatile industrial composite materials used for a wide range of applications because of their availability, machinability, strength to weight ratio, and their insulating characteristics. They can be used in many low temperature applications as an alternate to glass fiber-reinforced laminates (such as G-10 or G-11) where cost and ease of fabrication outweigh performance requirements. Low temperature applications that take advantage of their good mechanical and insulating properties are plentiful and often don't require an in depth knowledge of the mechanical properties. There are times that its use is prevented due to the lack of available data so engineers tend to select a glass fiber-reinforced laminate because data exist to support their material selection. Here we investigate tensile and compressive strength, elastic modulus, shear properties and thermal expansion characteristics of a few linen/phenolic and canvas/phenolic laminates.

CP614, *Advances in Cryogenic Engineering:*
Proceedings of the International Cryogenic Materials Conference - ICMC, Vol. 48,
edited by B. Balachandran et al.

TABLE 1. List of Test Materials

Material Identification	Thickness, mm	Description
(A) Canvas	4	Canvas/Phenolic Westinghouse Grade H11030
(D) Canvas	3	Canvas/Phenolic Acculam NEMA Grade CE
(C) Linen	1	Linen/Phenolic Acculam NEMA Grade LE
(F) Linen	6.3	Linen/Phenolic Acculam NEMA Grade LE

MATERIALS

Cotton/phenolic composites are cotton cloth-reinforced high-pressure industrial grade laminates. The cloth reinforcement is referred to as linen or canvas depending on the weave texture. Linen refers to a fine-weave, while canvas typically has medium weave. The cloth is produced with a plain weave that is unbalanced, having warp and fill directions that have slightly different fiber volumes and typically have slightly different properties in the two directions. The NEMA specifications for linen/phenolic (Grades L, LE) and canvas/phenolic (Grade C and CE) are specified to have a thread count of not more than 28.3 threads/cm in the fill direction and not more than 55 threads/cm in the combined warp and fill directions. Nema G-10 (glass fiber-reinforced epoxy laminate) products typically have a 43:32 warp to fill ratio, resulting in about a 30% difference in strength in the two directions [1]. Grades L and C are not recommended for electrical applications, grades CE and LE are better for electrical applications due to superior moisture resistance but are not recommended for electrical insulation involving power exceeding 600 volts.

Distinction between the warp and fill direction (cloth orientation) of the materials is not directly obvious. For the materials tested here (shown in Table 1) an arbitrary 0° and 90° orientation were initially assigned that correlated to the cloth texture. The orientation designation was maintained for sample preparation and testing. The assignment of the warp and fill designation was performed afterward based on the thermal expansion test results.

TEST PROCEDURE

The mechanical tests are conducted according to ASTM procedures whenever applicable. All the mechanical tests are conducted using a hydraulic test machine equipped with a test cryostat that enables the test specimen and associated fixture to be immersed in boiling liquid helium (4 K).

Tensile test are performed according to the procedures outlined in ASTM D-638 and ASTM D3039. The tabbed tensile specimen geometry was used to permit tests of the full laminate thickness. All tensile tests were performed in displacement control rate of 0.5 mm/min. Strain was measured either with a 25 mm clip-on extensometer or bondable resistance strain gages. Back-to-back strain gages are used to minimize error contribution due to bending strains or misalignment. The adhesive used to bond the tensile tabs is Stycast 2850 FT (w/ 24 LV catalyst).

TABLE 2: Cotton/Phenolic Reference Data

	Temp. K	Orientation	Canvas CE Ref.6	Canvas Gr. C Ref. 2,4	Linen Gr. L Ref.3	Linen LE Ref.6	Average* Properties
Elastic Modulus GPa	295	Warp	6.2	xx	7.9	6.9	6.48
		Fill	5.5	xx	xx	5.9	
		Normal	xx	xx	3.8	xx	
	4	Warp	xx	xx	11.4	xx	
		Normal	xx	xx	7.7	xx	
Compressive Strength MPa	295	Warp	xx	150	239	xx	182.5
		Fill	169	xx	xx	172	
		Normal	269	280	282	255	271.5
	77 or 4	Warp	xx	263	463	xx	363.0
		Normal	xx	528	479	xx	503.5
Tensile Strength MPa	295	Warp	62	92	xx	83	79.0
		Fill	48	62	xx	59	56.3
	77 or 4	Warp	xx	115	78	xx	89.7
		Fill	xx	76	xx	xx	

* Shaded values are the average of the warp and fill measurement

Two types of shear strength measurements were made. The through thickness shear strength was measured using the ASTM D 732 method. The Interlaminar Shear Strength (ILSS) of the materials is evaluated using ASTM D 2344 (Test Method for Apparent Interlaminar Shear Strength of Parallel Fiber Composites by Short Beam Method). This three-point bend method is good for comparison or screening but provides no real engineering data since tensile and compressive stresses are present. A span to thickness ratio of 5 was used for the interlaminar shear tests.

Compression tests are conducted according to the guidelines provided in ASTM D-695. The stress is introduced through end loading which can result in reduced strength measurements due to stress concentration at the end of the specimen.

The thermal contraction is measured on coupons of the materials using a strain gage method described previously [5]. The estimated accuracy of the measurements is +/- 3% of the reported values, and is based on calibration of the strain gages with OFHC Cu reference material. Strain gages are applied to measure contraction in the laminates principal orientations (Warp, Fill and Normal). Two gages are applied back to back and their outputs averaged to minimize contributions due to bending that may occur. The samples are thermally cycled from 295 K (ambient room temperature) to 77 K (liquid nitrogen) and on to 4 K (liquid helium). The through thickness (normal to the cloth) direction is only measured for the linen/phenolic Material F. The data that are reported are the average of at least five thermal cycle tests each.

TABLE 3:   NHMFL Test Results

Property	Temp. K	Orientation	A 4mm Canvas	D 3mm Canvas	C 1mm Linen	F 6 mm Linen	Average* Properties
Elastic Modulus GPa	295	Warp	6.7	x	x	7.55	6.84
		Fill	5.8	7.55	6.97	6.45	
	4	Warp	13.1	x	x	15.84	13.82
		Fill	13.5	13.7	12.7	14.1	
Compressive Strength MPa	295	Warp	148	x	x	xx	158.3
		Fill	x	170	x	157	
		Normal	243	x	x	x	243.0
	4	Warp	222	x	x	x	280.0
		Fill	x	331	x	287	
		Normal	458	x	x	x	458.0
Tensile Strength MPa	295	Warp	x	x	x	63.3	58.6
		Fill	53.1	58	59.8	53**	
	4	Warp	x	x	x	61.7**	64.1
		Fill	54.8	66	73.1	62.3	
Apparent Interlaminar Shear Strength MPa	295	Warp	10	x	x	x	10.9
		Fill	9.8	11.7	x	12.1	
	4	Warp	13.3	x	x	x	14.1
		Fill	12.3	15.9	x	14.8	
Thru Thickness Shear Strength MPa	295	Normal	94	102	68	95	89.8
	4	Normal	139	109	74	150	118.0

* Shaded values are the average of the warp and fill measurement

** Results of only one test

RESULTS

This program's main objective is to compare various cotton/phenolic materials to each other and to existing data in the literature as well as to generate data in some areas of need. The available data found in the literature and manufacturer's data sheets are shown in Table 2. The results for the mechanical property tests conducted at NHMFL are shown in Table 3, all values are the average of at least two tests unless otherwise noted. The comparative properties of G-10 CR are shown in Table 4. The graph in Figure 2 is a summary of the average tensile and compressive strength measurements made here. This plot also contains the average of the strengths of the data that are reported in the literature.

The reference data reports the 295 K elastic modulus of the warp and fill directions ranges from 5.9 to 7.9 GPa for linen and canvas phenolic materials. There is only one report of the normal direction modulus, which was measured to be 3.8 GPa [3].

TABLE 4: Reference Data for G-10CR

Material	Test Temp. K	Orientation	Elastic Modulus GPa	Compressive Strength MPa	Tensile Strength MPa	Apparent Interlaminar Shear Strength MPa
		Warp	28	375	415	60
	295	Fill	22	283	257	45
G-10CR		Normal	x	420	x	x
Ref. 1		Warp	36	862	862	x
	4	Fill	29	598	496	105
		Normal	x	749	x	x

Reference 3 also reports low temperature modulus measurements of 11.4 for the warp direction and 7.7 GPa for the normal direction. We tested two canvas/phenolic materials and two linen/phenolic materials. The 295 K warp and fill modulus measurements range from 5.8 GPa to 7.5 GPa, comparable to the reference data quoted above. The 4 K warp and fill modulus measurements range from 12.7 to 13.7 GPa, slightly higher than previously reported 11.4 GPa. Comparing these elastic properties to G-10CR [1] shown in Table 4, we see that the 295 K modulus for G-10's warp and fill direction ranges from 22 to 28 GPa. G-10 CR is about 3 to 4 times stiffer at 295 K than cotton phenolic, in the warp and fill directions. The G-10 modulus increases about 30 % upon cooling to 4 K, while the cotton phenolic modulus increases 80 to 100 % at 4 K. This stronger temperature dependence, in the cotton/phenolic, results in G-10 being about 2.5 to 3 times as stiff as a cotton/phenolic laminate at 4 K.

The NHMFL compressive strength measurements are slightly lower than the reference data found in the literature. The range of warp and fill compressive strengths reported in the literature (Table 2) for cotton/phenolic laminates at 295 K is from 150 MPa to 239 MPa with the average of the reference values being 182 MPa. The NHMFL 295 K warp and fill direction compressive measurements average 158 MPa with an error band of +/- 11 MPa. The average 295 K normal direction compressive strength (reference table 2) is 271 MPa with a small error band of about +/-13 MPa. The 295 K normal compressive strength (NHMFL tests, Table 3) is only measured on material A and is 243 MPa at 295 K. There are two references for the 4 K compressive strength in the warp direction (263 MPa and 463 MPa) and the 4 K normal direction compressive strength (528 MPa and 479 MPa). The corresponding 4 K values measured here are an average of 280 MPa (warp and fill direction) and a normal direction compressive strength of 458 MPa for material A only. Comparing the compressive strengths to G-10CR (Table 4), we see that at room temperature G-10 CR is essentially twice as strong for all three principal directions. At 4 K G-10 CR is two to three times stronger in the warp and fill directions than cotton/phenolic. The 4 K normal direction compressive strength of G-10CR is about 1.6 times greater than that of the canvas/phenolic measured here.

The tensile strengths of the cotton/phenolic from the literature appear to exhibit a trend of distinction between the warp and fill directions with the warp direction being the stronger. There is not a recognizable distinction between strengths of the linen/phenolic compared to canvas/phenolic. From the reference data (Table 2) the average 295 K strength for the warp and fill directions are 79 MPa and 56 MPa respectively. For the

TABLE 5: Thermal Contraction Data

Temp. K	Orientation	NHMFL Test Materials				Reference Data		
		% Contraction, Tref = 293 K				% Contraction, Tref = 293 K		
		A	D	C	F	Ref. 2	Ref. 3	Ref. 1
		Canvas	Canvas	Linen	Linen	Canvas	Linen	G-10 CR
77	Warp	0.28	0.29	0.27	0.28	0.244	0.3	0.215
	Fill	0.35	0.39	0.34	0.36		0.58	
	Normal				0.7	0.647	0.85	0.644
4	Warp	0.34	0.34	0.32	0.33	0.264		0.241
	Fill	0.42	0.45	0.4	0.42			
	Normal				0.8	0.73		0.706

NHMFL tests there is not enough data to support this trend. At 295 K the average tensile strength of 55 MPa for the fill direction compares well with the published data.

The shear test results are also shown in Table 3. The apparent interlaminar shear (ILSS) tests are done to compare if there are differences between the cotton/phenolic types and to compare to the G-10CR published values. The apparent ILSS of linen and canvas/phenolic are approximately the same. The apparent ILSS is very low compared to that of G-10CR (about a factor of 5 to 7 less). The through thickness shear test results are relatively consistent for the materials with the exception of the lower results for the 1 mm thick linen sheet.

The thermal contraction test results are shown in Table 5 along with the previously published thermal contraction of cotton/phenolic laminates and G-10CR for comparison. The thermal contraction of the cotton/phenolic material is sensitive to the material orientation. As with most laminated composite materials, the through thickness (or normal direction) exhibits the largest contraction since it is perpendicular to the fiber reinforcement. The two directions parallel to the cloth direction exhibit a consistent variation in contraction of about 30 % between the two orientations for all four materials. This difference was used as the basis for the determination and designation of the warp and fill direction as mentioned previously. The thermal contraction isn't dependent on material thickness or the type of cloth for the linen or canvas/phenolic.

CONCLUSIONS

Some important low temperature mechanical properties and thermal contraction data has been generated for industrial high-pressure cotton/phenolic laminates. The properties have been compared to the cryogenic industry standard (NEMA G-10CR) to assist in material selection issues. The G-10 and G11 laminates are obviously superior but this general-purpose cotton/phenolic material does have an attractive set of properties that allow it to be utilized in many cryogenic applications. The cotton/phenolic laminates have good compressive strength and through thickness shear properties but applications that place tensile or interlaminar shear stress on the material should be addressed carefully. Finally, with respect to the properties addressed here, there isn't much difference between linen/phenolic and canvas/phenolic laminates.

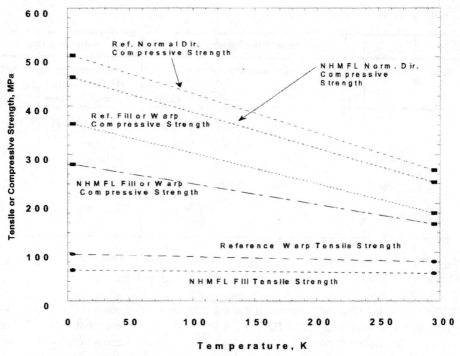

FIGURE 1. Graph of cotton/phenolic laminate tensile and compressive strengths vs. temperature.

ACKNOWLEDGEMENTS

This work is based upon research conducted at the NHMFL, which is supported by the National Science Foundation, under Award No. DMR-9527035. We also thank Mike Haslow for his strain gage and measurement instrumentation expertise.

REFERENCES

1. M.B. Kasen, et. al., "Mechanical, Electrical, and Thermal Characterization of G-10CR and G-11CR Glass-Cloth/Epoxy Laminates Between Room Temperature and 4K", Adv. Cryogenic Engineering, Vol. 26, Plenum Press, NY (1980).

2. "Materials at Low Temperatures" Reed and Clark, p. 96, p. 427 Clark 1968

3. Wang, S.T., et. al. "Low Temperature Measurement of the Thermal and Mechanical Properties of Phenolic Laminate, the Pultruded Polyester Fiberglass, and A & B Epoxy Putty" Advances in Cryogenic Engineering Vol 26 pp. 286-294 Plenum Press, NY

4. "LNG Materials and Fluids" (1977) ed. D. Mann, National Bureau of Standards, Boulder, CO

5. Walsh, R.P. "Use of Strain Gages for Low Temperature Thermal Expansion Measurements" 16th CEC/ICMC Conference, Kitakyushu, Japan, Elsevier Science, May 1996.

6. "Industrial Laminating Thermosetting Products" NEMA Standards Publication No. LI 1-1998, pp. 102 and 128, published by National Electrical Manufacturers Association, Copyright 1998

THERMAL PROPERTIES OF EPOXY RESINS AT CRYOGENIC TEMPERATURES

H.Nakane[1], S.Nishijima[2], H.Fujishiro[3], T.Yamaguchi[4], S.Yoshizawa[4], and S.Yamazaki[1]

[1]Kogakuin University
Shinjuku-ku, Tokyo, 163-8677, Japan
[2]Osaka University
Ibaraki-shi, Osaka, 567-0047, Japan
[3]Iwate University
Morioka-shi, Iwate, 020-8551, Japan
[4]Meisei University
Hino-shi, Tokyo, 191-8506, Japan

ABSTRACT

In order to establish the design technique of epoxy resin at cryogenic temperature, its thermal contraction coefficients and dynamic Young's modulus were measured from room to cryogenic temperatures when plasticizer was both present and absent. The disappearance of the effects of the plasticizer were confirmed by measuring its thermal expansion coefficient. The process in which the addition of plasticizer reduces the glass transition temperature was clarified by measuring its dynamic Young's modulus. It was also discovered that blunt peak is caused by addition of plasticizer. The data obtained by measuring the dynamic Young's modulus clearly indicate that this peak disappears at cryogenic temperature resulting in the disappearance of the effects of the plastizer. The conclusion is that when epoxy resin is to be used at cryogenic temperature it is desirable that the addition of plastizer is kept at the minimum level.

INTRODUCTION

Since epoxy resin is composed of a strong structure of three-dimensional chains, it has the advantage of having small thermal contraction [1][2][3] and high modulus of elasticity in comparison with other highly polymerized materials. On the other hand, it

CP614, *Advances in Cryogenic Engineering:*
Proceedings of the International Cryogenic Materials Conference - ICMC, Vol. 48,
edited by B. Balachandran et al.
© 2002 American Institute of Physics 0-7354-0060-1/02/$19.00

has the disadvantage of being brittle. Thus it is doped with plasticizers to improve its resistance to impact and heat when used at room temperature. At cryogenic temperature, however, it is well known that too much doping causes large heat shrinkage and decreases its resistance against thermal shock [4][5]. Investigation into its molecular design reveals that epoxy resin with a large molecular weight between its crosslinks is better suited to such purposes [5][6]. From a practical point of view, such resin is generally in solid state at room temperature and not quite suited to all purposes. As a practical plan to both improve its moldability and properties at cryogenic temperature, the proper ratio of doping is reported in this paper. Thermal contraction coefficients and dynamic Young's modulus at cryogenic temperature are also included when plasticizers are both added and not added, and the difference between them is discussed using measured data.

EXPERIMENT

Sample

The sample was made by curing the Bisphenol-A epoxy resins by using aliphatic amine. The both materials was mixed at the weight ratio of 2 to 1 and cured at 293 K (room temperature) and 353 K (80°C) for 12 hours. To discuss the effect of plasticizer, the samples which NBR polymer (Carboxyl Terminated Butadiene Nitril) were mixed on the weight ratio of 10, 20, 30 and 40% were also made. The samples for measuring the thermal expansion coefficient were cut out to the rectangular of $10 \times 10 \times 5$ mm. The disk samples of diameter of 30 mm and thickness of 0.5 mm for measuring the tan δ were made by pouring the liquid material into the pattern of silicon rubber.

Measurement of Thermal Expansion Coefficient

The thermal expandion of rectangular samples were measured by a clip type dilatometer. The clip gauge was calibrated by two copper reference specimens of different length. By calculating the moving average of output with the ascent and descent of temperature, this instrument can follow for the rapid change of thermal expansion with the transformation of samples. The thermal expansion coefficient was measured at the temperature range between 5 and 380 K.

Measurement of Dynamic Young's Modulus

The dynamic Young's modulus was measured by the tensile test. The rectangle samples were cut to the width of mm from the disk samples of diameter of 30 mm and thickness of 0.5 mm. The both sides of sample which is not included the center part of 9 mm is fixed by mechanical terminals. The tensile stress at frequency of 1 Hz and amplitude of 9 μm is added into the sample of length of 9 mm. The strain was within linear range of 0.1% (= 9 μm / 9 mm). The stress [Pa] was measured by the force rebalance transducer. Then $E^* = (stress / strain)$ is obtained, and $E' = (E^* \cos \delta)$ and $E'' =$

($E^{*}\sin\delta$) were still evaluated, where δ is a phase angle between the stress and strain.

RESULT AND DISCUSSION

The temperature dependence of thermal contraction is shown in Fig. 1. The epoxy resins without plasticizer were cured at 293 K (room temperature) and 353 K (80 °C). Since there are a certain amount of difference in thermal expansion coefficient, we think that the meaningful difference on the engineering does not exist. It is considered that such difference is caused on the difference of reactivity and inner stress during curing time. Generally, the effect is cancelled by carrying out the post cure process.

Next, let us consider the epoxy resin added plasticizer. The temperature which the thermal expansion coefficient rapidly increases was used as the glass transition temperature Tg. When plasticizer is added by 20% into epoxy resin, Tg reduces by about 30 K. When it were cured at 353 K, the thermal contraction is larger at cryogenic temperature by adding the plasticizer into it. Since the thermal contraction from a curing temperature to Tg is large, the amount of thermal contraction up to cryogenic temperature becomes larger for the material which Tg is lower. When the amount of contraction at the curing temperature shifts to zero in Fig. 1, it is actually understood that thermal contraction is larger as Tg is lower. When the plasticizer is used with other materials, the fabrication capability is improved by the plasticizer. However it is necessary to attend that the large thermal contraction difference at the cryogenic temperature is caused by it.

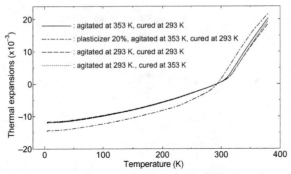

FIGURE 1. Temperature dependence of thermal expansion of epoxy resin without plasticizer.

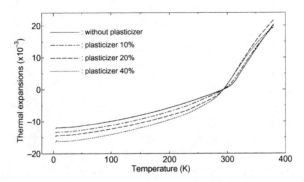

FIGURE 2. Temperature dependence of thermal expansion of epoxy resin with plasticizer.

When the amount of plasticizer was changed, the temperature dependence of thermal contraction was shown in Fig. 2. Tg can be perceived to decrease as the amount of the plasticizer increases. It means that thermal contraction up to cryogenic temperature increases as the amount of plasticizers increases. If heat stress becomes zero at curing temperature, it can be understood that the depression of Tg brings the increase of internal stress. Therefore, from this viewpoint, the material with high Tg is preferable as the material used at cryogenic temperature.

The thermal expansion coefficient which the thermal expansion in Fig. 2 was differentiated by temperature was shown in Fig. 3. It can clearly grasp from this figure that the bending point of a thermal expansion coefficient moves to low temperature according to the increase in the amount of plasticizers. As significant as it is the changes of thermal expansion coefficient around 230 K. It can be understood for the thermal expansion coefficient of all samples to decrease rapidly around this temperature, and for the thermal expansion coefficient of samples which contains plasticizer to approach that of the sample which does not add the plasticizer. It is considered that the molecular movement of the plasticizer freezes at this temperature region. It is suggested that the effect of plasticizer which used in this experiment disappears at this temperature region.

The results measured internal friction (tan δ) and dynamic Young's modulus was shown in Fig. 4 for the sample of epoxy resin and Fig. 5 for the epoxy including 40% plasticizer. The real and imaginary parts of dynamic Young's modulus are shown as E'

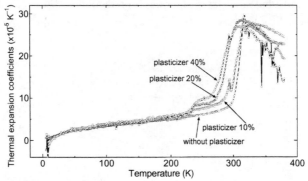

FIGURE 3. Temperature dependence of thermal expansion coefficient.

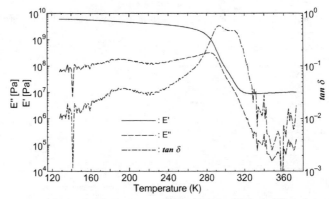

FIGURE 4. Temperature properties between complex elasticity (E' and E'') and tan δ for epoxy resin without plasticizer.

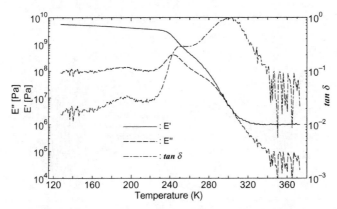

FIGURE 5. Temperature properties between complex elasticity (E' and E'') and tan δ for epoxy resin with 40% plasticizer.

and E'', respectively. In an epoxy resin, a real part E' in Fig. 4 decreases from 273 K, and it expresses a constant value (10^7 Pa) above 320 K. This temperature is equivalent to the glass transition temperature Tg of epoxy resin, and it corresponds to the peak of the imaginary part E''. In order to analyze the measuring result in detail, let us focus on the internal friction (tan δ). From the behavior of tan δ, although it seem to be simply a glass transition temperature, it can grasp that two relaxed mechanisms are actually intermingled. Moreover, the relaxed peak which exists around 200 K is a relaxation as seen in the epoxy resin.

On the other hand, the decrease of the glass transition temperature and the decrease of the Young's modulus at the room temperature region are seen in Fig. 5 for the material containing a plasticizer. Moreover, the temperature dependency of the imaginary part E'' of elasticity has different property from the pure epoxy resin and shows the peak around 240 K. The shape of the peak corresponding to the glass transition temperature of tan δ is also different. Although a detailed argument will be carried out later, an important problem is technologically pointed out. Though the Young's modulus decreases around room temperature, the Young's modulus hardly change below 230 K when the plasticizer is mixed. From this data, the effect of the plasticizer means the disappearance at cryogenic temperature. This corresponds to the disappearance of the effect of the plasticizer discussed in Fig. 3.

In order to clarify the effect of the plasticizer, the temperature dependence of tan δ of the material changed the amount of the plasticizer was shown in Fig. 6. The dispersion of tan δ which originates in the plasticizer in 230 – 240 K is plainly seen in this figure. This is an effect of the plasticizer. When the plasticizer is added, the sub-dispersion which originates in the epoxy resins is not influenced around 200 K. It is also able to understand that the size of main dispersion becomes large by adding the plasticizer. It can be understood by this behavior that the plasticizer added into the epoxy resin has softened.

In order to clarify the relation of dispersion, the imaginary part $E''$and real part E' of complex elasticity modulus are plotted on the vertical axis and horizontal axis, respectively, in Fig. 7. If the dispersion of the Debye type exists in the property, the semicircle will be drawn. In the epoxy which is not mixing the plasticizer, two circles are visible. It means that the two relaxed modes exist. The two circles correspond to the

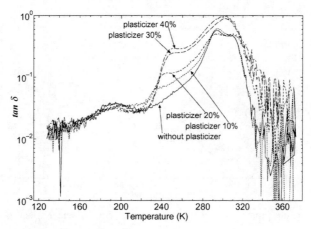

FIGURE 6. Temperature dependence of tan δ.

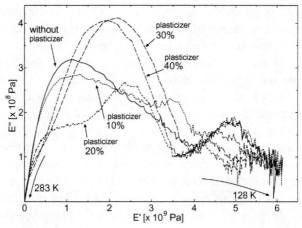

FIGURE 7. Relation between E' and E''.

main dispersion and the sub-dispersion. When a plasticizer is added, the peak of the main dispersion becomes small and the relaxed mode which newly originates in a plasticizer can be grasped. Especially, three relaxed modes is clearly detected for the materials which added 10% and 20% of the plasticizer. The relaxed peak produced at this middle region is the relaxation which originates in the plasticizer. When the plasticizer further increases, the peak of the main dispersion becomes further small, the peak is not visible at a glance and new peak which originates in the plasticizer remarkably grows up. The reason which the peak of the main dispersion becomes small is explained by the decrease of Tg with the obstruction of the construction of the three-dimensional chains of the epoxy resin by the plasticizer. Although the circle of the main dispersion for the sample mixed the 40% plasticizer is not clear in Fig. 7, it must be clarified if the axis of E'' in Fig. 5 is extended around 0 °C (273 K).

It is obvious in Fig. 7 that the position of the relaxed peak which originates in the plasticizer shifts to small value of E' as the amount of plasticizers increases. As the reason, it is considered to be equivalent to the cohesion of the plasticizer and the increase of the plasticizer region within an epoxy matrix.

CONCLUSION

In order to establish the design technique of the epoxy resin at the cryogenic temperature, when the plasticizer was systematically added to the epoxy resin, a heat expansion coefficient and dynamic Young's modulus were measured from room temperature to cryogenic temperature, and the following matter became clear.

1. The molecular movement of the plastizer was detected by measuring the thermal expansion coefficient, clarifying its freezing process at cryogenic temperatures. This explains the disappearance of the effects of the plasticizer. Addition of plasticizers increases the thermal expansion coefficient and hence execessive addition degrades the physical properties of epoxy resin at cryogenic temperatures.

2. From the measurement of dynamic Young's modulus, the process in which the plasticizer reduces the glass transition temperature Tg becomes obvious. Then the relaxed peak of temperature dependence of tan δ caused by the plasticizer was detected. It was shown that the relaxed peak caused by the plasticizer also disappears at cryogenic temperature, and the effect of a plasticizer is lost at cryogenic temperature. This phenomenon is corresponding to the measuring result of the heat expansion coefficient.

3. From the above-mentioned conclusion, it is shown that it is preferable to reduce the amount of the plasticizer as much as possible because the effect of the plasticizer disappears at cryogenic temperature. When the improvement of moldability with the room temperature neighborhood is necessary, it should be assumed a minimum amount of the plasticizer should be used for acceptable cryogenic temperature performance.

ACKNOWLEDGMENT

The authors would like to thank T. Nomura of Rheometric Scientific F. E. Ltd for his helpful experiment of dynamic Young's modulus.

REFERENCES

1. T.Okada, H.Okuyama, S.Nishijima, T.Hirokawa, J.Yasuda and T.Uemura, Development of Three Dimensional Fabric Reinforced Plastics for Cryogenic Application, Proc. Int. Cryog. Mater. Conf. Shenyang China, pp.771-776 (1988).

2. S.Nishijima, Y.Honda, S.Ueno, S.Tagawa and T.Okada,.Development of High Performance Composites for Cryogenic Use - Application of Positron Annihilation Method, Material Science Forum, Vol.255-275, pp.766-768 (1997).

3. T.Ueki, K.Nojima, K.Asano, S.Nishijima and T.Okada, Toughening of Epoxy Resin Systems for Cryogenic Use, Adv. Cryog. Eng. Vol.44, pp.277-283 (1998).

4. T.Ueki, K.Nojima, K.Asano, S.Nishijima, and T.Okada, Improvement of Fracture Toughness of Epoxy Resin for Cryogenic Use, Proc. ICEC16-ICMC, Kitakyushu, Japan pp.2061-2064 (1996).

5. S.Nishijima, Y.Honda, S.Tagawa, T.Okada, Study of Epoxy Resin for Cryogenic Use by Positron Annihilation Method, Journal of Radioanalytical and Nuclear Chemistry, vol.211, pp. 93-101 (1996).

6. F.Sawa, S.Nishijima, Y.Ohtani, K. Matsushita and T.kada, Fracture Toughness and Relaxation on Epoxy Resins at Cryogenic Temperatures, Adv. Cryog. Eng. Vol.40, pp. 1113-1119 (1994).

NON-METALLIC MATERIALS — TESTING AND EVALUATION

INFLUENCE OF SAMPLE AND TEST GEOMETRY ON THE INTERLAMINAR SHEAR STRENGTH OF FIBER REINFORCED PLASTICS UNDER STATIC AND DYNAMIC LOADING AT RT AND 77 K

P. Rosenkranz[1], D. H. Pahr[2], K. Humer[1], H. W. Weber[1]
and F. G. Rammerstorfer[2]

[1]Atominstitut der Österreichischen Universitäten, A-1020 Wien, Austria

[2]Institut für Leicht- und Flugzeugbau, TU Wien, A-1040 Wien, Austria

ABSTRACT

Fiber reinforced plastics are employed as insulation system for the superconducting magnet coils in fusion devices. In view of this application the material performance especially under interlaminar shear loading has to be assessed under conditions including the appropriate radiation environment at the magnet location. The established standards for their mechanical characterization involve sample sizes that are by far too large for reactor irradiation. As a consequence, the short-beam-shear (SBS) and the double-lap-shear (DLS) geometry were chosen for static and dynamic tests of the interlaminar shear strength at room temperature and at 77 K. Scaling experiments were made to investigate the influence of the sample and test geometry on the shear strength. In addition, finite element calculations were made for both geometries, in order to assess the stress distribution and the failure criteria of the specimens and to compare the theoretical results with the experimental data.

INTRODUCTION

Fiber-reinforced-plastics (FRP's) show excellent mechanical and electrical properties and are, therefore, candidate insulating systems for the superconducting magnet coils of fusion devices (i.e. ITER and future nuclear fusion reactors). Because of the pulsed operation of these devices, the high Lorentz forces lead to fatigue loading of the whole magnet system. As pointed out in the literature [i], not only the static but also the dynamic material performance of the insulation system is of special interest and has to be assessed accurately, in order to ensure safe operation over the plant lifetime. It turned out, that the interlaminar shear behavior of the laminate is the most sensitive and the weakest spot in the mechanical material performance of the insulation in the irradiated state.

CP614, *Advances in Cryogenic Engineering:*
Proceedings of the International Cryogenic Materials Conference - ICMC, Vol. 48,
edited by B. Balachandran et al.

In the past, most of the research was done under static loading conditions at room or cryogenic temperatures in the tensile, compressive or shear mode [2-4]. A few studies refer to fatigue studies mainly under tensile but also under shear loading of various FRP´s [5,6]. We reported on tensile fatigue tests of unirradiated and irradiated FRPs at room and low temperature as well as on scaling experiments to find a small sample geometry suitable for low temperature irradiation and testing [7,8].

Furthermore, it should be noted that the assessment of the interlaminar shear strength for FRP laminates is not trivial. The existing standards (i.e. the single lap shear or the short beam shear test) do not provide a pure and uniform interlaminar shear stress over a sizeable region of the test section in the specimen. This problem is discussed in detail in Ref. [9].

We chose for our investigations double lap shear (DLS) specimens (Figure 1) because of their axial symmetry and their applicability under simple tensile loading in the static and in the fatigue mode. Single and double lap shear specimens contain stress concentrations at the boundaries of the shear stress field [10]. The ratio of the shear length (L) to the sample thickness (t) of these sample geometries (cf. Figure 1) influences the results on the interlaminar shear strength [8-12]. Therefore, nine different sample geometries were selected and measured both at room temperature and at 77 K to assess the influence of the L/t ratio on the fatigue behavior of DLS specimens. In addition, the experimentally simple short beam shear (SBS) tests was investigated as well, in order to compare the results on the interlaminar shear strength (ILSS) obtained from both methods. Furthermore, finite element calculations were made for both geometries, in order to assess the stress distribution and the failure criteria of the loaded specimens, to compare the theoretical results with the experimental data and to find correlations between them.

EXPERIMENTAL DETAILS

The samples were prepared from 1x2 m^2 plates with a thickness of 4 mm (SBS test) and of 3, 4 and 5 mm (DLS test), respectively, of the laminate ISOVAL 10/E (ISOVOLTA AG, Wr. Neudorf, Austria). This prepreg consists of a two-dimensionally woven E-glass-

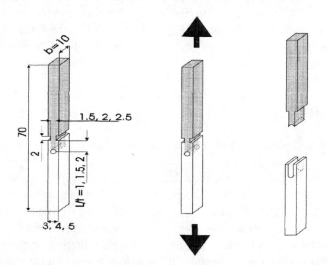

FIGURE 1. Sample geometry (width b, thickness t, and shear length L) and dimensions (mm) for the double lap shear specimen. The arrows indicate the loading direction. The interlaminar shear strength (ILSS) is calculated from the relation ILSS=F_{max}/2bL.

fiber reinforcement (0°/90°, orthotropic) in DGEBA epoxy. For the DLS specimens (Figure 1), the shear length (L) was varied at a constant shear width of 10 mm, resulting in various L/t-values for each thickness (t). The SBS tests (ASTM D2344) were carried out on 4 mm thick and 6, 8 and 10 mm wide samples at a span-to-thickness ratio of 3:1. This ratio was chosen as a consequence of our FEM calculations [13], which demonstrated that the critical global failure mode would be ply failure and not delamination failure for the span-to-thickness ratio of 5:1, suggested in the ASTM standard.

The fatigue tests were conducted with a MTS 810 TestStar II Material Testing System, which was modified for measurements in liquid nitrogen. The static interlaminar shear strength (ILSS) were conducted in displacement control using a rate of 0.5 mm/min^{-1} for the DLS test and of 1.3 mm/min^{-1} for the SBS test (according to ASTM D2344), respectively. All fatigue tests were run at 10 Hz under load control with sinusoidal loading at a minimum to peak stress ratio of R=0.1. For the shear fatigue tests (DLS specimen) various ratios of peak stress to the static ILSS were chosen and investigated up to 10^6 cycles to failure. Each load level was measured four or five times.

RESULTS AND DISCUSSION

Static interlaminar SBS and DLS properties

The ILSS (Fig. 2) for all DLS sample geometries are higher at 77 K than at room temperature, approximately by a factor of 1.5. In addition, the shear strength increases more rapidly with decreasing shear length at 77 K. Because of the low failure loads and the occasional occurrence of premature failure, we found it quite difficult to assess the ILSS for L/t<1, i.e. the standard deviation increases for decreasing L/t. Furthermore, for a sample thickness of 3 mm and L/t<1 it was often impossible to assess reliable ILSS data at room temperature. Evans et al. [12] also observed an increasing ILSS with decreasing shear length. Becker [10,11] suggested an extrapolation of the measured ILSS to L/t=0, in order to get the 'true' ILSS. This assumption may also hold for double lap shear specimens. Therefore, the results obtained with a specimen geometry of L/t=1 might be closest to the 'true' ILSS.

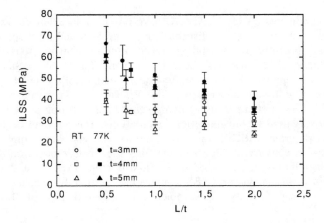

FIGURE 2. Interlaminar shear strength (ILSS) versus L/t for all DLS geometries and test temperatures.

223

TABLE 1. Static ILSS (MPa) of 4 mm thick SBS specimens obtained at 293 K and at 77 K. The ILSS is calculated from the relation ILSS=$3F_{max}/4bt$, where b and t are the sample width and thickness, respectively.

Temperature	Width 6 mm	Width 8 mm	Width 10 mm
293 K	52	44	43
77 K	70	60	60

On the other hand, as can be seen in Table 1, the ILSS of ISOVAL 10/E obtained by the SBS test is higher than the above values (~40 MPa). Therefore, the 'true' ILSS is underestimated by the double lap shear method. In order to clarify this situation, a finite element analysis of both test geometries was made in combination with additional acoustic emission experiments. As discussed in detail in the next section, the FE analysis confirms delamination (i. e. fracture between laminates layers) as the failure mode for both types of specimens and indicates, that the 'true' ILSS is underestimated by up to 50% at L/t=1 for DLS specimens and by up to 15% for SBS specimens.

Interlaminar DLS fatigue properties

The final results of the scaling experiments on ISOVAL-10/E under dynamic interlaminar shear loading at room temperature and at 77 K are presented in Figure 3, where the normalized shear stress versus cycle-to-failure curves (S/N-curves, Wöhler curves) are plotted for various sample thicknesses and L/t ratios. Compared to the S/N-curves obtained under dynamic load in tension [7,8], the S/N-curves for interlaminar shear (Fig. 3) show a smaller decrease of the strength (a decrease by ~25 to 50% at 10^6 cycles as compared to ~70% for the tensile tests). In general, higher stress levels are found for decreasing L/t ratios. While the S-N curves at room temperature are almost identical for L/t=2 and L/t=1.5, the material sustains about 5% higher relative load levels at L/t=1. However, this minor influence of the shear length is absent at 77 K. In addition, the S/N-curves obtained at 77 K and under shear load show a considerably smaller decrease of the strength (i.e. a decrease by 25% at 10^6 cycles) as compared to room temperature (~50%). In agreement with our previous work on the static and dynamic tensile scaling properties [7], no systematic influence of the geometry is found in the experimental results (overall effects within 5-10%), i.e. the small samples, developed in our program and required for irradiation experiments, are definitely suitable for this purpose.

Finite element analysis on SBS and DLS specimens

The scaling experiments on ISOVAL 10/E have provided an extended data base and input parameters to calculate the stress distributions in the SBS and DLS specimens with two-dimensional and three-dimensional linear elastic finite element methods (FEM). The assumption of a linear elastic material behavior is valid for DLS specimens, if first delamination loads are applied. Hence, the first delamination failure (FDF) loads were determined by acoustic emission experiments. The FE analyses are based on these FDF loads and the "real" ILSS is obtained by applying some kind of fitting procedure. The FE code MSC-Nastran was used for all numerical calculations. The pre- and post-processing was done by the program MSC-Patran. The failure models were implemented by FORTRAN routines. In contrast, the ILSS of the SBS specimen is based on the ultimate load, although local failure occurs first before the final fracture takes place. A comparison of linear and non-linear FE analyses on SBS samples has led to a difference of ~4% in the ILSS. Therefore, the linear FE analysis is sufficient for the investigation of SBS specimens too.

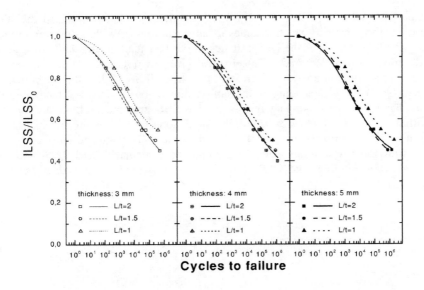

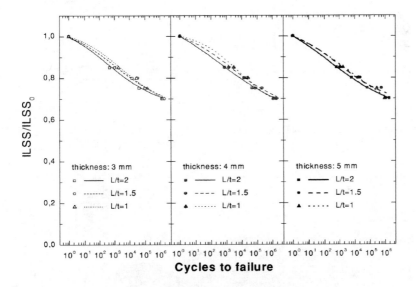

FIGURE 3. Influence of the shear-length-to-sample-thickness ratio L/t on the fatigue behaviour under interlaminar shear loading of 3, 4 and 5 mm thick ISOVAL-10/E at room temperature (upper panel) and at 77 K (lower panel).

Figure 4 shows the FE meshes for the SBS and DLS specimens. Figures 5 and 6 show the computed delamination risk parameter for both models for $F=F_{ult}$, where F_{ult} is the ultimate failure load (100% F_{ult}). For the DLS specimen the critical region occurs near the notch (not around the hole), i.e. crack initiation starts at the notch for $F<F_{ult}$. The evaluation of the average delamination risk parameter within the overlapping zone gives a value of

225

~50%. From this point of view the ILSS is significantly underestimated by the "standardized" analysis method which assumes mean shear stresses within the overlapping zone for F=F_{ult}. The problem of the DLS test lies in the fact that the standard evaluation procedure describes a complex fracture mechanical problem (crack initiation and crack growth) by a simple strength analysis. Therefore, the "true" ILSS of the DLS specimen is underestimated by ~40-50% depending on the sample dimensions and the test temperature. On the other hand, the corresponding results for the SBS specimen (cf. Figure 5) are "correct" within ~15%. Thus, stress distribution calculations with FEM are indispensable for an assessment of a "correlation factor" for the experimentally evaluated ultimate stress level for each individual type of laminate. Table 2 compares calculated "real ILSS" and the experimentally evaluated SBS and DLS ILSS by presenting the percentage difference between them.

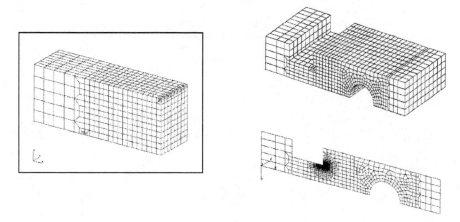

FIGURE 4. FEM meshes for the SBS and DLS specimens.

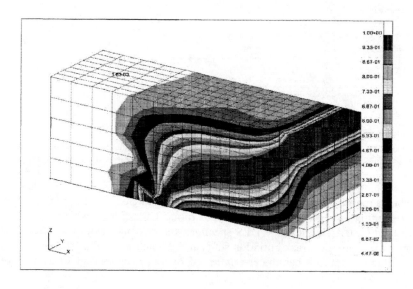

FIGURE 5. Delamination risk parameter of the short beam shear (SBS) specimen for 100% F_{ult}.

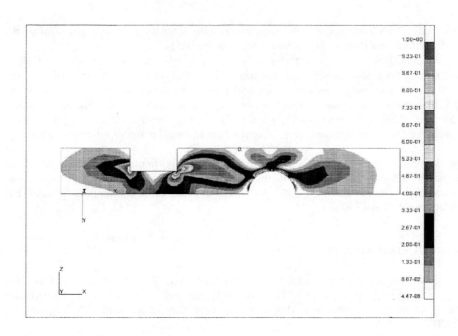

FIGURE 6. Delamination risk parameter of the double lap shear (DLS) specimen for 100% F_{ult}.

SUMMARY

In view of their applications superconducting magnet insulation materials in the ITER device, the interlaminar shear properties of fiber reinforced plastics are of special interest. Existing standard test procedures for such laminates do not provide a uniform interlaminar shear stress distribution over a sizeable region in the test section of the specimen.

For the double lap shear (DLS) specimen geometry, the ratio of the shear length to the sample thickness influences the measured interlaminar shear strength (ILSS). On the other hand, the span to thickness ratio of the short beam shear (SBS) test may influence the evaluated ILSS. Therefore, a scaling program was carried out to assess the influence of the specimen geometry on the ILSS under static and fatigue loading both at room temperature and at 77 K.

Furthermore, finite element calculations were made for both test geometries to assess the stress distribution and the failure criteria for the loaded specimens as well as to compare the theoretical results with the experimental data and to find a correlation between them. The results may be summarized as follows.

TABLE 2. Comparison of the calculated "real ILSS" R^{IL}_{13} and the experimental SBS and DLS ILSS R^{IL}_{13exp}

Test	Temperature (K)	Width b (mm)	R^{IL}_{13} (MPa)	R^{IL}_{13exp} (GPa)	Difference (%)
SBS	293 (77)	10	48 (69)	43 (60)	-10 (-13)
	293 (77)	8	48 (69)	44 (60)	-8 (-13)
DLS	293 (77)	8	48 (69)	28 (36)	-41 (-48)
	293 (77)	6	48 (69)	33 (44)	-31 (-36)
	293 (77)	4	48 (69)	33 (47)	-31 (-32)

- The static ILSS increases at both temperatures with decreasing shear length L, while the sample thickness t shows only minor effects on the ILSS.
- The dyamic ILSS does not vary in a systematic way with the shear length L as well as the sample thickness t at both temperatures. The overall effects are within 5-10%.
- A finite element analysis as well as accompanying acoustic emission experiments on the DLS and SBS specimens identified delamination as the failure mode. For the DLS test, the 'real' ILSS is underestimated by up to 50%, if the standard evaluation procedure is employed. The corresponding results for the SBS test are "correct" within ~10-15%.

These scaling results of the DLS specimen and, in particular, the FEM calculated "correlation factors" for the DLS and the SBS tests were used for investigations of the influence of reactor irradiation on the interlaminar shear fatigue behaviour of an ITER candidate material at 77 K.

ACKNOWLEDGEMENTS

The authors are greatly indebted to ISOVOLTA AG, Wiener Neudorf, Austria, for providing them with the samples of ISOVAL-10/E. Technical support by H. Niedermaier and E. Tischler is acknowledged. This work has been carried out within the Association EURATOM-OEAW.

REFERENCES

1. IAEA-TECDOC-417, "Insulators for fusion applications", IAEA, Vienna, Austria (1987).
2. Humer, K., Weber, H. W. and Tschegg, E. K. *Cryogenics* **35**, pp. 871-882 (1995).
3. Spiessberger, S. M., Humer, K., Tschegg, E. K., Weber, H. W. and Gerstenberg, H. *Cryogenics* **38**, pp. 79-83 (1998).
4. Rice, J. A., Hazelton, C.S. and Fabian, P. E., "Wrappable ceramic insulation for superconducting magnets", *Advances in Cryogenic Engineering* 46A, 2000, pp. 267-273.
5. Humer, K., Rosenkranz, P., Weber, H. W., Fabian, P. E. and Rice, J. A. *Journal Nuclear Materials* **283-287**, pp. 973-976 (2000).
6. Reed, R. P., Fabian P. E. and Schutz, J. B., "Development of U.S./ITER CS model coil turn insulation", *Advances in Cryogenic Engineering* 44A, 1998, pp. 175-182.
7. Rosenkranz, P., Humer, K. and Weber, H. W. *Cryogenics* **40**, pp. 155-158 (2000).
8. Rosenkranz, P., Humer, K. and Weber, H. W., "Influence of the test geometry on the fatigue behavior of fiber reinforced plastics in tension and interlaminar shear at RT and 77 K", *Advances in Cryogenic Engineering* 46A, 2000, pp. 181-187.
9. Rosenkranz, P., Humer, K., Weber, H. W., Pahr, D. H. and Rammerstorfer, F. G. *Cryogenics* **41**, pp. 21-25 (2001).
10. Becker, H., "Problems of cryogenic interlaminar shear strength testing", *Advances in Cryogenic Engineering* 30, 1983, pp. 33-40.
11. Becker, H. and Erez, E. A., "A study of interlaminar shear strength at cryogenic temperatures", *Advances in Cryogenic Engineering* 26, 1979, pp. 259-267.
12. Evans, D., Johnson, I. and Dew Hughes, D., "Shear testing of composite structures at low temperatures", *Advances in Cryogenic Engineering* 36B, 1990, pp. 819-826.
13. Pahr, D. H., Rammerstorfer, F. G., Rosenkranz, P., Humer, K. and Weber, H. W. Composites Part B, submitted (2001).

RESULTS OF A VAMAS ROUND ROBIN ON THE CRYOGENIC INTERLAMINAR SHEAR STRENGTH DETERMINATION OF G-10CR GLASS-CLOTH/EPOXY LAMINATES

Y. Shindo[1], K. Horiguchi[1] T. Ogata[2], A. Nyilas[3],
Z. Zhang[4], and K. Humer[5]

[1]Tohoku University,
  Sendai, Miyagi 980-8579, Japan
[2]National Institute for Materials and Science,
  Tsukuba, Ibaraki 305-0047, Japan
[3]Forschungszentrum Karlsruhe,
  D-76021 Karlsruhe, Germany
[4]Chinese Academy of Sciences,
  Beijing 100080, China
[5]Atominstitut der Österreichischen Universitäten,
  A-1020 Wien, Austria

ABSTRACT

This paper presents the results of a round-robin test program on cryogenic interlaminar shear strength of G-10CR glass-cloth/epoxy laminates. This program was organized under the auspices of the Technical Working Area 17 (TWA17) on Cryogenic Structural Materials of the Versailles Project on Advanced Materials and Standards (VAMAS), and performed to investigate the applicability of ASTM Standards D 3846-94 and D 2344-84 to woven fabric composite laminates at cryogenic temperatures. Specifically, two standard test specimen geometries (the double-notch shear specimen, and the short-beam shear specimen) were prepared. A total five laboratories in four different countries participated in this testing program. Each laboratory tested both specimens at cryogenic temperatures. A three-dimensional finite element analysis was also utilized to conduct a detailed study of the stress distributions within the specimens.

INTRODUCTION

Composites made from fiberglass fabrics have low through-thickness mechanical properties, poor impact damage tolerance, and low interlaminar fracture toughness.

CP614, *Advances in Cryogenic Engineering:*
Proceedings of the International Cryogenic Materials Conference - ICMC, Vol. 48,
edited by B. Balachandran et al.

The critical value of the interlaminar shear strength at failure, therefore, is a very important parameter in the design of composites structures. Several attempts have been made to determine the interlaminar shear strength for laminated composites at cryogenic temperatures. While the double-notch [1] and short-beam [2] shear tests for evaluating the cryogenic interlaminar shear strength of composite laminates have been used extensively, experimental results obtained using these different methods vary so much that the question remains as to which method is the most reliable [3, 4]. Shindo et al. [5] evaluated the cryogenic interlaminar shear strengths of G-10CR glass-cloth/epoxy laminates by both double-notch shear test and three-dimensional finite element method. The apparent interlaminar shear strength at liquid nitrogen temperature (77 K) and liquid helium temperature (4 K) depended on specimen geometry, and decreased with an increase in notch separation, but the maximum shear stress was nearly independent of notch separation variations. The maximum shear stress near the notch tip appeared to be a true indication of the interlaminar shear strength of the double-notch shear specimen at low temperatures. Shindo et al. [6] also discussed the cryogenic short-beam interlaminar shear behavior of G-10CR through theoretical and experimental characterizations. For the specimens failed in pure shear, the maximum shear stress at the mid-plane appeared to be a true indication of the interlaminar shear strength of the short-beam specimen at low temperatures. The best agreement occurred between the maximum shear stress of the double-notch shear specimen and the short-beam shear specimen.

Round-robin test programs to evaluate testing procedures for cryogenic interlaminar shear strength are performed under the auspices of the Technical Working Area 17 (TWA17) on Cryogenic Structural Materials of the Versailles Project on Advanced Materials and Standards (VAMAS). Ogata et al. [7] reported the results of the second VAMAS round-robin tests on interlaminar shear strength for G-10CR glass-cloth/epoxy laminates, and showed that the average 4 K interlaminar shear strength in short-beam was 128 MPa with a standard deviation of 14 MPa and in guillotine type was 106 MPa with a standard deviation of 7 MPa. This paper summarizes the results of the VAMAS round-robin effort aimed at validating the proposed cryogenic interlaminar shear strength test methods. Participating laboratories volunteered to perform both double-notch and short-beam shear tests at cryogenic temperatures. Each laboratory received several test specimens appropriate for their tests. The five laboratories listed in Table 1 were able to submit test results for analysis and comparison. A three-dimensional finite element analysis was also used to study the stress distributions within the test specimens and to interpret the experimental measurements. The double-notch and short-beam shear tests are compared as methods for determining cryogenic interlaminar shear strength of G-10CR.

MATERIALS AND TEST METHODS

The material in this program was G-10CR supplied in the form of 2.5 mm and 6.35 mm thick plates. These test coupons were cut with the length parallel to the fill

TABLE 1. Round robin participants listed in alphabetical order.

Atominstitut der Österreichischen Universitäten	K. Humer
Chinese Academy of Sciences	Z. Zhang
Forschungszentrum Karlsruhe	A. Nyilas
National Institute for Materials and Science	T. Ogata
Tohoku University	Y. Shindo

direction. Four laboratories performed interlaminar shear tests either at 77 K or at 4 K or both. Low temperature environments were achieved by immersing the loading fixture, specimen, and clip gage in liquid nitrogen (77 K) or liquid helium (4 K). One laboratory tested at 7 K. The temperature of 7 K was obtained by testing in helium flow cryostat. Loading rates between 0.06 and 0.5 mm/min were used by all groups.

Double-Notch Shear Test

The shape of a typical test specimen is depicted in Fig. 1. The specimens were notched with a cut on each of the opposite faces of the specimen across the entire width, of depth equal to half of the specimen thickness. The dimensions of a double-notch specimen are 79.5 mm in length (L), 12.7 mm in width (B), 6.35 mm in thickness (T), and the notch separation (S) is 1.6 mm. The procedure recommended by ASTM D3846-94 [1] is to load the double-notch shear specimen edgewise in compression with a supporting jig to prevent buckling. The interlaminar shear strength $ILSS$ was calculated from the solution for double-notch shear specimens:

$$ILSS = \frac{P_{max}}{SB} \qquad (1)$$

where P_{max} is the maximum load.

Short-Beam Shear Test

Figure 2 shows the three-point bend setup and specimen. The dimensions of a short-beam specimen are 15.0 mm in length (l), 10.0 mm in width (b), 2.5 mm in thickness (t), and the span-to-thickness ratio (s/t) is about 5.0. Each specimen was placed on two roller supports that allow lateral motion and a load was applied directly at the center of the specimen. Following the ASTM D 2344-84 [2], the apparent interlaminar shear strength S_H was calculated from the solution for short-beam shear specimens:

$$S_H = \frac{3P_B}{4bt} \qquad (2)$$

where P_B is the breaking load.

FINITE ELEMENT ANALYSIS

In order to evaluate the interlaminar shear strength of G-10CR at 77 K and 4 K, a three-dimensional finite element analysis was carried out. The boundary conditions for the finite element analysis are shown in Fig. 3. The uniformly distributed load p was applied as nodal forces. This approximately corresponded to the experimentally

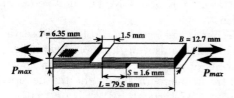

FIGURE 1. Specimen geometry for double-notch shear testing.

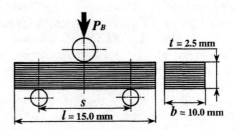

FIGURE 2. Experimental setup and geometry of short-beam shear specimen.

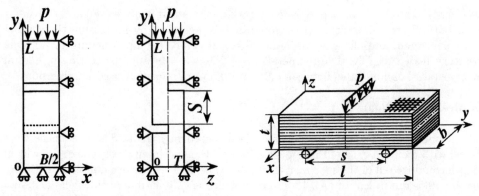

(a) Double-notch shear specimen   (b) Short-beam shear specimen

FIGURE 3. Boundary conditions assumed for finite element calculations.

TABLE 2. Predicted elastic moduli for G-10CR at 77 K and 4 K.

Temperature	E_x (GPa)	E_y (GPa)	E_z (GPa)	G_{xy} (GPa)	G_{yz} (GPa)	G_{zx} (GPa)	ν_{xy}	ν_{yz}	ν_{zx}
77 K	32.53 (33.7)	29.53 (37.0)	20.60	8.82	8.63	8.59 (0.19)	0.18	0.35	0.21
4 K	35.60 (35.9)	32.94 (29.1)	25.19	10.35	10.16	10.13 (0.21)	0.21	0.36	0.24

determined maximum load. Effective elastic moduli of G-10CR are determined under the assumption of uniform strain inside the representative volume element [6]. Predicted elastic moduli along with experimental data [8] are listed in Table 2, and (E_x, E_y, E_z) are the Young's moduli, (G_{xy}, G_{yz}, G_{zx}) are the shear moduli, $(\nu_{xy}, \nu_{yz}, \nu_{zx})$ are the Poisson's ratios and figures in brackets indicate the experimental data. The subscripts x, y and z will be used to refer to the coordinate directions, and the Poisson's ratio ν_{xy} reflects shrinkage (expansion) in the y-direction due to tensile (compressive) stress in the x-direction. The predicted elastic moduli agree well with the experimental data.

RESULTS AND DISCUSSION

A complete list of $ILSS$ values for the double-notch shear specimens tested in this program is recorded in Table 3. Laboratories 1, 2, 3 and 4 used the supporting jig and tested in compression. Laboratory 5 tested in tension without the supporting jig. These results are quite similar to the average $ILSS$. The average $ILSS$ value at 77 K is 125.9 MPa. The average $ILSS$ values at 7 K and 4 K range from 142.0 MPa to 135.1 MPa or +3.3% to -1.7% on the value of $ILSS$ from the average of all data, that is, 137.4 MPa. The $ILSS$ values increase between 77 K and 4 K. The double-notch shear specimens did not have significant problems.

All S_H data for the short-beam shear specimens are listed in Table 4. Observation of failed specimens showed a complex failure mechanism in the short-beam shear test, while the double-notch shear test specimens appeared to have failure in shear. The short-beam shear specimens had damage (resin microcracks) on the tensile side. No shear failure could be observed for specimens SB7 through SB12. Note that the specimens tested by laboratory 2 produce the high strength, but do not fail along the centerline of the specimen. With the exception of laboratories 2 and 5, the short-beam shear results are quite similar. The S_H values at 7 K and 4 K are higher than those at 77 K. Both tests show the same trends in strength, but provide significantly different quantities for the cryogenic interlaminar shear strength values.

Consider a Y_1-axis with the origin at the specimen midsection and Y_1 increases (decreases) along the fracture plane to a maximum of $Y_1 = 1$ (a minimum of $Y_1 = -1$) at the notch tip (i.e., $Y_1 = 2(y-L/2)/S$ for $L/2 - S/2 \leq y \leq L/2 + S/2$). Figure 4 shows the distributions of predicted shear stress σ_{zy}, normalized with respect to the average shear stress P_{max}/SB, along the centerline (-1 $\leq Y_1 \leq 1$, $x = B/2$, $z = T/2$) at 77 K and 4 K. The results of stress analysis demonstrate that the shear stress distribution is nonuniform on the fracture plane. As expected, large shear stress concentrations are

TABLE 3. Values of $ILSS$ for G-10CR.

Temperature	Lab	Specimen No.	T (mm)	B (mm)	S (mm)	P_{max} (kN)	$ILSS$ (MPa)	
77 K	1	DN1	6.35	12.7	1.6	2.53	124.3	
		DN2	6.35	12.7	1.6	2.40	122.0	
		DN3	6.35	12.7	1.6	2.43	123.7	
		DN4	6.35	12.7	1.6	2.69	136.8	
							Average	126.7
77 K	5	DN5	6.35	12.7	1.6	2.24	110	
		DN6	6.35	12.7	1.6	2.80	138	
		DN7	6.40	12.7	1.6	2.54	125	
		DN8	6.35	12.7	1.6	2.89	142	
		DN9	6.35	12.7	1.6	2.33	115	
		DN10	6.40	12.7	1.6	2.46	121	
							Average	125
4 K	1	DN11	6.35	12.7	1.6	2.74	134.8	
		DN12	6.35	12.7	1.6	2.67	131.6	
		DN13	6.35	12.7	1.6	2.83	139.3	
							Average	135.4
4 K	2	DN14	6.35	12.7	1.6	2.81	135.8	
		DN15	6.35	12.7	1.6	2.60	126.1	
		DN16	6.35	12.7	1.6	2.87	139.6	
		DN17	6.35	12.7	1.6	2.67	134.1	
		DN18	6.35	12.7	1.6	2.94	142.7	
		DN19	6.35	12.7	1.6	2.98	144.7	
							Average	137.2
4 K	3	DN20	6.35	12.7	1.6	2.89	142.4	
		DN21	6.35	12.7	1.6	2.79	137.0	
		DN22	6.41	12.7	1.6	2.56	125.9	
							Average	135.1
7 K	4	DN23	6.35	12.7	1.6	2.83	139.5	
		DN24	6.38	12.7	1.6	2.72	134.0	
		DN25	6.44	12.7	1.6	3.29	161.8	
		DN26	6.41	12.7	1.6	2.81	138.3	
		DN27	6.39	12.7	1.6	2.77	136.3	
							Average	142.0

found at each end of the fracture plane and the maximum shear stress occurs near the reentrant corners of the notches. The interlaminar shear failure will initiate here under the maximum shear stress.

Consider a Y_2-axis with the origin at the specimen midsection (i.e., $Y_2 = 2y/l$ for $0 \leq y \leq l/2$) and Y_2 increases (decreases) along the mid-plane to a maximum of $Y_2 = 1$ at the center of the loading nose (a minimum of $Y_2 = 0$). Figure 5 shows the distributions of predicted shear stress σ_{zy}, normalized with respect to the average shear stress $3P_B/4bh$, along the centerline ($0 \leq Y_2 \leq 1$, $x = 0$, $z = t/2$) at 77 K and 4 K. The maximum shear stress is found at $Y_2 = 0.32$. The maximum value of the shear stress is higher than the maximum value given by beam theory ($S_H = 3P_B/4bh$).

TABLE 4. Values of S_H for G-10CR.

Temperature	Lab	Specimen No.	t (mm)	b (mm)	s/t (mm)	P_B (kN)	S_H (MPa)	
77 K	1	SB1	2.5	10.0	5.0	3.02	89.4	
		SB2	2.5	10.0	5.0	3.16	91.3	
							Average	90.4
77 K	2	SB3	2.5	10.0	5.0	4.00	122.1	
		SB4	2.5	10.0	5.0	3.96	120.0	
		SB5	2.5	10.0	5.0	4.01	120.5	
							Average	120.9
77 K	4	SB6	2.5	10.0	5.2	2.93	89.2	
77 K	5	SB7	2.5	10.0	4.9	4.90	150	
		SB8	2.4	10.0	5.0	4.93	154	
		SB9	2.5	10.0	4.9	5.29	162	
		SB10	2.4	10.0	5.0	5.06	158	
		SB11	2.4	10.0	5.0	5.22	163	
		SB12	2.5	10.0	4.9	5.26	161	
							Average	158
4 K	1	SB13	2.5	10.0	5.0	3.36	98.7	
		SB14	2.5	10.0	5.0	3.37	99.1	
							Average	98.9
4 K	2	SB15	2.5	10.0	5.0	4.04	123.0	
		SB16	2.5	10.0	5.0	4.16	125.3	
		SB17	2.5	10.0	5.0	4.16	126.3	
		SB18	2.5	10.0	5.0	3.92	118.6	
		SB19	2.5	10.0	5.0	4.04	122.4	
							Average	123.1
4 K	3	SB20	2.5	10.0	5.1	3.25	99.6	
		SB21	2.5	10.0	5.1	3.31	100.0	
		SB22	2.5	10.0	5.1	3.11	94.8	
							Average	98.1
7 K	4	SB23	2.5	10.0	5.2	3.03	92.1	
		SB24	2.5	10.0	5.2	3.11	95.1	
		SB25	2.5	10.0	5.2	2.92	89.4	
		SB26	2.5	10.0	5.2	3.05	92.7	
		SB27	2.5	10.0	5.2	3.07	93.5	
							Average	92.6

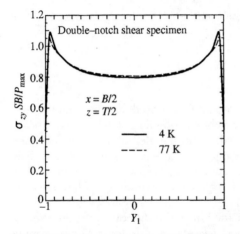

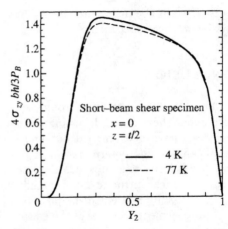

FIGURE 4. Cryogenic shear stress distributions through the length (double-notch shear specimen).

FIGURE 5. Cryogenic shear stress distributions through the length (short-beam shear specimen.

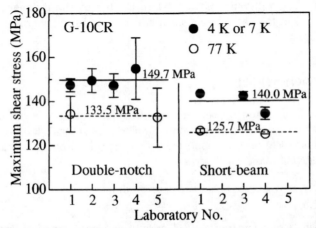

FIGURE 6. Comparison of maximum shear stress values for all laboratories testing specimens.

The expressions for the maximum shear stress of double notch shear specimens obtained by the finite element method are given below :

At 77 K :

$$\sigma_{max}^{DN} = 1.06\frac{P_{max}}{SB} \tag{3}$$

At 7 K and 4 K :

$$\sigma_{max}^{DN} = 1.09\frac{P_{max}}{SB} \tag{4}$$

Equations for short-beam shear specimens have similar forms :

At 77 K :

$$\sigma_{max}^{SB} = 1.40\frac{3P_B}{4bt} \tag{5}$$

At 7 K and 4 K :

$$\sigma_{max}^{SB} = 1.45\frac{3P_B}{4bt} \tag{6}$$

The maximum shear stresses obtained in the double-notch and short-beam shear tests are comparatively plotted for each laboratory in Fig. 6. The reasonably good agreement occurs between the maximum shear stress of the double-notch shear specimen and the short-beam shear specimen.

CONCLUSIONS

The results of the round-robin test program are presented in this paper. Two interlaminar shear strength testing methods, namely the double-notch and short-beam shear tests have been carried out and compared. The double-notch shear specimen is applicable to the determination of the cryogenic interlaminar shear strength. It is clear that failure occurs in shear along the longitudinal plane between the notches. The short-beam shear introduces uncertainly when it comes to determining the failure stress since it is heavily related to the position of failure.

Finite element analyses of the specimens loaded in the configurations, were used to approximate the state of shear stress at failure in the region of failure. The maximum shear stress for the specimens failed in shear appears to be a true indication of the cryogenic interlaminar shear strength. The 4 K interlaminar shear strength of G-10CR measured from the double-notch shear test is about 7% higher than that from the short-beam shear test. The difference is thought to be due to the influence of the damage (resin microcracks) in the short-beam shear specimens. Resin microcracks may lead to local delamination.

ACKNOWLEDGEMENTS

The contribution made by the participating laboratories, each of which carried out its tests at no cost in order to contribute to the international effort, is greatly appreciated. This research was supported by the Special Coordination of Science and Technology Agency, Japan.

REFERENCES

1. *ASTM D3846-94, Standard Test Method for In-Plane Shear Strength of Reinforced Plastics.*
2. *ASTM D2344-84, Standard Test Method for Apparent Interlaminar Shear Strength of Parallel Fiber Composites by Short-Beam Method.*
3. Kasen, M. B., "Current Status of Interlaminar Shear Testing of Composite Materials at Cryogenic Temperatures," in *Advances in Cryogenic Engineering* 36, edited by R. P. Reed and F. B. Fickett, Plenum Press, New York, 1990, pp.787-792.
4. Evans, D., Johnson, I. and Hughes, D. D., "Shear Testing of Composite Structures at Low Temperatures," in *Advances in Cryogenic Engineering* 36, edited by R. P. Reed and F. B. Fickett, Plenum Press, New York, 1990, pp.719-826.
5. Shindo, Y., Wang, R., Horiguchi, K. and Ueda, S., *ASME J. Eng. Mater. Technol.* **121**, pp.367-373(1999).
6. Shindo, Y., Wang, R. and Horiguchi, K., *ASME J. Eng. Mater. Technol.* **123**, pp.112-118(2001).
7. Ogata, T., Evans, D. and Nyilas, A., "VAMAS Round Robin Tests on Composite Material and Solder at Liquid Helium Temperature," in *Advances in Cryogenic Engineering* 44, edited by Balachandran et al., Plenum Press, New York, 1998, pp.269-276.
8. Kasen, M. B., MacDonald, G. R., Beekman, D. H., Jr. and Schramm, R. E., "Mechanical, Electrical, and Thermal Characterization of G-10CR and G-11CR Glass-Cloth/Epoxy Laminates Between Room Temperature and 4 K," in *Advances in Cryogenic Engineering* 26, Plenum Press, New York, 1980, pp.235-244.

REDUCTION IN THE THERMAL CONTRACTION OF HYBRID CERAMIC INSULATED COMPOSITE STACKS

J. A. Rice, C. S. Hazelton, and P. E. Fabian

Composite Technology Development, Inc.
Lafayette, CO 80026

ABSTRACT

Composite stacks of niobium tin superconductor insulated with traditional epoxy glass insulation and a wrappable hybrid ceramic insulation system (CTD 1102x) were fabricated. The effect of the insulation type on the thermal expansion was measured between room temperature and liquid nitrogen temperature. Two different thicknesses of the ceramic matrix were examined as well as the effect of not reacting the conductor. The results show that the ceramic insulation reduces the overall thermal expansion in the thickness direction by approximately 15% from the CTD 101K epoxy / S2 glass insulated composite. No effect of repeated cooling to liquid nitrogen temperatures for composite stacks insulated with the ceramic insulation was observed. CTE measurements on a laminate of the hybrid ceramic insulation show a close match with the CTE of copper. The lower CTE mismatch between the conductor and the insulation will reduce thermal stresses and improve the mechanical performance of ceramic insulated composites.

INTRODUCTION

New high performance magnet systems will require increased performance from the electrical insulation used to make them. In the past, the only way to achieve the high strengths required was to employ an epoxy fiberglass sleeve or wrap. While these epoxy systems met the mechanical requirements and were easy to process, they forced the manufacturers to work around their limitations. The high performance organic insulation systems suffered from temperature limits and high thermal contraction upon cooling to cryogenic temperatures. A15 conductor based magnets were forced to compromise on either react and wind processing or use S2 glass fabric that has deteriorated during the high temperature heat treatment.

This paper will describe another alternative for achieving the high performance necessary for advanced magnet systems. A new set of ceramic-based insulation systems have been developed at Composite Technology Development, Inc. (CTD) that can overcome these limitations. There are two general families of ceramic insulation. The first

CP614, *Advances in Cryogenic Engineering:*
Proceedings of the International Cryogenic Materials Conference - ICMC, Vol. 48,
edited by B. Balachandran et al.
© 2002 American Institute of Physics 0-7354-0060-1/02/$19.00

is a hybrid ceramic / organic insulation that contains a ceramic fiber and two interpenetrating matrices of a ceramic phase and an organic phase. The CTD 1102x insulation described below is one of these hybrid systems. Mechanical and electrical properties of the hybrid ceramic/organic insulation have been described previously in references [1] and [2]. The second is an all-ceramic system that does not employ any organic phase. The CTD 1005x insulation is in this family.

Lower thermal contraction of the insulation system will lower the thermal stresses placed upon the magnet coil during operation. By matching the CTE of the insulation to the copper or Cu-Nb$_3$Sn cable, further stress reductions can be achieved. In addition, a close match can enable stronger bonding between the insulation and the cable to further enhance performance and reduce cracking.

PROCEDURE

Materials

Both of the ceramic based insulation families process similarly to conventional organic insulation. The CTD 1002x ceramic matrix samples (equivalent to the CTD 1102x insulation prior to the addition of CTD 101K epoxy) are typically fabricated using a Vacuum Pressure Impregnation (VPI) process to fill the fabric preform. In contrast, the CTD 1005x samples are made using a pre-preg process that coats and fills each fabric layer before the laminate is pressed and set into shape. A magnet would be insulated by first wrapping the cable with the insulation (using either the VPI or pre-preg approach) and then wound into the coil shape. The magnet coil is heated to cure the ceramic matrix forming a solid completed form. The ceramic insulation is heat treated along with the superconducting cable at temperatures from 600°C to 800°C. After this heat treatment, the coil is still in a monolithic form.

The integrity after the heat treatment is another benefit of the ceramic insulation. Less costly tooling is required for handling and subsequent processing. The integrity also helps reduce the risk of damage and costly rework. At this stage, the coil can be impregnated or potted in a high performance organic to maximize the mechanical performance (for the hybrid ceramic / organic systems) or left as is for the all-ceramic insulation for highest radiation resistance.

Material Fabrication

The thermal expansion of the insulation materials was measured using flat plate laminates stacked together to give a total thickness of over 12 mm. Four specimens were made as listed in TABLE 1. Samples L1, L3, and L4 are all-ceramic and contain no organic phase. Sample L2 was impregnated with CTD 101K epoxy after finishing the heat treatment. Samples L1 and L2 were fabricated by a VPI process that filled the fabric preform with the ceramic matrix material. Samples L3 and L4 were fabricated using a pre-preg process using the CTD 1005x ceramic matrix applied to the individual ceramic fabric layers. These two samples vary in the fiber volume fraction, with sample L3 having a higher percentage of the CTD 1005x ceramic matrix. Sample L3 is nominally 50% fiber while L4 is closer to 42% ceramic fibers. The layers were then pressed together under heat to set the matrix. After fabrication of the laminates, they were heat treated according to the schedule listed in TABLE 2. In order to increase the density of sample L1, it was reimpregnated with the ceramic matrix partway through the heat treatment.

TABLE 1. Laminate samples fabricated to measure the effect of changing the ceramic matrix and adding epoxy.

Sample	Ceramic Matrix	Fibers	Epoxy
L1	CTD 1002x	CTD CF100	none
L2	CTD 1002x	CTD CF100	CTD 101K
L3	CTD 1005x	CTD CF100	None
L4	CTD 1005x	CTD CF100	None

TABLE 2. Heating Schedule for composite stack samples. Heating and cooling rate was 1.5°C/min between dwell steps.

Temperature (°C)	Dwell Time (hours)
210	60
340	48
650	180

Small samples, approximately 5 mm x 25 mm were cut from the laminates. For each specimen type, a single 12 mm thick specimen was fabricated by gluing four small specimens together using a minimum amount of strain gage adhesive. The final step mounted the cryogenic strain gage onto the side to measure the through thickness contraction.

Four sample composite stacks were fabricated to measure the combined thermal expansion characteristics of both the Nb_3Sn conductor cable and the insulation. TABLE 3 lists the variables examined in this study. These samples were fabricated from the same cable used to make the RD3 magnet at LBNL. One of the variables investigated was the effect of the compaction pressure on the stack properties. Two values were evaluated – a high-pressure stack at 2000 psi and a low-pressure stack at 500 psi. The forces were calculated from the compressibility of the cable and the insulation. The mold sizes were adjusted accordingly. There was a 0.26 mm height difference between the low-pressure and the high-pressure sample molds that translated into a lower (sample DE19-15) and a higher (DE19-17) volume fraction of insulation in the stack.

For each composite stack, short cable lengths approximately 0.2 m long were cut and the ends welded. Each length was covered with the fabric sleeve and stacked in a stainless steel mold fixture. Stacks DE19-15 and DE19-17 were vacuum impregnated with the CTD 1002x ceramic matrix using a flow through VPI method with the ceramic matrix slowly introduced into the evacuated fixture mold. The molds were cured at 150°C for 2 hours. Sample DE19-19 was set aside and the rest were placed in an inconel retort furnace for the heat treatment. The furnace was continuously purged with nitrogen gas. The furnace was heated at 1.5°C / minute according to the schedule listed in Table 2. The furnace vent plugged shortly after reaching 650°C and the cycle had to be interrupted for maintenance. The cycle was continued at the point where it was terminated. The appearance of samples looked normal after the cycle was finished.

TABLE 3. List of variables examined by the conductor/insulation composite stacks. All stacks were impregnated with epoxy as a finishing step.

ID	Conductor Cable	Ceramic Matrix CTD 1002x	Fabric	Compaction Pressure (psi)	Heat treatment
DE19-13	LBNL RD3	No	S2 glass sleeve	2000	Yes
DE19-15	LBNL RD3	Yes	S2 glass sleeve	500	Yes
DE19-17	LBNL RD3	Yes	S2 glass sleeve	2000	Yes
DE19-19	LBNL RD3	No	S2 glass sleeve	2000	No

The mold fixtures were then carefully disassembled to allow the inner surfaces to be mold released. The fixtures were heated to 80°C and impregnated with CTD 101K epoxy. The epoxy was cured at 135°C for 3 hours. After curing, the sample stacks were removed from the molds and cut to length with a low speed diamond wafer saw. The short lengths were then ground slightly to square and flatten each face and remove any resin rich layers from the surface. This composite stack procedure is similar to that reported by Chow and Millos [7]. The sample stacking procedure neglects the effect of winding tension in the coil and only matches the epoxy fraction of the coil straight section (coil ends are generally less tightly packed and have more epoxy after impregnation).

Testing

The thermal expansion (TE) and coefficient of thermal expansion (CTE) were measured using the comparative strain gage method. This test method is a modified version of other strain gage thermal expansion measurement methods used previously [3], [4], [5] and is intended primarily for use at cryogenic temperatures. The accuracy of the thermal expansion is approximately 1% and the reproducibility is ± 20 µε in the temperature range from 4K to 300K.

The machined samples were cleaned and prepared according to the standard procedures used for mounting strain gages at cryogenic temperatures. Two gages were mounted on one sample of each stack. Large strain gages, 9.5 mm long, were chosen because their length will average 4 to 6 conductor layers. One gage was mounted on each side as shown in FIGURE 1. The "H" direction measured the change in thickness of the conductor and insulation layers. The "L" direction measured the change in length along the conductor. The "H" direction should be more sensitive to the properties of the insulation.

A copper standard of known thermal expansion characteristics was also strain gaged with the same type and lot of strain gage as was used on the specimen to help eliminate any variation in strain measurement and to allow for the calculation of the strain gage thermal output as a function of temperature. The gages on the specimens and the standard are wired in order to transmit strain data to a strain conditioner, which in turn will transmit strain data into a PC-based data acquisition system. Thermocouples are attached to the specimen and to the standard.

The specimen and standard are placed inside an insulated copper specimen holder that is equipped with a heater on the outer surface. The specimen holder is lowered into a cryogenic dewar and the strain gage wires and thermocouples are attached to the strain conditioner. The strain gages are balanced so that the strain is zero at room temperature. The temperature of the specimen and standard is lowered using a liquid cryogen, either liquid helium (4.2 K) or nitrogen (76 K), until equilibrium at the desired temperature is reached. Once the temperature is stable, the data acquisition program is started and the

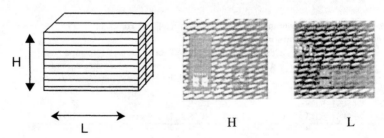

FIGURE 1. Location of the strain gages for the "H" or height direction and the "L" or length direction.

temperature and strain recorded at set intervals. The temperatures and strains are recorded at these intervals while allowing the temperature of the specimen and standard to slowly warm up to ambient temperature.

To analyze the data set, the known thermal expansion of copper was subtracted from the reference copper data to give the strain gage thermal output. The strain data were normalized to zero at 298K. The gage thermal response was then subtracted from each specimen's data set to calculate the thermal expansion (TE) of the specimen. Each data set was plotted as strain versus temperature.

RESULTS

Laminate data

The thermal expansion of samples L1 and L2 are given in TABLE 4 and shown in FIGURE 2. The thermal expansion (TE) of the all-ceramic sample L1 is very low and did not exceed -200 microstrain from 4K to room temperature. However, after impregnation with epoxy (sample L2), the TE is increased significantly to −3100 microstrain at approximately 77K. The expansion curve of sample L2 closely matches the calculated result of combining the all-ceramic L1 curve with the pure epoxy curve, according to the rule of mixtures. This "rule of mixtures" curve is also shown in FIGURE 2. The hybrid ceramic/organic insulation, as shown by sample L2, exhibits a thermal expansion that very closely matches that of pure copper. TABLE 4 lists the TE values of pure copper [6] and shows that the expansion matches within 5 to 8% above 50K. The value of the TE for sample L2 at 10K is anomalous and needs to be repeated to verify the behavior below 50K.

The thermal expansion curves of all-ceramic insulation samples L3 and L4 are shown in FIGURE 3. These two samples vary in the fiber volume fraction, with sample L3 having a higher percentage of the CTD 1005x ceramic matrix. Sample L3 is nominally 50% fiber while L4 is closer to 42% ceramic fibers. However, both of these specimens exhibit a lower expansion curve than the hybrid sample L2. Sample L3, in particular, closely matches the expansion of a typical copper sheathed Nb_3Sn cable listed in reference [6]. The lower thermal contraction of these ceramic insulation systems and their close CTE match to copper wire or $Cu-Nb_3Sn$ cable would reduce the thermal stress placed upon the copper during magnet cool down.

TABLE 4. Thermal expansion values measured on insulation components and a Rule of Mixtures Estimate.

	Typical Epoxy [6]	Sample L1 Ceramic Matrix + Fibers	Sample L2 Ceramic + Fibers + Epoxy	Rule Of Mixtures	Copper [6]
10 K	-11550	100	-3700	-3395	-3240
50 K	-10900	-100	-3400	-3340	-3180
80 K	-10200	-100	-3100	-3130	-3000
100 K	-9590	-100	-2900	-2947	-2820
150 K	-7780	-200	-2300	-2474	-2200
200 K	-5500	-200	-1600	-1790	-1480
250 K	-2765	-100	-800	-899.5	700
295 K	0	-100	0	-70	0

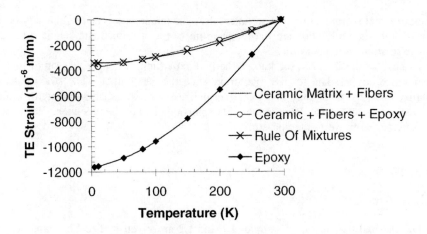

FIGURE 2. Comparison of the thermal expansion of the individual insulation components, the ceramic insulation, and the calculated expansion using the rule of mixtures.

Stack data

Each of the four composite stacks had two gages equaling eight data sets collected. One sample (DE19-15) was measured three additional times in the "H" direction to give a feel for the repeatability of the testing procedure and measure any effects due to cooling the sample to liquid nitrogen temperatures. TABLE 5 lists the thermal expansion variation for the four runs of stack sample DE19-15. The first three runs were within 0.3% of each other

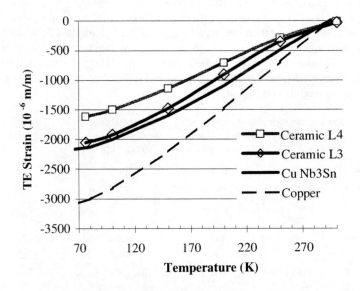

FIGURE 3. Thermal expansion of two all-ceramic insulation laminates compared to copper sheathed, niobium tin cable and copper wire. Copper and Cu Nb_3Sn data from [6]

TABLE 5. Thermal expansion values at 76K for each sample in the "H" direction.

	DE19-13	DE19-15	DE19-17	DE19-19
Thermal expansion	-2924	-2491	-2530	-3330
(10^{-6} m/m)		-2486		
		-2484		
		-2440		
Average	-2924	-2475	-2530	-3330
% Difference from DE19-13	0.0%	-15.4%	-13.5%	+13.9%

at 76K. The fourth run was 1.8% lower than the other three. Since the first three were nearly identical, it is concluded that there is no effect on the thermal expansion of the ceramic insulated samples due to repeated cooling to cryogenic temperatures. This finding is in contrast to the epoxy insulated specimens described in references [7] and [8] in which there was an almost 10% difference in TE after the first cool down cycle.

FIGURE 4 shows that the expansion of the RD3 cable ceramic insulated samples is lower than that of the standard epoxy insulated stack. The unreacted superconducting cable stack expansion is significantly higher. TABLE 5 lists the thermal expansion values at 76 K and the percentage change relative to Sample 1, DE19-13 (reacted cable with epoxy/S2 glass). The ceramic samples are approximately 15% lower and the unreacted stack is almost 14% higher.

There is a slight difference between the two pressure levels investigated for the ceramic insulation. The thermal expansion of the low-pressure stack was 15.4% lower than the standard and the high-pressure stack was 13.5% lower. The low-pressure stack does have a higher percentage of ceramic matrix when compared to the high pressure stack. However, this difference is slight. Assuming the difference in mold height is completely absorbed within the insulation layers, the difference in insulation content between the two reveals that there is approximately one volume percent more insulation in the low pressure sample. Image analysis confirms this volume fraction calculation.

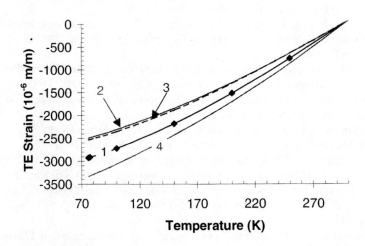

FIGURE 4. Thermal expansion of all of the composite stacks in the "H" direction. Sample 1 (DE19-13) conductor reacted with epoxy matrix; sample 2 (DE19-15) low pressure ceramic matrix and reacted; sample 3 (DE19-17) high pressure ceramic matrix and reacted; sample 4 (DE19-19) conductor not reacted. All samples were impregnated with CTD 101K epoxy.

CONCLUSIONS

The thermal expansion of the ceramic insulation reduces the overall thermal expansion in the thickness direction of a magnet coil by approximately 15% compared to standard S2 glass / CTD 101K epoxy insulation. The lower overall CTE will lower thermal stresses in the magnet sections and allow the designer more control over the desired pre-stress. In addition, the ceramic/organic hybrid insulation expansion nearly matches that of copper, again reducing shrinkage stress along the bond interface. The lowered stress levels should reduce the macro- and micro-cracking of the insulation translating into higher performance and reduced training.

No effect of repeated cooling to liquid nitrogen temperatures for composite stacks insulated with the hybrid ceramic insulation (CTD 1102x) was observed. Earlier work suggested that yielding in the copper sections was responsible for lower contraction during the first cool down. The ceramic matrix must be changing the residual stress field during the cure cycle or during cooling to prevent deformation in the copper.

The ceramic-based insulation systems have thermal performance advantages over conventional epoxy – S2 glass insulation systems in addition to their processing benefits.

ACKNOWLEDGEMENTS

This research was supported through US Department of Energy SBIR Grants DE-FG03-96ER82147 and DE-FG03-99ER82766. We thank Dr. Steve Gourlay of the Laurence Berkley National Laboratory for providing the RD3 superconducting cable.

REFERENCES

1   John A. Rice, Paul E. Fabian, Craig S. Hazelton, "Mechanical and Electrical Properties of Wrappable Ceramic Insulation," IEEE Transactions on Applied Superconductivity, Vol 9 No 2, June 1999, p 220-223.
2   John A. Rice, Paul E. Fabian, Craig S. Hazelton, "Wrappable Ceramic Insulation for Superconducting Magnets", Advances in Cryogenic Engineering (Materials), Vol. 46. (2000) p 267-273.
3   P. Fabian, T. Bauer-McDaniel, and R. Reed, "Low Temperature Thermal Properties of Composite Insulation Systems," in Cryogenics, V35, No. 11, 719, (1995).
4   R. Walsh, R. Reed, "Thermal Expansion Measurements of Resins (4 K - 300 K)," in Adv. Cryo. Eng. – Matererials 40: 1145, (1994).
5   "Measurement of Thermal Expansion Coefficient Using Strain Gages," Tech Note TN-513, Measurement Group, Inc., P.O. Box 27777, Raleigh, NC (1986).
6   R. P. Reed and A. F. Clark, eds., Materials at Low Temperatures, ASM, 1983.
7   Ken P. Chow and Gabriel A. Millos, "Measurements of Modulus of Elasticity and Thermal Contraction of Epoxy Impregnated Niobium-Tin and Niobium-Titanium Composites," IEEE Transactions on Applied Superconductivity, Vol 9, No 2, June 1999, pp 213-215
8   C. A. Swenson, I. R. Dixon, and Denis Markiewicz, "Measurement of Thermal Contraction Properties for NbTi and Nb_3Sn Composites," IEEE Transactions on Applied Superconductivity, Vol 7, No. 2, June 1997 pp 408-411.

MODE I AND MODE II INTERLAMINAR FRACTURE TOUGHNESS OF GLASS-CLOTH/EPOXY LAMINATES AT CRYOGENIC TEMPERATURES

Y. Shindo, K. Horiguchi, and S. Kumagai

Department of Materials Processing,
Graduate School of Engineering, Tohoku University,
Aoba-yama 02, Sendai 980-8579, Japan

ABSTRACT

Experiments and analysis on glass-cloth/epoxy laminate double cantilever (DCB) and end-notched flexure (ENF) specimens are presented. The Mode I and Mode II interlaminar fracture tests were conducted by the DCB and ENF test methods, respectively, at room temperature, liquid nitrogen temperature (77 K) and liquid helium temperature (4 K). The effects of temperature and geometrical variations on the interlaminar fracture toughness are shown graphically. The fracture surfaces were also examined by scanning electron microscopy to verify the fracture mechanisms. A finite element model was further used to perform the delamination crack analysis. Critical load levels, and the geometric and material properties of the test specimens were input data for the analysis which evaluated the Mode I and Mode II energy release rate at onset of delamination crack propagation. The results of the finite element analysis are utilized to supplement the experimental data.

INTRODUCTION

Glass-cloth/epoxy laminates are used mainly as thermal insulation, electric insulation, and permeability barrier, which provide minimal structural support in superconducting magnets. Although fiber-reinforced composite laminates have excellent mechanical properties along fiber directions, they lack through-thickness reinforcement. Hence, they have poor interlaminar properties and are susceptible to delamination. Depending on loading and structural conditions, such delamination may lead to either complete or partial failure in a composite structure.

It is relatively well accepted that delamination growth in laminated composite structures can be predicted by comparing the energy release rate, G, to its critical value, G_C. Most commonly, the double cantilever beam (DCB) test is used to determine the Mode I interlaminar fracture toughness. For Mode II, the end-notched flexure (ENF) test is perhaps the most commonly used method. In the present paper, the DCB and ENF tests were performed at room temperature (R.T.), liquid nitrogen temperature (77 K), and liquid helium temperature (4 K) to evaluate the Mode I and Mode II interlaminar fracture toughness of SL-E glass-cloth/epoxy laminates. The effects

CP614, *Advances in Cryogenic Engineering:*
Proceedings of the International Cryogenic Materials Conference - ICMC, Vol. 48,
edited by B. Balachandran et al.

of temperature and geometrical variations on the interlaminar fracture toughness are experimentally investigated. The fracture mechanisms were also analyzed by means of post-failure scanning electron microscopy (SEM). A three-dimensional finite element analysis was further used to investigate the behavior of strain energy release rate at the delamination crack front. The results of the finite element analysis were compared to the interlaminar fracture toughness values obtained from DCB and ENF tests.

MATERIALS AND TEST SPECIMENS

Glass-cloth/epoxy laminates were used as specimen material, namely SL-E, supplied by Nitto Shinko Corporation, Japan. The basis for SL-E is the glass fabric of E glass. E glass has a round cross-sectional shape (9 μm diameter), a tensile modulus of 72.5 GPa and a tensile strength of 1.4 GPa; the strain-to-failure is about 4.0 %. The E glass plain weave is produced by interlacing warp (length) threads (44 per 25.4 mm) and fill (width) threads (33 per 25.4 mm). Each reinforcing ply has a thickness of 0.175 mm. The epoxy resin used is a bisphenol-A with an acid anhydride curing agent. The glass-cloth/epoxy prepreg stacks were laminated as flat panels and placed in a hot press at 470 K and 4.9 MPa for 90 minutes. Woven prepregs of 20-ply stack resulted in a 3.5 mm laminate. The resulting fiber volume fraction was found to be about 56 %. A 0.05 mm thick nonadhesive Teflon film is inserted along one edge of the panel prior to processing to provide the initial midplane delamination crack.

These test coupons were cut with the length parallel to the warp direction. DCB and ENF specimens were made in accordance with JIS (Japanese Industrial Standard)K 7086 [1]. The DCB specimen geometry is shown in Fig. 1. In the figure, a_p and a_0 denote the initial (Teflon film) crack and precrack lengths. The specimen width ($B = 22$ mm), thickness ($2H = 3.5$ mm), distance from the load-line to the specimen end ($c = 6$ mm), and specimen length to precrack length ratio ($l/a_0 = 3.5$) were held constant, but the specimen length was varied ($l = 142, 120, 95$ and 70 mm). The precracking was performed by pushing a wedged razor blade into a clamped DCB specimen. Aluminum end blocks measuring 22 mm × 12 mm × 12 mm with a loading hole of 5 mm diameter were adhesively bonded to the specimens to enable load application. Figure 2 shows the three-point bend setup and ENF specimen. The specimen width ($B = 22$ mm), thickness ($2H = 3.5$ mm), and span to precrack length ratio ($2L/a_0 = 3.75$) were held constant, but the specimen length and span were varied ($l = 140, 110, 90, 70$ mm, $2L = 100, 80, 65, 50$ mm, respectively). For the load nose apparatus used in the tests, the loading nose cylinder diameter, d_1, and the support roller diameters, d_2, were 10 and 4 mm in diameter.

EXPERIMENTAL PROCEDURES

The procedures outlined in the JIS for testing method for interlaminar fracture

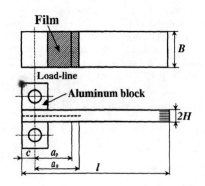

FIGURE 1. Geometry of DCB specimen.

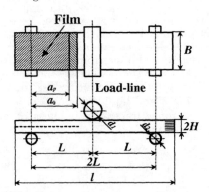

FIGURE 2. Experimental setup and geometry of ENF specimen.

toughness of carbon fiber reinforced plastics [1] were used for Mode I DCB and Mode II ENF tests. Both tests were conducted in stroke control (crosshead speed = 0.5 mm/min) on a 10 kN capacity screw-driven testing machine at room, liquid nitrogen and liquid helium temperatures. Low temperature environments were achieved by immersing the loading fixture and specimen in liquid nitrogen (77 K) or liquid helium (4 K).

Mode I DCB

The loads were applied to the specimen via pins through universal joints and aluminum blocks that were bonded on the specimen. The crack opening displacement (COD) on the load-line was measured from the displacement of the testing machine. Each specimen was loaded at room temperature, 77 K and 4 K until the crack propagated. After the crack growth arrested completely, the specimen was unloaded and loaded again. The crack extension process was then repeated. This procedure yields several interlaminar fracture toughness values for each specimen. During the test the load (P) and COD were continuously recorded by a computer. The crack lengths were calculated from the compliance calibration curves.

The interlaminar fracture toughness was determined in terms of the Mode I critical strain energy release rate using the modified compliance calibration (MCC) method. The compliance (C), defined as the ratio COD/P, was first measured for specimens with a wide range of delamination crack lengths. The relation between the compliance of the specimen and normalized delamination crack length ($a/2H$) may be written as [1]

$$\frac{a}{2H} = \alpha_0 + \alpha_1 (BC)^{1/3} \tag{1}$$

where a is the delamination crack length and the constants α_0 and α_1 are parameters fitted to the experimental results. Denoted as G_I for Mode I fracture, the strain energy release rate is defined as the energy released per unit increase in crack surface area, and can be related to the elastic compliance as [1]

$$G_I = \frac{3}{2(2H)} \left(\frac{P}{B}\right)^2 \frac{(BC)^{2/3}}{\alpha_1} \tag{2}$$

The critical conditions occur when G_I reaches its critical value that occurs at the instant of crack extension, i.e., when $P = P_C$:

$$G_{IC} = \frac{3}{2(2H)} \left(\frac{P_C}{B}\right)^2 \frac{(BC)^{2/3}}{\alpha_1} \tag{3}$$

where P_C is the critical load. The G_{IC} is also referred to as Mode I interlaminar fracture toughness. In this study, the area method was also used to determine the energy release rate. The applied load is increased until a load of P_i is reached, which causes an existing delamination crack of length a_i to extend to length a_{i+1}. Due to the delamination crack extension, the load drops to P_{i+1}, and the load-line displacement increases to δ_{i+1}. The applied load is then removed. The area between the loading and unloading curves represents the decrease in stored strain energy caused by the delamination crack extension. The critical strain energy release rate is obtained by dividing this released energy $W_i = (P_i \delta_{i+1} - P_{i+1} \delta_i)/2$ with the crack area $A_i = B(a_{i+1} - a_i)$ created. The equation describing this method is [2]

$$G_{IC} = W_i/A_i = \frac{(P_i \delta_{i+1} - P_{i+1} \delta_i)/2}{B(a_{i+1} - a_i)} \tag{4}$$

Mode II ENF

The ENF specimens were loaded in a three-point bend-test fixture. The mid-point deflection in the ENF specimens was monitored by means of crosshead displacement only. A load-deflection $(P-\delta)$ curve was plotted for each test. The interlaminar fracture toughness was determined in terms of the Mode II critical strain energy release rate using the beam analysis method. The Mode II interlaminar fracture toughness G_{IIC} is given by [1]

$$G_{IIC} = \frac{9a_1^2 P_C^2 C_1}{2B(2L^3 + 3a_1^2)} \tag{5}$$

$$a_1 = \left[\frac{C_1}{C_0}a_0^3 + \frac{2}{3}(\frac{C_1}{C_0} - 1)L^3 \right]^{1/3} \tag{6}$$

where a_1 is the instantaneous crack length corresponding to critical load P_C, C_1 is the compliance at critical load, and C_0 is the initial compliance. The ENF tests exhibit unstable delamination growth for $2L/a_0 > 2.86$ [3]; therefore, the crack initiation energy only for G_{IIC} was obtained for $2L/a_0 = 3.75$.

FINITE ELEMENT ANALYSIS

Effective elastic moduli were determined under the assumption of uniform strain inside the representative volume element [4]. The 77 K and 4 K effective predicted elastic moduli are listed in Table 1, and (E_x, E_y, E_z) are the Young's moduli, (G_{xy}, G_{yz}, G_{zx}) are the shear moduli and $(\nu_{xy}, \nu_{yz}, \nu_{zx})$ are the Poisson's ratios. The subscript $x, y,$ and z will be used to refer to the coordinate directions, and the Poisson's ratio ν_{xy} reflects shrinkage (expansion) in the y-direction due to tensile (compressive) stress in the x-direction.

In order to evaluate the Mode I and Mode II interlaminar fracture toughness of SL-E glass-cloth/epoxy laminates at 77 K and 4 K, a three-dimensional finite element analysis was carried out. A global energy method was used to calculate the strain energy release rate at the crack front [5]. Since the values of G depend on the choice of assumed crack extension, we used values of crack extension where G is insensitive to the variation of crack extension.

DCB Model Generation

Owing to symmetry, only a quarter of the specimen needs to be modeled. The uniformly distributed load p_i was applied as nodal forces. Analyses of sensitivity were carried out for the effects of mesh size. The elements immediately ahead of the crack tip in the ligament direction were $0.1 \times 0.35 \times 1.1$ mm. The finite element grid consisted of 3000 nodes and 1815 three-dimensional, eight-node isoparametric elements in ANSYS 5.3 [6]. This mesh was the finest mesh possible taking account of limitations of available computer CPU time and memory limitations. The value of crack extension used in the present study was 0.2 mm.

ENF Model Generation

Owing to symmetry, only a half of the specimen needs to be modeled. The elements immediately ahead of the crack tip in the ligament direction were $0.125 \times 0.375 \times 1$ mm. The finite element grid consisted of 1780 nodes and 10464 three-dimensional, eight-node isoparametric elements. The Mode II crack propagation causes sliding of the crack surfaces. To prevent overlapping of elements on opposite crack surfaces and to allow the transmission of contact forces, contact elements were used along the crack surfaces. The coefficients of friction between the crack surfaces, μ, were assumed to be 0, 0.4, and 0.8. The value of crack extension used in the present study was 0.25 mm.

TABLE 1. Predicted elastic moduli for SL-E at 77 K and 4 K.

Temp.	E_x (GPa)	E_y (GPa)	E_z (GPa)	G_{xy} (GPa)	G_{yz} (GPa)	G_{zx} (GPa)	ν_{xy}	ν_{yz}	ν_{zx}
77 K	41.89	38.87	31.87	13.62	13.25	13.30	0.22	0.32	0.23
4 K	45.09	42.58	36.82	15.24	14.93	14.98	0.23	0.31	0.24

RESULTS AND DISCUSSION

Mode I Delamination (DCB)

Compliance of the DCB specimen was evaluated from experimental load - COD curves at various delamination crack lengths. Plots of $(BC)^{1/3}$ versus $a/2H$ at room temperature and 77 K showed good linearity. From equation (1), linear segments representing two equations were chosen to approximate the data. Thus, at room temperature

$$\frac{a}{2H} = -0.740 + 7.252(BC)^{1/3} \tag{7}$$

and at 77 K

$$\frac{a}{2H} = -0.461 + 7.840(BC)^{1/3} \tag{8}$$

Equation (8) would be adopted for applications at 4 K.

The P-COD response at room temperature exhibited a gradual decrease of load with crack opening, indicating a steady crack growth rather than sudden crack advances. The critical loads were taken at the onset of unloading. The P-COD curve at 77 K showed several critical points with sudden load drops corresponding to crack propagation. With decreasing temperature the polymeric matrix becomes stiffer and stronger but also less ductile [7]. The critical loads were taken at the onset of fracture leading to a load drop. The Mode I interlaminar fracture toughness G_{IC} was determined as a function of crack length a. Figure 3 shows R-curve profiles for $l = 142$ mm specimens at room temperature and 77 K. No dependence of G_{IC} values on crack length a is observed. There is good agreement between the MCC and area methods at room temperature. The area method result averages approximately 20 % lower than the MCC method result at 77 K. This is due to the fact that the area method provides a measure of the average fracture toughness as the total area between the loading and unloading curves is divided by the crack extension area. The mean values of the Mode I interlaminar fracture toughness $\overline{G_{IC}}$ can be obtained by averaging G_{IC} over five or more steps of crack propagation.

Because of space limitation in existing helium cryostat, a test specimen much smaller than the standard DCB specimen would be extremely desirable at 4 K. The effects of test temperature and geometrical variations on G_{IC} values obtained by the MCC and area methods can be clearly observed in Table 2 where the final values of measured and predicted crack lengths, and $\overline{G_{IC}}$ are shown. The final value of crack length a_{fm} was measured with an optical microscope. The dimension a_{fm} was taken as the average of three physical measurements taken at equally spaced points across the width direction between the load-line and the final crack front after testing. The predicted crack length a_f was determined from the last compliance using the equations (7) and (8). The predicted crack length values are overall in very good agreement with those determined experimentally. The mean values of interlaminar fracture toughness $\overline{G_{IC}}$ at room temperature and 77 K were consistent among specimens of the different sizes. The interlaminar fracture toughness at 4 K was obtained using the 70-mm-long specimens. The P-COD curve for the 4 K test was similar to the 77 K results. Figure 4 shows R-curve profiles for $l = 70$ mm specimens at room temperature, 77 K and 4 K. The interlaminar fracture toughness G_{IC} is virtually constant and independent of the crack length a. For the 4 K test, the crack extension step during load drop is smaller than for the 77 K test over the whole range of a. $\overline{G_{IC}}$ increases with decreasing temperature over the interval room temperature to 77 K as a result of matrix stiffening. The $\overline{G_{IC}}$ value at 4 K is lower than that at 77 K.

After the DCB tests, the fracture surfaces were examined by scanning electron microscopy (SEM) to identify the fracture characteristics. Particles of the matrix resin were observed in the fracture surface of specimen tested at room temperature, indicating significant multiple failure of the matrix material. However, the fracture surfaces of specimens tested at 77 K and 4 K showed relative clean fiber surfaces with less matrix materials stuck on them. The dominant failure mode at cryogenic temperatures was

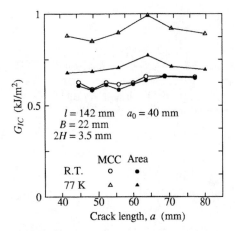

FIGURE 3. Delamination resistance curves for $l = 142$ mm specimens at R.T. and 77 K.

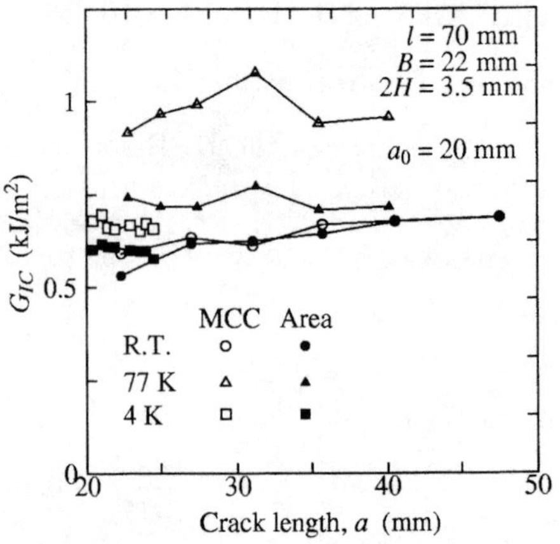

FIGURE 4. Delamination resistance curves for $l = 70$ mm specimens at R.T., 77 K and 4 K.

TABLE 2. DCB test results.

Temp.	l (mm)	a_0 (mm)	a_{fm} (mm)	a_f (mm)	$\lvert(a_f - a_{fm})/a_{fm}\rvert$ $\times 100$ (%)	$\overline{G_{IC}}$ (kJ/m^2) MCC	Area
	142	40	77.20	75.10	-2.7	0.632	0.619
R.T.	120	34	75.00	72.00	-4.0	0.660	0.642
	95	26	66.00	64.10	-2.9	0.638	0.625
	70	20	48.20	47.50	-1.5	0.646	0.631
	142	40	97.15	96.05	1.1	0.906	0.708
77 K	120	34	81.50	81.05	-0.6	0.947	0.722
	95	26	62.50	63.70	1.9	0.899	0.708
	70	20	44.50	45.16	1.5	0.975	0.728
4 K	70	20	25.20	25.23	0.1	0.664	0.598

interfacial failure rather than matrix failure. Using higher magnification, resin fracture normal to the main fracture plane due to property mismatch between the fibers and matrix during the cryogenic-temperature exposure were observed visually on the fracture surface of specimen tested at 4 K. For the 77 K tests, higher resistance exists at the fiber-matrix interface because of relative stronger fiber-matrix adhesion. Cooling further to 4 K embrittles the resin and creates residual stresses at the fiber-matrix interface, adversely affecting the interlaminar fracture performance [7]. Such a mechanisms is proposed as the explanation for the effect of temperature on the interlaminar fracture toughness.

Table 3 presents a comparison of the results of average Mode I interlaminar fracture toughness $\overline{G_{IC}}$ from the MCC and area methods with the finite element method for $l = 70$ mm specimens tested at room temperature, 77 K and 4 K. No significant difference is observed between the values obtained by the MCC and area methods at room temperature. The area method result averages approximately 25 % lower than the MCC method result at 77 K and about 10 % lower at 4 K when specimens are characterized by load peaks, at which unstable fracture occurs. The finite element method produced similar shapes of R-curves as the MCC and area methods. The finite element method result averages about 25 % lower than the area method result at room temperature. The specimen tested at room temperature responds in a ductile stable manner with similar nonlinear behavior just prior to crack growth. Due to this nonlinear behavior, linear finite element analysis is not strictly applicable to the specimens tested at room temperature. Since nonlinear fracture mechanics methods are considered outside the scope of this paper, the fracture toughness values were determined from the critical load recorded at each crack length. This procedure thus considers only the updated

TABLE 3. Comparison of $\overline{G_{IC}}$ for DCB with $l = 70$ mm.

Temp.	$\overline{G_{IC}}$ (kJ/m^2)		
	MCC method	Area method	Finite element method
R.T.	0.646	0.631	0.469
77 K	0.975	0.728	0.731
4 K	0.664	0.598	0.521

linear response at each crack length and neglects accumulated plasticity effects. At cryogenic temperatures, MCC method values overestimate $\overline{G_{IC}}$, but there is reasonably good agreement between the area and finite element methods although the finite element method predicts less 4-K toughness than obtained by area method. Such a discrepancy may possibly be due to matrix cracking in the flexure loaded arms.

Mode II Delamination (ENF)

At room temperature (R.T.) test for $2L = 100$ mm specimens, the load increased linearly with the deflection at the mid-span to a certain point and then the load-deflection curve became nonlinear. With the further increase of the load to a maximum value the load slowly decreased. In contrast, at liquid nitrogen temperature tests, 77 K, for $2L = 100, 80, 65, 50$ mm specimens, unstable crack growth occurred which was shown by the sudden drop in the load. The effects of geometrical variations on Mode II interlaminar fracture toughness G_{IIC} values obtained by the beam analysis method can be clearly observed in Table 4. Equation (5) was used to calculate the G_{IIC}. The critical load, P_C, is the intersection of the load-deflection curve with a line denoting a 5 % increase in the initial elastic compliance. The $\overline{G_{IIC}}$ value was taken for an average of the G_{IIC} values. The interlaminar fracture toughness $\overline{G_{IIC}}$ at 77 K did not show much dependence on the geometrical variations. The interlaminar fracture toughness at 4 K was obtained using the 70-mm-long specimens ($2L = 50$ mm). The P-δ curve for the 4-K test was similar to the 77 K results. For all the ENF specimens tested at 77 K and 4 K, damage under the loading nose cylinder was observed. The effect of temperature on interlaminar fracture toughness showed similar trends to the Mode I DCB test.

The Mode II fracture surface exhibited similar characteristics to the Mode I fracture surface. A detailed examination of the fracture surface using SEM highlighted another interesting point. Hackles were observed on the fracture surface of specimens tested at 77 K and 4 K. The fracture surface micrograph at 4 K showed lower hackle mark density in the matrix resin than at 77 K. Since hackle mark density indicates the shear stress intensity in the matrix, the low hackle mark density suggests that the load carried by the matrix was reduced.

The finite element analysis, as an alternative to the beam analysis method, can also be used to obtain the Mode II interlaminar fracture toughness based on the critical load and instantaneous crack length. To evaluate the friction effect, a series of calculations were performed for $2L = 100$ mm specimen tested at 77 K. Calculation results for $\mu = 0, 0.4$, and 0.8, were nearly identical, and hence moderate friction had negligible effect on strain energy release rate. For a real ENF specimen, the friction effect can therefore be neglected since the Teflon film insert has a very low friction coefficient value [8]. In all the specimens the coefficient of friction was set equal to zero. Table 5 presents a comparison of the results of average Mode II interlaminar fracture toughness $\overline{G_{IIC}}$ from the beam analysis method with the finite element method for ENF specimens tested at 77 K and 4 K. The finite element method result averages approximately 20 % lower than the beam analysis method result at 77 K and about 25 % lower at 4 K. Such a discrepancy may possibly be due to damage under the loading nose cylinder. In the cryogenic ENF test, high compressive bending and transverse stresses cause damage under the cylinder. After the damage, the capability of resisting deformation for the specimen will not be as good as it was previously. Therefore, the existence of damage can be expected to increase the compliance. The compliance obtained with the experiment is higher than that from the finite element analysis. Consequently, $\overline{G_{IIC}} > \overline{G_{IIC}}$(FEM).

TABLE 4. ENF test results.

Temp.	2L (mm)	G_{IIC} (kJ/m²)	G_{IIC} (kJ/m²)
R.T.	100	1.288	1.288
	100	1.186	
	100	1.390	
	100	2.664	2.812
	100	2.927	
	100	2.845	
	80	2.711	2.598
77 K	80	2.484	
	65	2.696	2.701
	65	2.706	
	50	2.699	2.667
	50	2.634	
	50	2.912	2.527
4 K	50	2.541	
	50	2.129	

TABLE 5. Comparison of Mode II toughnesses from two data reduction methods.

Temp.	2L (mm)	G_{IIC} (kJ/m²)	G_{IIC}(FEM) (kJ/m²)
77 K	100	2.812	2.367
	80	2.598	2.066
	65	2.701	2.108
	50	2.667	2.015
4 K	50	2.527	1.880

CONCLUSIONS

This paper has presented experiments and finite element analysis on glass-cloth/epoxy laminate DCB and ENF specimens. The following conclusions can be drawn from this study.

1. The specimen tested at room temperature was characterized by stable crack propagation. In contrast, the specimens tested at 77 K and 4 K were characterized by load peaks, at which unstable fracture occurred.

2. No dependence of interlaminar fracture toughness values at 77 K on geometrical variations was observed.

3. The interlaminar fracture toughness increased between room temperature and 77 K, further cooling to 4 K produced a toughness decrease.

4. For the delamination cracks in cryogenic Mode I DCB tests the finite element analysis or area method must be used. In Mode II, the finite element analysis provided the conservative estimate.

REFERENCES

1. *JIS K 7086, Testing Method for Interlaminar Fracture Toughness of Carbon Fiber Reinforced Plastics*

2. Hashemi, S., Kinloch, A. J. and Williams, J. G., *Compos. Sci. Technol.* **37**, pp.429-462(1990).

3. Cowley, K. D. and Beaumont, P. W. R., *Compos. Sci. Technol.* **57**, pp.1433-1444(1997).

4. Hahn, H. T. and Pandey, R., *ASME J. Eng. Mater. Technol.* **116**, pp.517-523(1994).

5. Buchholz, F. -G., Rikards, R. and Wang, H., *Int. J. Fract.* **86**, pp.37-57(1997).

6. *ANSYS Revision 5.3*, ANSYS, Inc., Houston, PA, 1996.

7. Hartwig, G. and Knaak, S., *Cryogenics* **24**, pp.639-647(1984).

8. Yang, Z. and Sun, C. T., *ASME J. Eng. Mater. Technol.* **122**, pp.428-433(2000).

NON-METALLIC MATERIALS — INSULATION

ANODIZED INSULATION FOR CICC COILS

A. F. Zeller

National Superconducting Cyclotron Lab
Michigan State University
E. Lansing, MI 48824 USA

ABSTRACT

Anodization techniques for production of CICC for radiation resistant magnets are described. Aluminum conduit can be anodized on the inside to provide electrical isolation while leaving the outside of the conductor available for structural use. This technique should work well for NbTi conductor, but other materials are necessary for Nb_3Sn. Titanium alloys are suggested for production of wind-and-react Nb_3Sn coils.

INTRODUCTION

Planned large accelerator projects such as the Muon Collider, and its precursor the Neutrino Factory, and the Rare Isotope Accelerator will also be prodigious producers of high-energy neutrons around the target areas. These projects require that magnets be in close association with the target area, particularly the Muon Collider, where the target resides inside a 20 T solenoid. The intense radiation fields require the magnets be radiation resistant to insure sufficient operational life times. Since machine down time is costly and replacement of failed components difficult and hazardous, materials must be selected that maximize service life.

The target of the Muon Collider [1] is inside a 20 T solenoid. The solenoid is a hybrid consisting of a resistive inner part and a superconducting outer part. The outsert will generate 14 T and consists of eight separate sections. The inner sections will of necessity be Nb_3Sn. In addition, the heat deposition in the coil windings, 1 kW/m^3, requires Cable-in-Conduit-Conductor (CICC) to deal with this 4 K load.

At proton intensities of 1 MW on the target, the energy deposition in the superconducting solenoid is 6×10^6 Gy per year of operation (65% beam on). In addition, the coils are subjected to a high-energy neutron flux of 4×10^{17} n/cm^2 per year. Therefore, conventional organic insulations and epoxies cannot be used, as lifetimes would be on the order of months. The desired lifetime is approximately twenty years, requiring an all-

CP614, *Advances in Cryogenic Engineering:*
Proceedings of the International Cryogenic Materials Conference - ICMC, Vol. 48,
edited by B. Balachandran et al.
© 2002 American Institute of Physics 0-7354-0060-1/02/$19.00

inorganic system. Several inorganic insulations, such as aluminum oxide, silicon dioxide, magnesium oxide and spinel, are sufficiently radiation resistant so that the coil lifetimes will be limited by the sensitivity of the superconductor itself [2,3]. The brittleness, thermal contraction issues and lack of strength of these materials is well known, however. Additionally, there isn't a good substitute for epoxy to provide restraint against wire motion that has been shown to work at cryogenic temperatures.

Providing insulation for the conductor isn't a very difficult problem. Techniques such as plasma spray, mineral cloth and anodization can be used, although there are drawbacks to each of these processes. They are either too porous or too rigid or both. The major difficulty is providing the inorganic equivalent of epoxy. With the problems associated with differential thermal contractions, simply clamping the wires together doesn't work. Conductors move around during temperature cycling and during energizing. This quickly causes abrasion and shorts. Another possible technique is to pot the coil in a conventional, organic epoxy and tightly encase it in container. Even though the epoxy is destroyed by the radiation, it is kept in place because there is no way for it to leave the coil volume. The problem with this is the added bulk of the containment vessel, which adds to the radial build of the magnet – increasing the difficulty of the outer coils. Therefore, a different approach is needed.

ANODIZING

External

One approach to insulating CICC is to use an aluminum conduit and to anodize it. The problems associated with other inorganic insulation are here, too. The conductor cannot be bent after it has been anodized, as the layer is disrupted when placed in compression. This is not too serious a problem, because the coil could be loosely wound, then anodized. Such an approach has been used for resistive coils [4]. The problem of clamping and lack of an inorganic epoxy remains, however. Anodic layers are good dielectrics and relatively durable, but repeated rubbing against each other will quickly remove the insulating layer from a conductor. Stainless steel banding around the coil would not help slow down the disruption because the thermal contraction of aluminum is so much higher than stainless that the coil would become very loose during cool down. Aluminum banding would help, but irregularities in the coil package result in individual conductors sliding during ramping and subsequent erosion of the anodic layer.

Internal

A different method of coil construction is proposed, then. If the interior hole of the conduit is anodized, then the outside of the conductor can be clamped or welded without fear of the insulating layer being removed. A schematic drawing of such a coil is shown in FIG 1. Note the anodic layer is exaggerated for clarity. In actual practice the anodic layer is only a few micrometers thick. The coil is shown welded together as a single unit, but banding could also be used as well or in addition. The aluminum alloy can be chosen for structural or protection purposes since the anodic layer is relatively independent of the alloy. In addition, there are several anodizing techniques that produce different mechanical or dielectric properties and thickness.

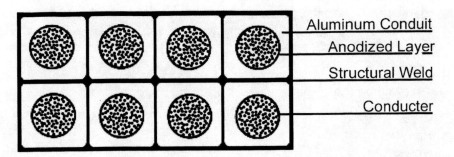

FIGURE 1. Coil cross section with CICC insulated with internal anodization. The anodic layer has been exaggerated for illustrative purposes.

CONDUCTOR CONSTRUCTION

Normally, CICC is constructed by wrapping the conduit around the cable and then welding it shut. This technique will not work with an anodized conduit because winding the coil will destroy the anodic layer on the inner radius. Therefore, it is necessary to wind the coil into the right shape, or close to it, and then anodize it. The next problem is getting the conductor into the conduit without scrapping off the anodic layer.

Winding the coil into the right shape when the interior channel is to be anodized is easier than when exterior insulation needs to be applied. The coil is wound with whatever support structure is required and then the conductor can be welded to form a very rigid structure. There is no need to wind the coil loose in order to accommodate insulation or epoxy. However, it is not a simple matter to anodize the interior of a long narrow tube. The simple technique of pumping the anodizing solution through the tube with voltage applied to the outside doesn't work. It is necessary to insert a wire cathode into the interior and electrically isolate it from the tube. This can be done with small o-rings, although it is necessary to reposition the o-rings to get the interior completely coated.

Getting the conductor through the conduit without damaging the anodic layer is more of a challenge. Putting a slip plane of some organic material defeats the purpose of having an all-inorganic system since such material would be damaged by the radiation. Small particles of former-organic materials would seriously contaminate the liquid helium system. It might be possible to use a low friction Teflon® sleeve to insert the conductor, and then remove it. Tests on short, 1 m, pieces were successful, but it is unlikely to work on 30 m parts.

PROTOTYPE COILS

Test coils of 9.52 mm diameter 6061 aluminum tubes with an internal opening of 6.35 mm were coiled into 300 mm diameter sections of varying lengths. Round conductor was used in the program because of availability. The cable was a 147-strand construct of 0.25 mm diameter NbTi wires with a copper-to-superconductor ratio of 2:1 having a cable diameter of 4.76 mm. Attempts to simply insert the conductor is short lengths of tubes (all that could be anodized on the inside with just immersing the tube in the anodizing bath) resulted in removal of the anodic layer.

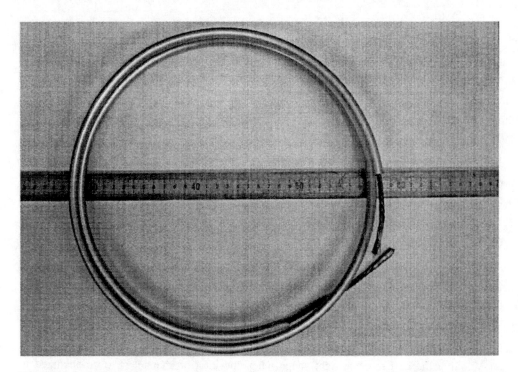

FIGURE 2. Test CICC coil.

A material was needed to provide the slip plane that was tough enough to withstand the cable insertion, yet be removable. It was discovered that Polyvinyl Alcohol (PVA) met our requirements. The material is a strong film, but has the property that it completely dissolves in 95 C water. The PVA film is 0.015 mm thick and is spiral-wrapped around the cable and the cable is pushed-pulled through the conduit. FIG 2 shows a test coil after insertion of the PVA insulated cable. The conduit is not anodized on the inside because the purpose of the test is to determine if the PVA is abraded during the insertion. FIGURE 3 shows one of the ends of the test coil with the PVA in place. Pulling the PVA wrapped cable through an approximately 1.5 m coil did not result in a short between the cable and the conduit. For longer coils, more layers or thicker film would be used. The coil was then connected to a steam cleaner and the PVA was completely removed.

PROSPECTS FOR Nb₃Sn CICC

Aluminum tubing appears to work for conductors like NbTi, but high field magnets, as in the Muon Collider application, requires Nb_3Sn. The CICC is formed from unreacted conductor and heat treatment done after winding. This is clearly unacceptable with an aluminum conduit since the melting point of aluminum is lower than the reaction temperature. A way around this problem would be to use titanium or a titanium alloy. The preferred choice would be one the standard formulations that are used in cryogenic applications: Ti-6Al-4V or Ti-5Al-2.5Sn. Unfortunately, there is little literature available on anodizing the various titanium alloys. Significant amounts of titanium are anodized each year for jewelry purposes, but there isn't much demand for these two alloys in that area.

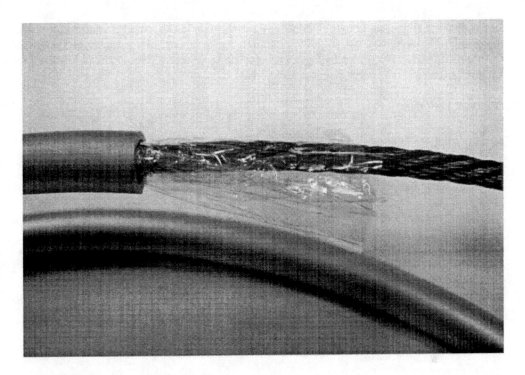

FIGURE 3. Close-up of the PVA film around the cable.

Sample plates of the 6-4 alloy were anodized for dielectric strength testing. The anodic layer is not very thick and is much less resistant to abrasion than the aluminum oxide layer on aluminum. The layer was capable of holding 100 V DC when applied with a 4 cm^2 electrode. This would likely be sufficient for a CICC magnet assuming the layer remains intact during cable insertion. The anodizing of the interior has the same difficulty that aluminum does – it requires an internal electrode. It is also clear that significant work would have to be done on the anodizing process, as well as determining the best alloy.

FUTURE DEVELOPMENTS

The direction towards using aluminum for NbTi coils is relatively straightforward. The major developments needed are the anodizing of the inside of long pieces of cable. Discussions with anodizers suggest this is not a very difficult task for solenoid coils. Racetrack coils would be significantly more difficult. The cathode could easily be put in the conduit before forming, but removal might be difficult unless a mechanical or chemical means are found. The part of the manufacturing process requiring the most work is cable insertion. The test coils produced so far have a relatively low fill factor – approximately 33%. Because of cooling requirements, a fill factor of greater than 50% would be unlikely, so the increase may be possible. Getting more area of conductor in increases the possibility of damage to the anodic layer. Making usable lengths of conductor will be the area of most research. Once acceptable lengths of conductor are produced, the cable will have to be tested to insure its safety during a quench and its operational reliability.

Prospects for titanium-clad conductor are less likely than aluminum clad. There are the same problems as with aluminum conduit, but the anodic layer is less known. Determining the right anodizing conditions and alloys will be the major tasks. Another question that will have to be answered is the suitability of titanium as a conduit in terms of conductor operation. This is likely not to be a problem since conduits have been constructed of materials such as stainless steel and nickel alloys.

CONCLUSIONS

Preliminary conductor construction development with an internal anodic layer shows promise for production of radiation resistant coils with NbTi superconductor. Possible conduit for Nb_3Sn was also discussed, but much work needs to be done before actual conductor development could begin.

ACKNOWLEDGEMENTS

This work is supported in part by the National Science Foundation through grant PHY-9528844 and by BNL through contract DE-AC02-98CH10886. Helpful discussions with M. A. Green (LBNL), John Miller (NHMFL), P. Mantica (MSU) and D. Kraft (Anacote Corporation). The author wished to thank Nigel Flynn of Isolyser Co., Inc. for supplying the PVA film.

REFERENCES

1. S. Ozaki, R. Palmer, M. Zisman and J. Gallardo, "Feasibility Study-II of a Muon-Based Neutrino Source", BNL-52623 (2001).
2. M. E. Sawan and P. L. Walstrom, "Superconducting Magnet Radiation Effects in Fusion Reactors", *Fusion Tech.* **10** (1986), p741.
3. R. Reed and D. Evans, "Insulation Systems for Muon Collider", (2000), unpublished.
4. W. J. Leonhardt, "A Radiation Hard Dipole Magnet Coils Using Aluminum Clad Copper Conductors", *Proc. 1989 IEEE Particle Accelerator Conf,* (1989), p. 366.

CHARACTERIZATION OF REACTOR IRRADIATED ORGANIC AND INORGANIC HYBRID INSULATION SYSTEMS FOR FUSION MAGNETS

K. Bittner-Rohrhofer[1], P. Rosenkranz[1], K. Humer[1], H.W. Weber[1], J.A. Rice[2], P.E. Fabian[2] and N.A. Munshi[2]

[1]Atominstitut der Österreichischen Universitäten, A-1020 Wien, Austria

[2]Composite Technology Development, Inc., Lafayette, CO, USA

ABSTRACT

Because of their excellent electrical and mechanical material performance, organic and organic/inorganic hybrid materials are candidate insulation systems for fusion magnets. Wrappable inorganic insulation systems combine the high strength and modulus of a ceramic composite with the simple processing of an organic insulation system. Various types of organic as well as organic/inorganic hybrid material compositions containing a ceramic fiber fabric as well as a ceramic and an epoxy matrix were irradiated at ambient temperature in the TRIGA reactor (Vienna) up to neutron fluences of 10^{21}, 10^{22} and 5×10^{22} m^{-2} (E>0.1 MeV). Tensile, short-beam-shear and double-lap-shear tests were performed at 77 K. The influence of reactor irradiation on the mechanical material performance will be presented for different compositions of these materials.

INTRODUCTION

Fiber composites offer a broad spectrum of technical applications (e. g. for space technology or mechanical engineering) due to their excellent material properties. The scientific application of fiber reinforced materials depends on different combinations of special fiber and matrix materials needed for high material performance. Thus, the improvement and modification of compositions led to new wrappable organic and organic/inorganic hybrid material compounds. Their high strength and stiffness as well as the simple processing meet all requirements for the insulation of the windings of superconducting magnet systems of ITER and future fusion reactors. In particular, inorganic compositions would simplify the insulation procedure of the superconducting coils due to their good high temperature performance. Therefore, the radiation resistance of the mechanical properties is of special interest for nuclear technology.

CP614, *Advances in Cryogenic Engineering:*
Proceedings of the International Cryogenic Materials Conference - ICMC, Vol. 48,
edited by B. Balachandran et al.
© 2002 American Institute of Physics 0-7354-0060-1/02/$19.00

A few years ago, a test program was started to investigate the mechanical behavior of organic and organic/inorganic composites for long-term operation of superconducting fusion magnets.

The ultimate tensile strength (UTS) and the interlaminar shear strength (ILSS) of ceramic-fiber reinforced epoxy composites were measured at room temperature, 77 and 4 K by Bruzzone et al. [1]. Furthermore, this paper reports on the ILSS data obtained on plasma sprayed ceramic coatings. Some extended mechanical studies were also done by Schutz et al. [2], Dienst et al. [3] and Clinard et al. [4] investigating the ultimate strength and the fracture thoughness of various ceramic systems suggested for fusion applications under a radiation environment (gamma and fast neutrons). Rice et al. [5] tested the mechanical and electrical properties of inorganic insulation systems, which were developed for high temperature applications (up to 900 °C) and thus represent an ideal candidate for insulating the complex magnet coil systems. Investigations of the influence of radiation damage on glass fiber reinforced plastics with different types of organic matrix material (e. g. epoxy or bismaleiminde resins) were done by Humer et al. [6] . Mechanical tests in tension as well as the inter- and intralaminar shear mode were carried out at liquid nitrogen temperature (77 K).

Recently, Humer et al. [7] characterized various ITER relevant insulation systems containing a ceramic fiber reinforcement as well as a ceramic and an epoxy matrix system. The UTS and the ILSS of the laminates were measured at 77 K before and after irradiation. They found that the tensile strength of compounds with a higher organic content of epoxy degraded more rapidly than systems with a lower content. However, no systematic influence of the organic content could be observed for the ILSS. Furthermore, this paper presented results on swelling and weight loss due to irradiation to different neutron fluences.

The following study presents a series of irradiation experiments carried out to investigate the tension and interlaminar shear behavior of various newly developed organic and organic/inorganic hybrid insulation systems under cryogenic conditions.

MATERIALS AND TEST PROCEDURES

Different types of fiber reinforced composites were included in the test program in order to assess the influence of reactor irradiation on their mechanical performance. A detailed summary of the compositions of these insulation systems is given in Table 1.

TABLE 1. Material compositions and designations of the investigated insulation systems.

Sample ID	Matrix	Matrix	Reinforcement Type	Fiber Volume Fraction
CTD-403	CTD-403	Organic	S2-Glass	0.538
CTD-406	CTD-406	Organic	S2-Glass	0.533
CTD-410	CTD-410	Organic	S2-Glass	0.537
CTD-1005x, g	CTD-1005x	Ceramic	CTD-CF 100	0.463
CTD-1005x	CTD-1005x	Ceramic	CTD-CF 100	0.485
CTD-1025x	CTD-1005x+CTD-1002x	Ceramic	CTD-CF 100	0.497
CTD-1102x	CTD-1002x+CTD-101K	Ceramic/Organic	CTD-CF 100	0.484
CTD-1105x	CTD-1005x+CTD-101K	Ceramic/Organic	CTD-CF 100	0.495
CTD-1402x-1	CTD-1002x+CTD-403	Ceramic/Organic	CTD-CF 100	0.506
CTD-1402x-2	CTD-1002x+CTD-403	Ceramic/Organic	CTD-CF 100	0.466

As can be seen in Table 1, three main groups of fiber composites were investigated: organic, ceramic and ceramic/organic compounds. Each organic system (CTD-403, CTD-406 and CTD-410) is a specially designed epoxy resin combined with S2 glass fibers, which offer high mechanical strength. Furthermore, two different ceramic series (CTD-1005x, g and CTD-1005x) with the same matrix material, but with a different fiber volume fraction, were investigated as well as a combination (CTD-1025x) of two different ceramic matrix materials. The organic/inorganic insulation systems consist of an industrial standard CTD-101K DGEBA-epoxy resin system in combination with the CTD-1002x and CTD-1005x ceramic system with different percentages of woven fabrics. For CTD-1402x-1 and CTD-1402x-2 a combination of the CTD-1002x ceramic matrix and the CTD-403 organic matrix was used, which was employed before as the main matrix material for the organic CTD-403 system. All ceramic and ceramic/organic insulation systems were reinforced with the CTD-CF 100 ceramic fabric. The percentage of the fiber content is about 54 % for two-dimensionally woven S-glass fabric/epoxies and varies for inorganic and organic/inorganic systems between 46-50%.

All materials were irradiated at ambient temperature (~340 K) in the TRIGA reactor (Vienna) to neutron fluences of 10^{21}, 10^{22} and 5×10^{22} m^{-2} (E>0.1 MeV), which correspond approximately to a total absorbed dose of 5, 50 and 250 MGy. The reactor is operating at a γ-dose rate of 1×10^{6} Gyh^{-1}, a fast neutron flux density of 7.6×10^{16} m^{-2}s^{-1} (E>0.1 MeV), and a total neutron flux density of 2.1×10^{17} m^{-2}s^{-1}, respectively.

The UTS as well as the ILSS were measured before and after irradiation. All experiments were performed in liquid nitrogen (77 K) using a servo-hydraulic Material Test System MTS 810. The tensile tests were carried out according to the standards DIN 53455 and ASTM D638. However, smaller sample geometries were taken because of the limited space in the irradiation facility of the TRIGA-reactor. The reduced sample size is based on earlier scaling experiments and cryogenic testing. Thus, the dimensions of the flat tensile specimens were 45x10x3 mm (length x width x thickness), the cross section in the test area was 3x3 mm (width x thickness). A crosshead speed of 1.5 mm min^{-1} was used for the tensile tests.

For the ILSS, the short-beam-shear (SBS) according to ASTM D2344 as well as the double-lap-shear (DLS) test were used. The SBS-tests were carried out at a crosshead speed of 1.3 mm min^{-1} on 3 mm thick samples with a span-to-thickness ratio of 4:1 for the organic systems and of 5:1 for the ceramic and ceramic/organic matrices. As for the tensile test, the dimensions of the DLS specimens were 45x10x3 mm (length x width x thickness). Further, all specimens were loaded in their strongest direction. Therefore, the warp direction of the laminates was chosen to be parallel to the longer axis of the samples. More details about the experimental set-up , the SBS testing device and the evaluation procedure were reported by Humer et al. [6].

RESULTS

The results on the UTS and ILSS (DLS and SBS test) on the organic and organic/inorganic insulation systems before and after irradiation and testing at 77 K are summarized in Table 2 and plotted in Figures 1 to 5.

In the case of the unirradiated organic insulation systems containing the S2-glass fiber reinforcement (Figure 1), the UTS is approximately 900 MPa for CTD-403 and CTD-406 and about 950 MPa for CTD-410. After irradiation to neutron fluences of 10^{21} and 10^{22}m^{-2} (E>0.1 MeV), the UTS of all organic materials degrades by 2-4% and finally by 10-15% at the highest dose level.

TABLE 2. Results on the insulation systems irradiated at ~340 K and tested at 77 K.

Sample ID CTD-	Neutron fluence (E>0.1 MeV) (m^{-2})	UTS (MPa)	ILSS DLS-test (MPa)	ILSS SBS-test (MPa)
403	-	894 ± 27	41 ± 9	108 ± 3
406	-	892 ± 26	45 ± 8	104 ± 3
410	-	950 ± 50	38 ± 7	96 ± 3
1005x, g	-	-	6 ± 0	21 ± 1
1005x	-	-	7 ± 1	16 ± 2
1025x	-	288 ± 0	10 ± 1	24 ± 2
1102x	-	-	38 ± 2	80 ± 4
1105x	-	-	15 ± 1	38 ± 4
1402x-1	-	-	30 ± 7	69 ± 9
1402x-2	-	450 ± 0	31 ± 0	-
403	1 x 10^{21}	874 ± 13	39 ± 3	98 ± 3
406	1 x 10^{21}	888 ± 15	42 ± 2	100 ± 3
410	1 x 10^{21}	941 ± 49	35 ± 2	95 ± 2
1005x, g	1 x 10^{21}	-	5 ± 1	21 ± 2
1005x	1 x 10^{21}	-	6 ± 0	17 ± 1
1025x	1 x 10^{21}	278 ± 11	9 ± 1	23 ± 2
1102x	1 x 10^{21}	-	20 ± 2	48 ± 1
1105x	1 x 10^{21}	-	14 ± 2	41 ± 4
1402x-1	1 x 10^{21}	-	16 ± 3	42 ± 10
1402x-2	1 x 10^{21}	417 ± 0	28 ± 0	-
403	1 x 10^{22}	868 ± 12	32 ± 2	97 ± 3
406	1 x 10^{22}	870 ± 25	35 ± 4	98 ± 2
410	1 x 10^{22}	921 ± 13	30 ± 4	94 ± 2
1005x, g	1 x 10^{22}	-	6 ± 1	19 ± 1
1005x	1 x 10^{22}	-	6 ± 1	16 ± 1
1025x	1 x 10^{22}	251 ± 31	7 ± 1	22 ± 1
1102x	1 x 10^{22}	-	14 ± 0	34 ± 2
1105x	1 x 10^{22}	-	11 ± 2	31 ± 2
1402x-1	1 x 10^{22}	-	15 ± 3	31 ± 2
1402x-2	1 x 10^{22}	386 ± 0	17 ± 0	-
403	5 x 10^{22}	760 ± 32	12 ± 1	48 ± 6
406	5 x 10^{22}	785 ± 24	15 ± 3	56 ± 7
410	5 x 10^{22}	858 ± 30	12 ± 2	64 ± 5
1005x, g	5 x 10^{22}	-	5 ± 1	19 ± 2
1005x	5 x 10^{22}	-	6 ± 1	18 ± 1
1025x	5 x 10^{22}	246 ± 25	8 ± 1	27 ± 3
1102x	5 x 10^{22}	-	6 ± 1	28 ± 3
1105x	5 x 10^{22}	-	12 ± 2	30 ± 1
1402x-1	5 x 10^{22}	-	16 ± 3	28 ± 3
1402x-2	5 x 10^{22}	388 ± 0	28 ± 0	-

In comparison to the CTD-400 series, a dramatically lower UTS (on average by 70%) was measured for the CTD-1025x ceramic fabric, and an approximately 50% lower UTS for the CTD-1402x-2 ceramic/organic system. On the other hand, no significant influence of the reactor irradiation was observed for both systems, which are reinforced by the CTD-CF 100 ceramic fabric. Only a slight decrease (~15%) of the UTS is found at the highest neutron fluence.

The results on the ILSS, measured by the DLS-test (Figures 2 and 3), show that for organic materials the ILSS starts at 40- 45 MPa and degrades continuously with irradiation to 12-15 MPa (degradation by ~70%). As for the UTS, rather low ILSS values were measured for composites consisting of ceramic CTD-CF 100 fibers and ceramic matrices. After irradiation, the ILSS of CTD-1005x, g and CTD-1005x remains almost constant and decreases slightly (by 20%) for CTD-1025x.

The ILSS results on the unirradiated CTD-1402x-1 and CTD-1402x-2 systems containing an organic/inorganic matrix combination are lower by 25% than those for the CTD-400 series, except CTD-1102x, which shows the same result for the ILSS as CTD-410. However, the ILSS of the unirradiated CTD-1105x material is very small (~15 MPa). As can be seen in Figure 3, almost all of these CTD-series degrade considerably by radiation (degradation by up to 80%).

In addition, the ILSS was measured with the SBS-test. We find the same general trends as with the DLS-test. The organic composites show again the highest ILSS before and after irradiation (Figure 4), but a lower degradation (50-70%). Further, an almost constant behavior is observed for the ceramic compounds CTD-1005x, g and CTD-1005x and a slight decrease (8%) of the ILSS for CTD-1025x (Figure 5). For systems containing ceramic/organic matrix materials, the radiation damage leads to a reduction of the ILSS by up to ~65%.

As for the organic and inorganic CTD-series, the ILSS of the hybrid systems is approximately two times higher than the results of the DLS-test. This significant difference (by up to 50%) in the ILSS corresponds well to earlier numerical investigations by Pahr et al. [8]. The standardized ILSS evaluation procedure assumes a mean shear stress over the whole shear area and describes a complex fracture mechanical problem by a simple strength analysis. However, the numerical evaluation of the average delamination risk parameter within the overlapping zone leads to approximately 50%. Therefore, the FE-simulations show that the standard evaluation method, especially for the DLS-test, underestimates the real ILSS. On the other hand, results for the SBS-test underestimate the real ILSS by only ~10-15%.

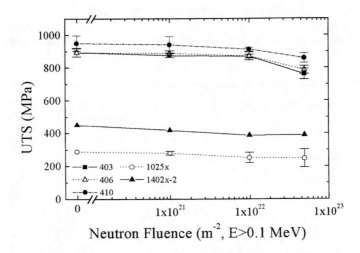

FIGURE 1. Ultimate tensile strength (UTS) of various insulation systems as a function of neutron fluence measured at 77 K.

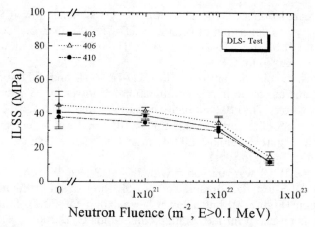

FIGURE 2. Interlaminar shear strength (ILSS) of organic insulation systems as a function of neutron fluence measured at 77 K with the double-lap-shear (DLS) test.

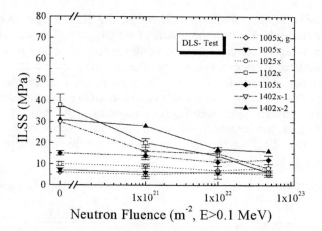

FIGURE 3. Interlaminar shear strength (ILSS) of ceramic and ceramic/organic insulation systems as a function of neutron fluence measured at 77 K with the double-lap-shear (DLS) test.

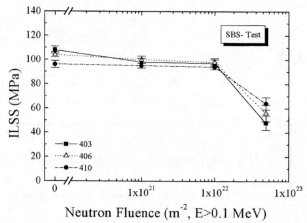

FIGURE 4. Interlaminar shear strength (ILSS) of organic composites as a function of neutron fluence measured at 77 K with the short-beam-shear (SBS) test.

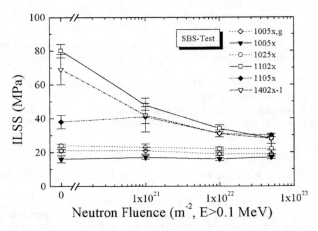

FIGURE 5. Interlaminar shear strength (ILSS) of ceramic and ceramic/organic composites as a function of neutron fluence measured at 77 K with the short-beam-shear (SBS) test.

SUMMARY

The present paper reports on the influence of reactor irradiation on the ultimate tensile strength and the interlaminar shear strength of newly developed organic and organic/inorganic hybrid materials, which could be used as future insulation materials for superconducting magnet coils in fusion technology.

Different compounds containing organic, ceramic and ceramic/organic matrix materials as well as S2-glass- and ceramic fiber reinforcements were irradiated at ambient temperature ($\sim$350 K) in the TRIGA reactor (Vienna) to fast neutron fluences of 10^{21}, 10^{22} and 5×10^{22} m^{-2} (E>0.1 MeV). The mechanical properties were investigated before and after irradiation by the tensile test, the double-lap-shear (DLS) test and the short-beam-shear (SBS) test. The main results are the following:

- Systems with organic matrices and S2-glass fiber reinforcement show the highest ultimate tensile strength (UTS) followed by the organic/inorganic compounds (lower by approximately a factor of two). The inorganic systems with woven ceramic CTD-CF 100 fabrics show a much lower UTS (by $\sim$70%). All materials show a continuous degradation of the UTS by up to $\sim$15% after irradiation.
- The nature of the matrix and of the fiber materials leads to a systematic influence on the interlaminar shear strength (ILSS). Both the DLS and the SBS-test show a similar trend before and after irradiation. The radiation damage degrades the ILSS (measured with the DLS-test) by up to 70% for organic and by up to 80% for organic/inorganic systems. The lowest ILSS is measured again for ceramic compounds, but it remains almost constant from the lowest to the highest dose level. This radiation hardness is observed again for the SBS-test. Furthermore, the results obtained by the SBS-test show a reduction of the ILSS by $\sim$50-70% for organic systems and by $\sim$65% for the organic/inorganic hybrid systems.

ACKNOWLEDGEMENT

Technical assistance by Mr. H. Niedermaier and Mr. H. Hartmann is acknowledged. This work has been carried out within the Association EURATOM-OEAW.

REFERENCES

1. Bruzzone, P., Nylund, K. and Muster, W.J., " Elecrical insulation system for superconducting magnets according to the wind and react technique", *Advances in Cryogenic Engineering* 36B, 1990, pp. 999- 1006.
2. Schutz, J.B. and Reed, R.P., "Inorganic and hybrid insulation materials for ITER", *Advances in Cryogenic Engineering* 40B, 1994, pp. 985-992.
3. Dienst, W. and Zimmermann, H., *Journal of Nuclear Materials* 1091, pp. 212-215 (1994).
4. Clinard, F.W., Jr., Dienst, W. and Farnum, E.H., *Journal of Nuclear Materials* 1075, pp. 212-215 (1994).
5. Rice, J.A., Fabian, P.E. and Hazelton, C.S., *Applied Superconductivity*, IEEE Trans-on, **9**, pp. 220-223 (1999).
6. Humer, K., Spießberger, S., Weber, H.W., Tschegg, E.K. and Gerstenberg, H., *Cryogenics* **36**, pp. 611- 617 (1996).
7. Humer, K., Rosenkranz, P., Weber, H.W., Rice, J.A. and Hazelton, C.S., "Mechanical strength, swelling and weight loss of inorganic fusion magnet insulation systems following reactor irradiation", *Advances in Cryogenic Engineering* 46A, 2000, pp. 135- 141.
8. Pahr, D.H., Rammersdorfer, F.G., Rosenkranz, P., Humer, K. and Weber, H.W., Composites Part B, submitted (2001).

TENSILE TESTING AND DAMAGE ANALYSIS OF WOVEN GLASS-CLOTH/EPOXY LAMINATES AT LOW TEMPERATURE

S. Kumagai, Y. Shindo, and K. Horiguchi

Department of Materials Processing,
Graduate School of Engineering, Tohoku University,
Aoba-yama 02, Sendai 980-8579, Japan

ABSTRACT

In order to evaluate the tensile properties of SL-ES30 glass-cloth/epoxy laminates for superconducting magnets in fusion energy systems, tensile tests were examined both experimentalty and analytically. The tensile tests were conducted in accordance with JIS K 7054 at room temperature and liquid nitrogen temperature (77 K). The general specimen geometry was a rectangular dog-bone shape with constant gage length, but with each specimen size having a different specimen width. The experimental finding provides the data for analytical modeling. The model utilizes two damage variables which are determined from experimental data. A finite element method coupled with damage was adopted for the extensional analysis. The effects of temperature, specimen geometry and gripping method on the tensile properties are examined.

INTRODUCTION

Composite materials are being increasingly employed to manufacture structural components which are exposed to low temperature environment in aerospace and superconducting applications. An understanding of the mechanical behavior of these composites at low temperatures will provide a basis for systematic development of material variants and of new materials to meet the needs of future designs. Woven glass-cloth/epoxy laminates, such as G-10CR and G-11CR [1], are used mainly as electrical, thermal and permeability barriers, which provide minimal structural support in superconducting magnets at cryogenic temperatures. SL-ES30 woven glass-cloth/epoxy laminates having high glass content are candidate materials for the insulation of superconducting magnet windings in the International Thermonuclear Experimental Reactor(ITER).

The most difficult mechanical test to run for composite materials has been the tensile test. The difficulty of testing composite specimens in tension is due to the high strength, stiffness, and anisotrophy of the resin matrix composite materials being tested. For the most part, nearly all of the standard test methods that are currently accepted and used are designed for room temperature testing. The test procedure JIS(Japan Industrial Standard) K 7054 [2] was written to generate resin matrix composite tensile

CP614, *Advances in Cryogenic Engineering:*
Proceedings of the International Cryogenic Materials Conference - ICMC, Vol. 48,
edited by B. Balachandran et al.
© 2002 American Institute of Physics 0-7354-0060-1/02/$19.00

properties under room temperature conditions. However, when this procedure was applied to glass-cloth/epoxy laminates at cryogenic temperatures problems arose [3]. Tschegg et al. [4] performed tensile tests for woven glass-cloth/epoxy laminates at cryogenic temperatures and examined the effect of specimen geometry on the tensile properties.

The purpose of this study is to investigate the effects of temperature and specimen geometry on the tensile properties of SL-ES30 glass-cloth/epoxy laminates. Tensile tests were performed with dog-bone shape specimens at room temperature and 77 K in accordance with JIS K 7054. A two-dimensional finite element analysis was also used to study the stress and damage distributions within the test specimens and to interpret the experimental measurements.

EXPERIMENTAL PROCEDURE

Materials and Specimens

SL-ES30 woven glass-cloth/epoxy laminates were used as specimen materials. It has a weight fraction of 86% of basket-weave E-glass. A shematic of a basket-weave fabric reinforced lamina is shown in Fig. 1. The basket weave is a variation of the plain weave. The reinforcing fabric has a ratio of 44 threads per 25.4 mm in the warp direction to 64 in the fill direction and is 0.185 mm thick. The epoxy resin used is a bisphenol-A with a acid anhydride curing agent.

Tensile specimens were made in accordance with JIS K 7054 [2]. The specimens were cut from the SL-ES30 laminate with the axis parallel to either the warp or fill threads. The specimen geometry and the dimensions are shown in Fig. 2. The width was varied (B=10, 15, 25 mm).

Testing Method

Tensile tests were conducted using a 100 kN capacity servo-hydraulic testing machine at room and liquid nitrogen temperatures. The specimen strains were measured using resistance strain gages or clip-on type extensometer(clip gage). The ultimate strength was also determined. Low temperature environment was achieved by immersing the loading fixture, specimen and strain gage in liquid nitrogen (77 K). Tensile tests of SL-ES30 were in accordance with JIS K 7054. Two types of test fixtures were prepared to examine the effect of loading procedures on tensile properties of SL-ES30. Tensile test fixtures are schematically shown in Fig. 3. The loads were continuous with constant crosshead speeds of 0.5 and 5.0 mm/min.

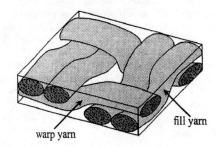

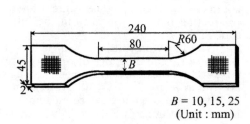

FIGURE 1. Shematic of basket-weave lamninate cross-section.

FIGURE 2. Specimen geometry for tensile test.

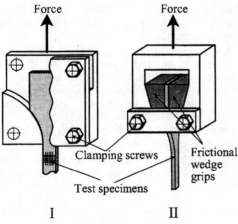

FIGURE 3. Tensile test fixtures.

TABLE 1. Tensile properties of SL-ES30 at room temperature(R.T.) and 77 K.

	R.T.	77 K
E_W^I (GPa)	42.1	55.0
ν_W^I	0.208	0.282
σ_W^I (MPa)	480	930
E_W^{II} (GPa)	41.7	47.7
ν_W^{II}	0.191	0.219
σ_W^{II} (MPa)	484	——
E_F^I (GPa)	41.3	51.8
ν_F^I	0.178	0.239
σ_F^I (MPa)	421	816
E_F^{II} (GPa)	40.0	44.5
ν_F^{II}	0.180	0.220
σ_F^{II} (MPa)	411	——

EXPERIMENTAL RESULTS AND DISCUSSION

Tensile Properties of SL-ES30

Specimens with $B=15$ mm were tested to analyze the effect of temperature on the elastic constants and tensile strength for SL-ES30 specimens. The results of these studies are tabulated in Table 1. The tensile strengths σ_W^I, σ_F^I, Young's moduli E_W^I, E_F^I, and Poisson's ratios ν_W^I, ν_F^I were obtained by using the test fixture I at room temperature and 77 K. The tensile strengths σ_W^{II}, σ_F^{II}, Young's moduli E_W^{II}, E_F^{II}, and Poisson's ratios ν_W^{II}, ν_F^{II} were also obtained for the test fixture II. The subscripts W and F denote the warp and fill directions of the laminates, respectively. At room temperature, there is no definite indication of a dependence of tensile properties on test fixture I or II. The tensile strengths σ_W^{II}, σ_F^{II} at 77 K are invalid values because the specimen slips out of test grips during testing. The values of σ_W^I and σ_F^I are about 2 times higher at 77 K than at room temperature, while Young's modulus and Poisson's ratio are slightly higher at 77 K than at room temperature. A strength anisotropy can be expected for woven glass-cloth/epoxy laminates depending on the warp/fill ratio of the fabric used in their construction [1]. The important fact of these tests is that the

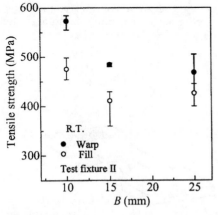

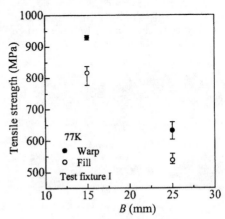

FIGURE 4. Effect of specimen width on tensile strength of SL-ES30 at room temperature (test fixture II).

FIGURE 5. Effect of specimen width on tensile strength of SL-ES30 at 77 K (test fixture I).

271

strength in the warp direction is higher than that in the fill direction. As seen in the Fig. 1, the warp yarns are less crimped than the fill yarns. This fact indicates that fiber waviness has an important effect on tensile strength.

The influence of specimen width on the tensile strengths σ_W^{II}, σ_F^{II} at room temperature is shown in Fig. 4. The tensile strengths in both the warp and fill direction depend on specimen width. The larger stress concentrations give rise to lower tensile strength. The JIS K 7054 test specimen is designed to be tested using friction-type wedge grips (test fixture II). There are inherent problems, however, with utilizing these friction wedge grips at cryogenic temperarures. The primary problem is that the specimen slips out of test grips during testing. This problem is avoided by using test fixture I. Fig. 5 also shows the variation of tensile strengths σ_W^I, σ_F^I with specimen width at 77 K. The tensile strengths σ_W^I and σ_F^I are strongly influenced by specimen width. Even if the test fixture I is used, wider specimens may give lower to strengths than narrow ones because of increased stress concentrations.

Fractography

Some interesting observations can be made regarding failure of the specimens in test fixture I at 77 K. Fig. 6 shows the appearance of specimens loaded to failure in the warp direction. All specimens fail in a fillet (specimen radius), presumably because of a stress concentration there. Failure of the specimen with $B=15$ mm is preceded by matrix microcracking at gage section while the specimen with $B=25$ mm shows no such features. The specimen with $B=25$ mm typically fails at the transition region ends.

DAMAGE MECHANICS ANALYSIS

Damage Variables

We consider a tensile specimen as shown in Fig. 7. The coordinate axes x and y are in the middle plane of the specimen and the z-axis is perpendicular to this plane. The coordinate axes x and y also correspond to the warp and fill directions. Consider an elementary volume and a stress σ_{x0} in the direction x. In a tensile test, tensile stress σ_{x0} induces longitudinal strain ε_{x0} and transverse strain ε_{y0} and ε_{z0}. The subscripts x, y, and z will be used to refer to the coordinate directions. The indicators of damage are defined as

$$E = \frac{\sigma_{x0}}{\varepsilon_{x0}} = E_x = E_y, \quad K = \frac{\sigma_{x0}}{\varepsilon_{x0} + \varepsilon_{y0} + \varepsilon_{z0}}, \tag{1}$$

where E is the Young's modulus, and K can be viewed as a volumetric modulus. Assuming that the elastic behavior of SL-ES30 woven glass-cloth/epoxy laminates can

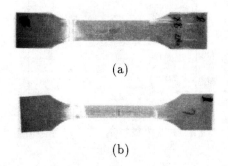

(a)

(b)

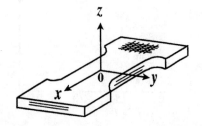

FIGURE 6. Photographs of typical fractured specimens (Test fixture I, Warp) with (a)$B=25$ mm, and (b)$B=15$ mm.

FIGURE 7. Coordinate system for tensile specimen.

be considered as quasi-isotropic in the plane of the layers and neglecting the influence of the strain ε_{z0} in the thickness direction on the damage development, K becomes

$$K = \frac{\sigma_{x0}}{\varepsilon_{x0} + \varepsilon_{y0}} = \frac{E}{1 - \nu}, \tag{2}$$

where ν represents the Poisson's ratio of the quasi-isotropic material. The damage variables [5] are given by

$$d = 1 - \frac{E}{E^0}, \quad \delta = 1 - \frac{K}{K^0}, \tag{3}$$

where the superscript 0 denotes the initial characteristics of materials. Then, the effective elastic properties of degraded material can be expressed as follows :

$$E = E^0(1 - d), \quad K = K^0(1 - \delta). \tag{4}$$

We assume that the laminate is in the plane stress state in the z direction. The strain energy of the material can be written as

$$U = \frac{1 + \nu}{2E}\left\{\frac{3}{4}(\sigma_{xx}^2 + \sigma_{yy}^2) + 2\sigma_{xy}^2 - \frac{\sigma_{xx}\sigma_{yy}}{2}\right\} + \frac{1 - 3\nu}{8E}(\sigma_{xx} + \sigma_{yy})^2, \tag{5}$$

where σ_{xx}, σ_{yy}, $\sigma_{xy} = \sigma_{yx}$ are the stress components. The degree of degradation of the mechanical properties due to damage have the following form :

$$\frac{E}{1 + \nu} = \frac{E^0}{1 + \nu^0}(1 - d_1), \quad \frac{E}{1 - 3\nu} = \frac{E^0}{1 - 3\nu^0}(1 - d_2), \tag{6}$$

where d_1 and d_2 are damage parameters. By introducing two damage parameters d_1 and d_2, the strain energy of the damaged material can be written as

$$U_d = \frac{1 + \nu^0}{2E^0(1 - d_1)}\left\{\frac{3}{4}(\sigma_{xx}^2 + \sigma_{yy}^2) + 2\sigma_{xy}^2 - \frac{\sigma_{xx}\sigma_{yy}}{2}\right\}$$
$$+ \frac{1 - 3\nu^0}{8E^0(1 - d_2)}(\sigma_{xx} + \sigma_{yy})^2. \tag{7}$$

Using Eq.(2), it is found that

$$\frac{1 + \nu}{E} = \frac{2}{E} - \frac{1}{K}, \quad \frac{1 - 3\nu}{E} = -\left(\frac{2}{E} - \frac{3}{K}\right). \tag{8}$$

Substituting Eqs.(4) and (6) into Eqs.(8) yields

$$\frac{1 + \nu^0}{1 - d_1} = \frac{2}{1 - d} - \frac{1 - \nu^0}{1 - \delta}, \quad \frac{1 - 3\nu^0}{1 - d_2} = -\frac{2}{1 - d} + \frac{3(1 - \nu^0)}{1 - \delta}. \tag{9}$$

Therefore, Eq.(7) can be rewritten as

$$U_d = \frac{1}{2}\left\{\frac{2}{E^0(1 - d)} - \frac{1 - \nu^0}{E^0(1 - \delta)}\right\}\left\{\frac{3}{4}(\sigma_{xx}^2 + \sigma_{yy}^2) + 2\sigma_{xy}^2 - \frac{\sigma_{xx}\sigma_{yy}}{2}\right\}$$
$$- \frac{1}{8}\left\{\frac{2}{E^0(1 - d)} - \frac{3(1 - \nu^0)}{E^0(1 - \delta)}\right\}(\sigma_{xx} + \sigma_{yy})^2. \tag{10}$$

From the consideration of thermodynamic potential, the conjugate variables Y_1 and Y_2 associated with the damage parameters d_1 and d_2 are defined by

$$Y_1 = (1 - d_1)^2 \frac{\partial U_d}{\partial d_1} = \frac{1 + \nu^0}{2E^0} \left\{ \frac{3}{4}(\sigma_{xx}^2 + \sigma_{yy}^2) + 2\sigma_{xy}^2 - \frac{\sigma_{xx}\sigma_{yy}}{2} \right\},$$

$$Y_2 = (1 - d_2)^2 \frac{\partial U_d}{\partial d_2} = \frac{1 - 3\nu^0}{8E^0}(\sigma_{xx} + \sigma_{yy})^2. \tag{11}$$

It is obvious that Y_1 and Y_2 are constants which are determined by the state of stress and the initial elastic properties.

At a given state of damage, the strains and stresses are related through the elastic energy as

$$\varepsilon_{xx} = \frac{\partial U_d}{\partial \sigma_{xx}}, \quad \varepsilon_{yy} = \frac{\partial U_d}{\partial \sigma_{yy}}, \quad 2\varepsilon_{xy} = \frac{\partial U_d}{\partial \sigma_{xy}}. \tag{12}$$

Substituting Eq.(10) into Eqs.(12), the strain-stress relationship is given by

$$\begin{Bmatrix} \varepsilon_{xx} \\ \varepsilon_{yy} \\ 2\varepsilon_{xy} \end{Bmatrix} = \begin{bmatrix} a & -a+b & 0 \\ -a+b & a & 0 \\ 0 & 0 & 4a-2b \end{bmatrix} \begin{Bmatrix} \sigma_{xx} \\ \sigma_{yy} \\ \sigma_{xy} \end{Bmatrix}, \tag{13}$$

where

$$a = \frac{1}{E^0(1 - d)}, \quad b = \frac{1 - \nu^0}{E^0(1 - \delta)}.$$

The stress (σ_{xx}) and strains ($\varepsilon_{xx}, \varepsilon_{yy}$) are recorded from the uniaxial tensile test along a principal material direction (warp direction). The elastic properties of the undamaged material can be obtained for the initial state : E^0=45.0 GPa(R.T.), 55.5 GPa(77 K), and ν_{xy}^0=0.190(R.T.), 0.263(77 K). The elastic properties E, ν and K of the damaged material are also recorded as damage progressed. Then the damage variables d and δ can be identified as a function of applied stress. The damage parameters d_1 and d_2 are readily obtained from Eqs.(9), respectively. Furthermore, using Eqs.(11),

$$Y_1 = \frac{3}{8}(1 + \nu^0)\frac{\sigma_{xx}^2}{E^0}, \quad Y_2 = \frac{1}{8}(1 - 3\nu^0)\frac{\sigma_{xx}^2}{E^0}. \tag{14}$$

Finally, the relationship between Y_1 and d_1, as well as Y_2 and d_2 for any given state of applied stress can be established. The data at 77 K (test fixture I) and corresponding curve fits are plotted together in Figs.8 and 9. The fitting functions are:

$$\begin{aligned} d_1 &= 0 & (0 \leq Y_1 < 0.489), \\ d_1 &= 0.0749Y_1 - 0.0366 & (0.489 \leq Y_1 < 2.22), \\ d_1 &= 0.00893Y_1 + 0.110 & (2.22 \leq Y_1), \end{aligned}$$

$$\begin{aligned} d_2 &= 0 & (0 \leq Y_2 < 0.0440), \\ d_2 &= 9.44Y_2 - 0.415 & (0.0440 \leq Y_2 < 0.115), \\ d_2 &= 0.435Y_2 + 0.625 & (0.115 \leq Y_2). \end{aligned} \tag{15}$$

Eqs.(15) are used to predict damage distribution in a tensile specimen. The fitting functions used to fit the data at room temperature (test fixture II) are:

$$\begin{aligned} d_1 &= 0.0207Y_1^3 - 0.093Y_1^2 + 0.107Y_1, \\ d_2 &= -256Y_2^4 + 177Y_2^3 - 45.9Y_2^2 + 6.33Y_2. \end{aligned} \tag{16}$$

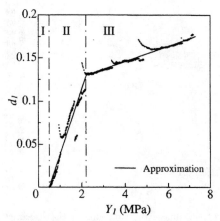
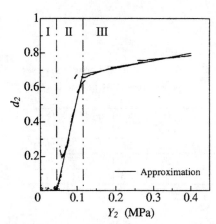

FIGURE 8. Plot of damage parameters (d_1 - Y_1 for SL-ES30 at 77 K (Test fixture I).

FIGURE 9. Plot of damage parameters (d_2 - Y_2 for SL-ES30 at 77 K (Test fixture I).

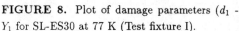
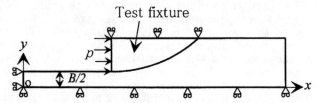

FIGURE 10. Finite element model of the tensile test.

Finite Element Analysis

The proposed damage mechanics model is implemented numerically using the finite element method. The model is used to analyze the dog-bone shaped specimens subjected to inplane tensile forces. Owing to symmetry, only quarter of the specimen needs to be modeled. The loading conditions used for the finite element analysis were dictated by the loadings applied by the actual test fixtures. These force and displacement boundary conditions for the model of a specimen are illustrated in Fig. 10, for test fixture I. In the model, specimen width B was 15 and 25 mm, with specimen radius R (60 mm) held constant. The finite element grid consisted of 779 nodes and 232 two-dimensional, eight-node isoparametric elements.

RESULTS AND DISCUSSION

Stress and Damage Distributions

Damage variables d and δ contour plots at 77 K for the model (B=15 mm), i.e., the tensile specimen tested in the test fixture I, are presented in Figs.11(a) and (b). Applied nodal force approximately corresponded to the experimentally determined breaking load; 31 kN. The predicted damage pattern is consistent with the photograph of the damaged specimen.

The finite element analysis results for stress concentration factor($\alpha = \sigma_b/\sigma_0$) and damage concentration factors($d_b/d_0, \delta_b/\delta_0$) are listed in Table 2. Here, subscript b indicates values at transition region ends, while subscript 0 indicates values at the midline of the specimen. Applied nodal forces approximately corresponded to the experimentally determined breaking loads;15 kN at R.T. and 31 kN at 77 K. At room temperature, it is suspected that the radius zone may be significantly affected by a stress concentration factor which would cause premature failure. The α values of specimen with $B = 15$

TABLE 2. FEM results for tensile stresses and damage parameters (test fixture I).

Temp.	B (mm)	α	d_b/d_0	δ_b/δ_0
R.T.	15	1.36	1.09	1.20
77 K	15	1.03	1.97	1.76
	25	1.35	1.84	1.81

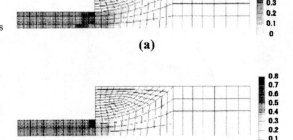

(a)

(b)

FIGURE 11. Damage distributions in the specimen with (a) d and (b) δ.

mm is smaller at 77 K than at room temperature, while the values of d_b/d_0 and δ_b/δ_0 are larger at 77 K than at room temperature. The damage analysis shows considerable stress reduction due to the damage at transition region ends. The narrow specimen (B=15 mm) has damage (resin microcracks) on the gage region of the specimen at 77 K. Primarily due to the stress reduction at transition region ends, the stress is therefore transferred to the gage region. At 77 K, stress concentrations would be expected to have a greater effect on the wider specimens than on the narrow ones, and it could be argued that this might explain some of the difference in strength. In summary, the use of a narrow specimen for cryogenic tensile testing(test fixture I) of woven glass-epoxy laminates is recommended.

CONCLUSION

This paper has presented experiments and finite element analysis on woven glass-cloth/epoxy laminate tensile specimens. The following conclusions can be drawn from this study.

- In the dog-bone shape specimens failure is always in the fillet, presumably because of a stress concentration there.
- The tensile strength increases as specimen width B is reduced.
- The damage parameters of SL-ES30 are determined from stress-strain curves at room temperature (test fixture II) and 77 K (test fixture I).
- The damage contours show that damage is qualitatively in line with the experimental observations.
- For cryogenic tension testing (test fixture I) of woven glass-epoxy laminates the narrow specimen must be used to obtain adequate test results.

REFERENCES

1. Kasen,M.B., *Cryogenics* **21**,pp.323-340 (1981).
2. *JIS K 7054, Testing method for tensile properties of glass fiber reinforced plastics*
3. Eisenreich,T.J. and Cox,D.S., "Modification of the ASTM D 3039 Tensile Specimen for Cryogenic Applications," in *Advances in Cryogenic Engineering* 38A, edited by F.R.Fickett and R.P.Reed, Plenum, New York, 1992, pp.437-444.
4. Tschegg,E., Hummer,K. and Weber,H.W., *Cryogenics* **31**, pp.312-318(1991).
5. Shan,H.Z., Pulvinage,P., Parvizi-Majidi,A. and Chou,T.W., ASME *J.Eng.Mater.Technol.* **116**, pp.403-407(1994).

HIGH VOLTAGE BREAKDOWN CAPABILITIES OF HIGH TEMPERATURE INSULATION COATINGS FOR HTS AND LTS CONDUCTORS

E. Celik[1], H. I. Mutlu[1], Y. Akin[1,2], H.Okuyucu[1],W. Sigmund[2], and Y. S. Hascicek[1]

[1] National High Magnetic Field Laboratory,
1800 E. Paul Dirac Dr., Tallahassee, FL, 32310, USA
[2] Department of Materials Science and Engineering,
The University of Florida, Gainesville, FL, 32611, USA

ABSTRACT

High temperature ZrO_2 based coatings were deposited on Ag and Ag/AgMg sheathed Bi-2212 tapes from solutions derived from alkoxide-based precursors using a reel-to-reel, continuous sol-gel technique. The insulation coatings were annealed at 850°C for 20 hours under O_2 flow. The surface morphology and structure of coatings were characterized by SEM and XRD. High voltage breakdown of insulation on tapes was measured by a standard high voltage breakdown power supply. It has been found that high voltage breakdown values of these insulations strongly depend on number of dipping, thickness, coating type, annealing conditions, and dopant content in ZrO_2. 20% Y_2O_3-ZrO_2 coatings showed the best high voltage breakdown value, 2.05 kV at 1.5 mA.

INTRODUCTION

High temperature superconducting (HTS) tapes are regarded as promising conductors for applications in high field superconducting magnets. To date, most developments have focused on oxide powder-in-tube (PIT) conductors with a wind-and-react (W&R) technique for making coils and magnets [1]. Despite the recent progress around the world in Ag/AgMg sheathed Bi-2212 and Bi-2223 HTS coils and magnets, the feasibility of these materials being wound into coils and magnets is limited by (1) the inherent 'weak-link' problem, i.e. poor connectivity at grain boundaries, (2) inhomogenity of phase formation over the entire length of the superconductor, (3) the low mechanical strength, and (4) the difficulty of the handling the material during winding of the conductor. To minimize the problem of the handling in the fabrication process and to improve the mechanical integrity of the magnet system, a W&R procedure is adopted together with the use of high temperature insulation coatings [2,3]. High temperature insulation coatings are necessary for turn-to-turn insulation in any wind and react (W&R) magnet built from the high temperature superconductors. Insulating materials surrounding the conductors prevent

CP614, *Advances in Cryogenic Engineering:*
Proceedings of the International Cryogenic Materials Conference - ICMC, Vol. 48,
edited by B. Balachandran et al.
© 2002 American Institute of Physics 0-7354-0060-1/02/$19.00

electrical short circuits within the winding of a coil [4,5]. The NHMFL sol-gel ZrO_2 insulation coating has been developed for this application and used since 1997 [6-12].

In this study, the high temperature insulation coatings on Ag/AgMg sheathed Bi-2212 superconducting tapes were produced by reel-to-reel, continuous sol-gel technique.

EXPERIMENTAL PROCEDURE

Zirconia and stabilized zirconia coatings were produced on Ag/AgMg sheathed Bi-2212 superconducting tapes using reel-to-reel, continuous sol-gel technique. A detailed discussion of solution preparation can be found in Ref.[5-13] Ag/AgMg on Bi-2212 tapes were cleaned with acetone. Ag tapes were first dipped into a solution of HNO_3 and water, and then rinsed in the acetone. The coatings were deposited onto tapes using the sol-gel dip-coating technique at a withdrawal rate of 2 cm/sec. The as-deposited films were dried at 100°C for 30 sec before burning out carbon based materials in amorphous layers at 300°C for 30 sec. They were then oxidized at temperatures in the range of 450°C and 600°C for 120 sec in air. The process was repeated 5 to 20 times depending on coating types. Finally, the coatings were annealed at 850°C for 20 hours under an oxygen atmosphere.

The surface topography and cross-section of insulation coatings were examined by Scanning Electron Microscopy (SEM). X-ray diffraction (XRD) patterns of coatings were carried out with a Philips diffractometer with a Cu K_α irradiation.

High voltage breakdown tests were performed at 1.5 mA using a standard high voltage power supply. After the annealing process, two Cu contacts were applied on insulation coatings as seen in Fig. 1 to measure high voltage breakdown values.

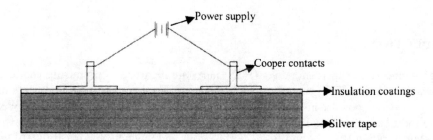

FIGURE 1. Schematic of the measurement of high voltage breakdown values of insulation coatings on Ag tapes.

RESULTS AND DISCUSSION

The fabrication of insulation on Ag or AgMg sheathed Bi-2212 tapes by solution deposition of ZrO_2 based insulation coatings involves four basic steps: (1) synthesis of the precursor solutions; (2) deposition of coatings by a reel-to-reel, continuous sol-gel set up, where drying processes usually begin depending on the solvent; (3) low temperature heat treatment for drying, combustion of organic species (typically 300-400°C), formation of amorphous structures, and oxidation process (450-600°C); and (4) annealing for densification and crystallization of coatings at 850°C for 20 hours under O_2 flow [14]. Reaction kinetics in the insulation system also affect microstructural development. The reaction kinetics are related to processing temperature and time during heat treatment and annealing. Our previous work [13] explained microstructural evaluations of insulation coatings on Ag/AgMg sheathed Bi-2212 tapes in the temperature range of 500 and 650°C for 2 min in air. The surface topographies of ZrO_2, MgO-ZrO_2, Y_2O_3-ZrO_2, CeO_2-ZrO_2, Sm_2O_3-ZrO_2, Er_2O_3-ZrO_2, In_2O_3-ZrO_2, and SnO_2-ZrO_2 coatings on Ag/AgMg tapes which were annealed at 850°C for 20 hours under oxygen flow, are seen in Fig. 2. The compositions of all coatings which are used in this study are 20% of dopant and the rest is ZrO_2.

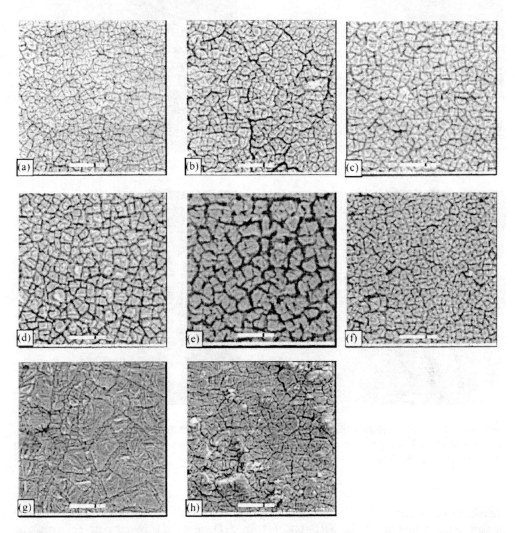

FIGURE 2. SEM micrographs of the surface of a) ZrO_2, b) MgO-ZrO_2, c) Y_2O_3-ZrO_2, d) CeO_2-ZrO_2, e) Sm_2O_3-ZrO_2, f) Er_2O_3-ZrO_2, g) In_2O_3-ZrO_2, and h) SnO_2-ZrO_2 coatings on Ag tapes annealed at 850°C for 20 hours under oxygen flow. The number of dipping was 15. The white scale bars are 20 μm.

The surfaces of the insulations have continuous long cracks. Not only annealing but also the thicknesses of insulations influence the surface quality and microstructural developments. Number of dipping for coatings, which are shown in Fig. 2, is 15 and the average thickness of the coatings is 12 μm (Fig. 3). Furthermore, the cracks in sol-gel dipped coatings of zirconia depend on number of dipping, structure and content of molecular precursor, viscosity, withdrawal speed, surface tension and contact angle with substrate. Especially, these cracks and coating thickness influence the high voltage breakdown values of the insulations on Ag/AgMg sheathed Bi-2212 tapes.

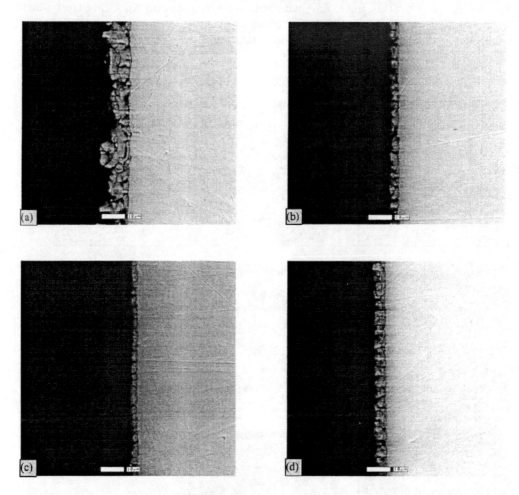

FIGURE 3. SEM micrographs of the cross-section of a) ZrO_2, b) MgO-ZrO_2, c) Y_2O_3-ZrO_2, d) CeO_2-ZrO_2, on AgMg sheathed Bi-2212 tapes annealed at 850°C for 20 hours under oxygen flow. Number of dipping is between 8 and 13. The scale bars are 10 μm.

XRD patterns of sol-gel coatings taken after the annealing process showed the characteristic peaks of insulation materials (Fig. 4). Cubic, orthorhombic and tetragonal phases were observed on the XRD patterns for ZrO_2 and ZrO_2 based coatings prepared from $Zr[O(CH_2)_3CH_3]_4$ and Mg, Y, Ce, Sm, Er, In and Sn based organometallic precursors. Although MgO-ZrO_2 exhibits hexagonal $Mg_2Zr_5O_{12}$, cubic ZrO_2 and MgO, cubic $Y_{0.15}Zr_{0.85}O_{1.93}$, cubic and orthorhombic ZrO_2 and monolithic Y_2O_3 phases were found in Y_2O_3-ZrO_2. Cubic $Ce_2Zr_2O_7$ and cubic, orthorhombic and hexagonal ZrO_2 phases were also determined in CeO_2-ZrO_2. Cubic ZrO_2 and Sm_2O_3, and cubic ZrO_2 and Er_2O_3 were phases of Sm_2O_3-ZrO_2 and Er_2O_3-ZrO_2 coatings on Ag/AgMg tapes, respectively. Furthermore,

In$_2$O$_3$-ZrO$_2$ exhibited the same structure with ZrO$_2$ except cubic In$_2$O$_3$ phase. An increase in the crystalline size of coatings was obtained with increasing heat treatment temperatures.

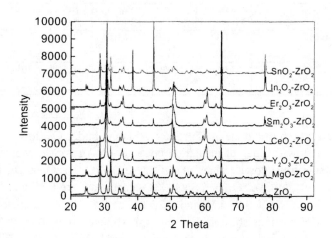

FIGURE 4. XRD patterns of insulation coatings on Ag tapes.

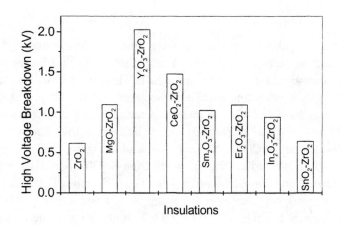

FIGURE 5. The high voltage breakdown values of insulation coatings on AgMg sheathed Bi-2212 tapes at 1.5 mA

The high voltage breakdown measurements were preformed for several ZrO$_2$ based sol-gel coatings on Ag/AgMg sheathed Bi-2212 tapes. The high voltage breakdown values of these insulations seems to depend on number of dipping, thickness, coating type, annealing conditions, dopant content in ZrO$_2$ [13]. In table 1 breakdown voltage values and concentrations of coatins are given. When the coating thickness was increased, the high voltage breakdown value increased. The annealing temperature and time influenced the

281

high voltage breakdown voltage values of insulations. The coatings showed sintering characterizations [9-13] depending on the temperature and the time. The coating thickness decreased with sintering. The optimum coating thickness was between 7 and 10 µm after

TABLE 1. High voltage breakdown voltage values and dopant ratios for insulation coatings.

Insulation	ZrO_2	MgO-ZrO_2	Y_2O_3-ZrO_2	CeO_2-ZrO_2	Sm_2O_3-ZrO_2	Er_2O_3-ZrO_2	In_2O_3-ZrO_2	SnO_2-ZrO_2
HVB values (kV)	0.617	1.1	2.032	1.479	1.028	1.096	0.943	0.643
Dopant ratio	20%	20%	20%	20%	20%	20%	20%	20%

the annealing process which was done at 850°C for 20 hours under O_2 flow. Fig. 5 depicts the high voltage breakdown values of insulation coatings on Ag/AgMg sheathed Bi-2212 tapes. The high voltage breakdown values of ZrO_2, MgO-ZrO_2, Y_2O_3-ZrO_2, CeO_2-ZrO_2, Sm_2O_3-ZrO_2, Er_2O_3-ZrO_2, In_2O_3-ZrO_2, and SnO_2-ZrO_2 coatings were 0.67 kV, 1.1 kV, 2.05 kV, 1.48 kV, 1.06 kV, 1.21 kV 0.98 kV, and 0.72 kV at 1.5 mA of breakdown current, respectively. Y_2O_3-ZrO_2 coatings showed the best high voltage breakdown value which is 2.05 kV at 1.5 mA. ZrO_2 based insulation coatings on Ag/AgMg sheathed Bi-2212 tapes have been in use in the insulation process for the HTS coil inserts for NMR program at the NHMFL.

CONCLUSIONS

Zirconia and stabilized zirconia coatings were produced on Ag and AgMg sheathed Bi-2212 superconducting tapes from Zr, Mg, Y, Ce, Sm, Er, In and Sn based precursors using reel-to-reel, continuous sol-gel technique. The fabrication of insulation on Ag/AgMg sheathed Bi-2212 tapes by this approach involves synthesis of the precursor solutions, deposition of coatings by a reel-to-reel, low temperature heat treatment, and annealing. Annealing conditions and the coating thickness influenced the surface quality and microstructural developments. Cubic, orthorhombic and tetragonal phases were observed by XRD for ZrO_2 and ZrO_2 based coatings prepared from $Zr[O(CH_2)_3CH_3]_4$ and Mg, Y, Ce, Sm, Er, In and Sn based organometallic precursors. The high voltage breakdown values of these insulations depend on number of dipping, thickness, coating type, annealing conditions, dopant content in ZrO_2. The Y_2O_3-ZrO_2 coatings showed the best high voltage breakdown value, which is 2.05 kV at 1.5 mA. ZrO2 based insulation coatings on Ag/AgMg sheathed Bi-2212 tapes have been in use for the insulation for the HTS high field insert coils for NMR program at NHMFL.

ACKNOWLEDGEMENT

We would like to thank R.B. Goddard for ESEM assistance, and K. Marken for kindly supliying Ag/AgMg sheathed Bi-2212 tapes. This work is based upon research carried out at the National High Magnetic Field Laboratory (NHMFL), which is supported by the National Science Foundation, under Award No. DMR-9527035.

REFERENCES

1. L.Y. Xiao, D.K. Hilton, Y.S. Hascicek, and S.W. Van Sciver, *Cryogenics,* 37:837-841 (1997).
2. N.V. Vo, *J. of Magnetism and Magnetic Materials,* 188:145-152 (1998).
3. K. Bauer, S. Fink, G. Friesinger, A. Ulbricht and F. Wuchner, *Cryogenics,* 38:1123-1134 (1991).
4. H.W. Weijers, Q.Y. Hu, Y.S. Hascicek, A. Godeke, Y. Viouchkov, E. Celik, K. Marken, W. Dai And J. Parrell, *IEEE Transactions on Applied Superconductivity,* 9:563-566 (1999).
5. E. Celik, H.I. Mutlu, and Y.S. Hascicek, *IEEE Transactions on Applied Superconductivity,* 10:1341-1345 (2000).
6. I.H. Mutlu and Y.S. Hascicek, *Advances in Cryogenic Engineering,* 34:233-237 (1997).
7. E. Celik, J. Schwartz, E. Avci and Y.S. Hascicek, *IEEE Transactions on Applied Superconductivity,* 9:1916-1919 (1998).
8. E. Celik, I. H. Mutlu and Y.S. Hascicek, US patent application pending, May, 1998.
9. E. Celik, E. Avci and Y.S. Hascicek, *2nd Advanced Technologies Symposium,* March 8-10 (1999).
10. E. Celik, E. Avci and Y.S. Hascicek, *Advances in Cryogenic Engineering* (Materials), 46:291-295 (2000).
11. E. Celik, H.I. Mutlu, E. Avci and Y.S. Hascicek, *IEEE Transactions on Applied Superconductivity,* 10:1329-1333 (2000).
12. E. Celik, E. Avci and Y.S. Hascicek, *Physica C: Superconductivity,* Volume 340, Issues 2-3, 1 December 2000, Pages 193-202
13. E. Celik, H.I. Mutlu, H. Okuyucu, and Y.S. Hascicek, *IEEE Trans. Appl. Supercond.,* 11 (1), 2881-2884 (2001)
14. R.W. Schwartz, *Chemistry of Materials,* 9:2325-2340 (1997).

NON-METALLIC MATERIALS —
PROCESS AND DEVELOPMENT

A DEFINITION OF THE PARAMETERS THAT INFLUENCE AND CONTROL THE FLOW OF RESIN DURING VACUUM IMPREGNATION OF MAGNETS AND OTHER STRUCTURES

S.J. Canfer, D. Evans, R.J.S. Greenhalgh and D. Morrow

CCLRC, Rutherford Appleton Laboratory, Chilton, Oxon. OX110QX UK

ABSTRACT

After winding and termination, the final stage in the preparation of many superconducting and conventional magnets is vacuum impregnation with an epoxide resin. Given the importance of the vacuum impregnation process in the overall scheme of magnet manufacture, it is surprising that there are many factors that effect the success and economics of the process that are still not quantified. This paper defines the pressure range, outgassing rates and moisture content that are considered necessary to achieve complete resin impregnation of tightly wound or other compacted, high impedance structures.

INTRODUCTION

High energy particle physics experiments such as ATLAS at CERN require large superconducting magnets. Future experiments such as neutrino factories will require higher magnetic fields with consequently higher stresses on materials.

Vacuum impregnation is commonly used to produce large superconducting magnets, which operate in thermally and mechanically demanding environments. The impregnation of magnets is often the last stage in a lengthy and costly process of magnet building. A good understanding of the effect of many process variables is necessary to optimise production conditions. In addition, over-engineering and unnecessary costs may be avoided. Failure of the impregnation may mean failure and scrapping of the complete system. It is therefore essential that the impregnation process is fully understood and the risk of failure is minimised.

RESIN DISTRIBUTION AND FLOW – EVALUATION PROGRAMME

There are a number of factors that will influence the flow of resin into a mould during the impregnation process, including the nature of the mould tool and the vacuum impregnation (often referred to as VPI – vacuum – pressure impregnation) process being used[1].

CP614, *Advances in Cryogenic Engineering:*
Proceedings of the International Cryogenic Materials Conference - ICMC, Vol. 48,
edited by B. Balachandran et al.
© 2002 American Institute of Physics 0-7354-0060-1/02/$19.00

The experimental programme was designed to quantify the following parameters of the VPI process;

- Effect of residual pressure at the time of impregnation
- Out-gassing rate of component at impregnation
- Relationship of out-gassing rate and water vapour content
- Effect of resin de-gassing efficiency

Effect of Residual Pressure and Resin Viscosity

For this aspect of the work, a small but practical tool was required. The component to be impregnated should allow uniform impregnation and allow the progress of the impregnation to be accurately monitored. In addition, the resin flow pattern and the resistance to flow should be entirely reproducible.

Figure 1(a) shows the glass plate rig. The glass plates measured 500 mm x 500mm. Eighteen layers of 0.25mm thick E-glass fabric were cut into squares of 400mm x 400mm and sandwiched between the glass plates. The assembly was clamped using 4mm shims between the glass plates. Impregnation was through the open edges of the tool. After impregnation, each composite panel contained 46 to 48 volume % glass fibres. The position of the moving resin front was recorded using a digital camera.

Impregnation of this package could be completed in two ways. In the horizontal mode, Figure 1(b), the glass fabric was covered with resin relatively quickly, no further out-gassing of the glass was possible and all vapours within the glass fibres were trapped. In the vertical mode, Figure 1(c), the filling rate was more easily controlled. This mode was used to investigate the effects of slower filling rates and speed of capillary action impregnation without necessarily trapping vapour volumes. It should be stressed that this ideal situation is unlikely to occur in practice.

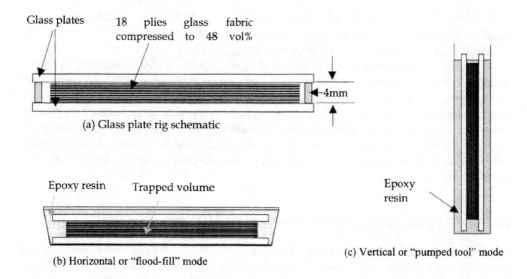

FIGURE 1. Schematic cross-sections of glass plate trial system.

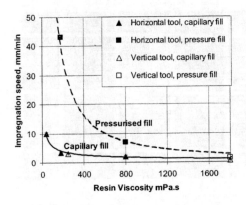
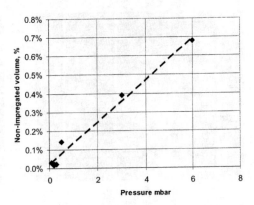

FIGURE 2. Speed of impregnation in horizontal and vertical modes, by capillary and pressurised impregnation.

FIGURE 3. Non-impregnated volume as a function of tank pressure before impregnation.

The rate of impregnation of the two modes of filling is presented in Figure 2. Also shown is the rate of filling of the tool by capillary action, and the rate of impregnation once the pressure is returned to atmospheric. In each case, the rate of impregnation is shown as a function of resin viscosity. For most practical purposes, a limiting viscosity of 200 mPa.s is indicated as being suitable for impregnation.

For each impregnation that is recorded in Figure 2, the 'tool' was evacuated to a pressure of <0.15 mbar at 50°C and once these conditions had been established, they were maintained for at least 72 hours before impregnation. There was no significant difference in the speed of impregnation by capillary action in the horizontal or vertical modes. However in the horizontal, "trapped volume" mode, the distance impregnated was limited.

A number of experiments were performed to examine the effect of the tank pressure before impregnation on the extent of impregnation. For each experiment, the panel was prepared, evacuated and conditioned as described above. After a minimum of 72 hours at 50°C and a pressure of <0.15 mbar, dry nitrogen was admitted to the VPI vessel to raise the pressure to a pre-determined value. These precautions insured that the system contained no residual water vapour and that pressure resulting from gas or vapour trapped within the tool was due only to dry nitrogen. Experiments were performed using pressures from 0.08 to 6 mbar. Following resin impregnation and when resin flow had stopped, the non-impregnated area (volume) was recorded. The results are presented in Figure 3.

The non-impregnated volumes shown in Figure 3 relate closely to the 'trapped' volumes predicted from the residual pressure at impregnation using simple 'gas law' relationships.

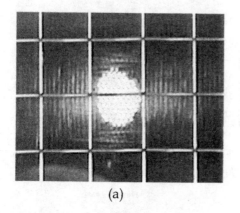

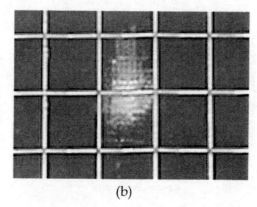

(a)

(b)

FIGURE 4(a). Void remaining at end of pressure-impregnation.
NB Grid squares are at 1cm intervals.

FIGURE 4(b). The same area after 12 hours at atmospheric pressure.

Images recorded during impregnation are presented in Figures 4(a) and 4(b). Figure 4(a) shows the non-impregnated area when the resin had stopped flowing and impregnation was nominally complete. Figure 4(b) shows the same component after a time lapse of 12 hours at atmospheric pressure. The previously dry, non-impregnated fibres appear to have been wetted as a result of capillary action. Resin appears to have been drawn along the fibre bundle (but not filling the inter-spaces in the woven fabric), causing some re-distribution of the available resin and void volumes.

Outgassing from glass cloth – Theory

Consider a volume of glass cloth trapped by encroaching resin. During the time it takes the resin to fill the volume, the cloth will continue to out-gas into the volume. Assume that as the resin covers the cloth, the covered cloth no longer out-gasses into the volume. Thus the out-gassing rate in to the void is proportional to the void volume which becomes smaller with time. The gas left in the final void will have two components, one being the gas that was present at the moment the volume was enclosed by the resin, the second being the additional contribution from out-gassing in the void as it was being filled. It can be shown[2] that under most circumstances:

$$\text{void volume} = \frac{V_0}{P_A}\left(P_0 + {r}/{2k}\right)$$

(1)

where
V_0 is the initial trapped volume (litres)
P_A is the final pressure in the void (atmospheric?) (mbar)
P_o is the evacuated pressure, before filling (mbar)
r is the outgassing rate of the component at the time of filling (mbar litres/sec)
k is the rate of resin filling (litres/sec)

Providing the outgassing rate of the VPI vessel is low, the net outgassing rate of the component may be measured. The leak rate of the empty VPI chamber at the temperature of impregnation is estimated by measuring the pressure rise, in mbar per second, multiplied by the chamber volume, to give leak rate in mbar litres per second. This is compared with the outgassing rate of the chamber plus component, also at the impregnation temperature.

Outgassing from glass cloth – Practice

The evacuation time necessary to achieve a low out-gassing rate is not a constant but will vary with a number of parameters, including the evacuated volume, surface area and degree of compaction of glass to be evacuated and pumping speed. Rates of out-gassing have been measured for the system of glass plates described above (area of glass fabric approximately 2.9 m²). Additional out-gassing measurements have been made on rolls of E-glass tape, with a total area of around 24 m². Data obtained from both sets of measurements using the same VPI chamber and evacuation system are presented in Figure 5.

When flowing through a mould tool, the resin will follow the path of lowest impedance. This invariably means that impregnation does not occur uniformly, because in most situations for impregnation, compaction of the component is not uniform and those regions that are most heavily compacted will be the last to impregnate. If the component is not fully outgassed, flow of resin through the component will trap gas or vapour in some regions with consequent lack of impregnation of the finished component.

Vacuum impregnation of superconducting magnets is likely to involve trapped volumes. These volumes are isolated by the advancing resin front as impregnation proceeds. It is difficult to envisage a situation where the resin will advance in a perfectly even fashion from inlet port to outlet or pumping port. The resin will take the path of least resistance, typically around the outside of the tool between ground plane insulation and the conductor, or through resin-rich areas in the volumes between the corners of the turns.

Impregnation may proceed by two mechanisms:
- pressure, where the resin is forced through the dry glass by pressure differential ·
- wicking impregnation, where capillary action draws the resin between fibres
- a combination of the above.

Using the data from Figure 5 in combination with Equation 1, it is possible to calculate the effect of various outgassing rates.

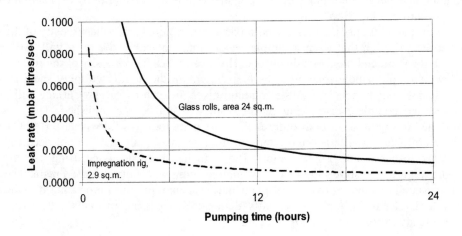

FIGURE 5. Out-gassing rates as a function of pumping time.

TABLE 1. Calculated Void Volume from Out-Gassing Rates

Leak Rate (mbar.l/sec)	Calculated Void Volume (mls)
0.2	60 (19%)
0.15	45 (14%)
0.10	30 (9%)
0.01	3.0 (1.0 %)
0.005	1.5 (0.5 %)

Figures in brackets represent void volume as a percentage of total volume of the glass plate tool

Using the glass plate tool described earlier, and a resin with a viscosity of around 200 mPa.s at the processing temperature, it was found that resin flow stopped approximately 10 minutes after applying atmospheric pressure. If the free volume within the tool (50 volume percent glass) was 0.32 litres (0.32 litres filling in 10 minutes is equivalent to a volume flow rate of 5.3×10^{-4} litres / second), the pressure at impregnation was 0.15 mbar , then the effect of out-gassing rate at the time of impregnation is illustrated in Table 1.

In practice, some difficulty was experienced in examining the effects of high out-gassing rates because of the relative small surface area of glass involved and the time taken to pump the system to the required impregnation pressure. However, the measured void volume from practical experiments was around a factor of two smaller than calculated, possibly due to dissolution of gasses and vapours within the resin.

Effect of resin degassing efficiency

Resin should be degassed to a pressure below that in the structure to be impregnated, to avoid further degassing in the tool. Further degassing is likely, because during impregnation the resin front is divided into a large surface area, which will further promote degassing. The gas or vapour so released will pool in areas cut off from the vacuum and result in "void-rich" areas. In such areas, wicking, or capillary action, is the only mechanism able to impregnate the fibres.

A hydrostatic pressure exists in a container of resin, so that every 1 millimetre of resin depth adds 0.1 mbar pressure. This is a significant pressure in a tool evacuated to a similar level. A degassing arrangement should be used that exposes all of the resin to a vacuum, whether by vigorously circulating the resin in a container or by exposing the resin to a vacuum in a thin film.

For many purposes, a high-speed mixer is not required since the viscosity of the resin is sufficiently high that even a simple paddle rotating within the resin will sufficiently disturb the resin and create fresh surface. However, when the viscosity of the resin system is low, a simple paddle rotating at high speed within the liquid, will only shear material that is immediately adjacent to the rotating mixer. In such cases new surface will not be created and complete de-gassing of the resin will not be accomplished. This effect is illustrated in Figure 6. In both Figures the glass plate tool was fully degassed to 0.1mbar. Figure 6(a) shows the large void resulting from gas or vapour that had not been removed from the resin prior to impregnation. The resin had been degassed using a high-speed paddle. Figure 6(b) shows the results of impregnating with fully degassed resin. The resin was degassed by pouring over a flat plate under vacuum, prior to introducing it to the glass plate tool. The resin bubbled violently during this degassing procedure, despite having been already exposed to vacuum in a can.

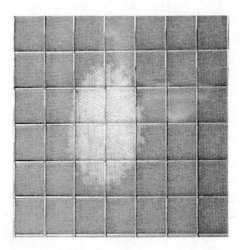 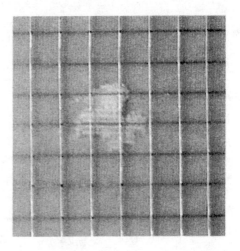

FIGURE 6(a). Large trapped void, a result of impregnating with poorly degassed, low viscosity resin. Void exists through all 18 layers of glass fabric. Impregnation pressure for both figures 0.1mbar.

FIGURE 6(b). Result of impregnating with low viscosity resin, degassed in a thin film. Void only present between top glass plate and top ply of glass fabric.

Moisture content

Moisture content is considered to be of interest because water will be held by the alkaline surface of the glass fibres. Other gases or vapours are not so tightly bound to the glass and may be pumped away far more easily. When fully out-gassed, the moisture content of the system to be impregnated should be low. At low pressures the water vapour pressure becomes a significant part of the pressure measured in a tool. The water vapour comes from four sources:

1. Air trapped in the tool;
2. Water molecules on the alkaline surface of glass fibres. A large surface area will be presented due to the small diameter of the glass fibres; typically 9 microns. $1m^2$ of glass cloth has a surface area of approximately $47m^2$.
3. Leaks of outside air into the tool or vacuum chamber.
4. On impregnation, from vapour dissolved in the resin and hardener mixture.

Figure 7 illustrates the relationship between out-gassing rate and moisture content. These data were collected by evacuating a number of rolls of glass fabric and measuring the water content and the out-gassing rate as a function of time. The water content is presented as the water vapour pressure as a percentage of the tank pressure. At higher pressures, the partial pressure of water vapour (as a percentage of the total pressure) is small. As the pressure is reduced, the water fraction increases, then slowly falls even after pressure has reached a steady value.

When out-gassing a component prior to impregnating, the moisture content will slowly fall but may take some days to reach an equilibrium value. Repeated flushing of the tool with dry nitrogen, followed by pumping, was found to reduce moisture content more quickly than by pumping alone.

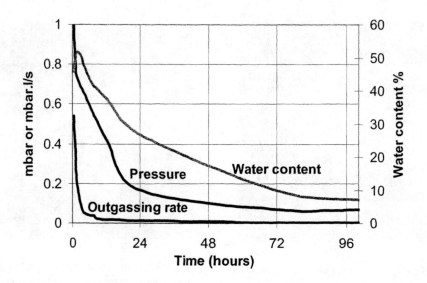

FIGURE 7. Pressure, outgassing rate and water content measured with time for 24m² glass tape.

CONCLUSIONS

The parameter that had the greatest effect on impregnation was outgassing of the low viscosity material. It is essential to degas resin thoroughly. Low viscosity resins, in particular, were found to be more difficult to degas. Hydrostatic pressure in the resin container was found to prevent full degassing.

A high vacuum, of the order of 0.15mbar, is required to avoid voids in the flood-fill process. The trapped volume of gases is directly proportional to the pressure at impregnation.

A resin viscosity of 200 mPa.s or lower is recommended to reduce impregnation times.

Flushing a tool with dry nitrogen gas will help reduce evacuation times and residual moisture content.

Evacuation times may be relatively short. The net outgassing rate of the structure may be used as a guide to a suitable time. An acceptable maximum outgassing rate is 0.01mbarls⁻¹. This corresponds to a moisture content of around 25% at a vacuum of 0.15mbar. Present work has not determined whether a lower moisture content would be beneficial.

Capillary action represents only a small contribution to the total impregnation process, in the trapped-volume case.

REFERENCES

1. Ten Kate H.J., Private communication
2. Greenhalgh, J.,Voids, outgassing, and residual pressure, Advanced Materials Group Technical Note (Internal), Rutherford Appleton Laboratory (2001)

HIGHLY RADIATION-RESISTANT VACUUM IMPREGNATION RESIN SYSTEMS FOR FUSION MAGNET INSULATION

P. E. Fabian, N. A. Munshi, and R. J. Denis

Composite Technology Development, Inc.
Lafayette, Colorado, 80026, USA

ABSTRACT

Magnets built for fusion devices such as the newly proposed Fusion Ignition Research Experiment (FIRE) need to be highly reliable, especially in a high radiation environment. Insulation materials are often the weak link in the design of superconducting magnets due to their sensitivity to high radiation doses, embrittlement at cryogenic temperatures, and the limitations on their fabricability. An insulation system capable of being vacuum impregnated with desirable properties such as a long pot-life, high strength, and excellent electrical integrity and which also provides high resistance to radiation would greatly improve magnet performance and reduce the manufacturing costs. A new class of insulation materials has been developed utilizing cyanate ester chemistries combined with other known radiation-resistant resins, such as bismaleimides and polyimides. These materials have been shown to meet the demanding requirements of the next generation of devices, such as FIRE. Post-irradiation testing to levels that exceed those required for FIRE showed no degradation in mechanical properties. In addition, the cyanate ester-based systems showed excellent performance at cryogenic temperatures and possess a wide range of processing variables, which will enable cost-effective fabrication of new magnets. This paper details the processing parameters, mechanical properties at 76 K and 4 K, as well as post-irradiation testing to dose levels surpassing 10^8 Gy.

INTRODUCTION

Large, capital-intensive, superconducting or resistive magnets are essential components of most current and planned fusion devices. Magnets for these applications must be reliable, have a long mean-time-between-failure, and be able to be manufactured using cost-effective materials and fabrication processes. Electrical insulation is often the weak link in magnet design, due to sensitivity to high radiation doses, embrittlement at cryogenic temperatures, and fabrication limitations.

CP614, *Advances in Cryogenic Engineering:*
Proceedings of the International Cryogenic Materials Conference - ICMC, Vol. 48,
edited by B. Balachandran et al.

Current magnet designs typically utilize a glass/epoxy composite insulation. This type of material provides sufficient electrical insulation, suitable mechanical properties at magnet operating temperatures, flexible processing for cost-effective coil fabrication and assembly, and reasonable cost. However, the epoxy vacuum-pressure impregnation (VPI) resins degrade to unacceptable levels of performance when exposed to high levels of radiation, particularly at the dose levels being considered for next-generation fusion devices [1]. The epoxy resins also do not withstand exposure to higher temperatures such as those to be encountered in these new magnet systems.

A future fusion device, the Fusion Ignition Research Experiment (FIRE) is part of the Next-Step-Option (NSO) devices in the US magnetic fusion energy program. FIRE will place great demands on the insulation due to the high expected radiation doses and mechanical stresses. According to the FIRE Engineering Status Report for Fiscal Year 2000, the expected lifetime neutron dose to the TF coil insulation is approximately 1.5×10^8 Gy (1.5×10^{10} Rad), an order of magnitude higher than the International Thermonuclear Experimental Reactor (ITER) requirements [2]. The highest mechanical stresses on the insulation are expected to be 520 MPa [3] in compression and 50 to 60 MPa in shear [4]. An additional consideration in FIRE is the ability of the insulation to perform at both cryogenic and elevated temperatures up to 373 K [4]. Current insulation systems have not been tested for operation and performance at these elevated temperatures.

Other fusion programs requiring high performance insulation are the National Compact Stellarator Experiment (NCSX) and the Quasi-Omnigenous Stellarator (QCS). Both systems are planned to demonstrate the advantages of the compact stellarator for magnetic fusion energy confinement. Because of the complex coil geometry in NCSX it is necessary to have a very low viscosity vacuum-impregnation resin system that is able to travel the tortuous path of the coil and wick into the very small voids of the conductor [5]. This insulation system must also be capable of withstanding elevated temperatures up to 100°C and preferably have a low temperature cure.

While epoxy insulation materials developed and tested for ITER have been exposed to no more than 23 MGy (2.3×10^9 Rad) combined neutron and gamma radiation, [1] glass reinforced composites using polyimide (PI) and bismaleimide (BMI) matrix materials have been tested to greater than 100 MGy (1×10^{10} Rad) with little or no degradation in shear strength [6]. It is known, though, that these materials are difficult and expensive to process, and have initial mechanical properties lower than those achievable with the epoxies. Therefore, new VPI resin systems, based on cyanate esters combined with bismaleimides and polyimides, were developed to both improve the insulation system's ability to withstand high levels of radiation and to improve the resin system's overall processibility.

MATERIALS AND PROCESSING PROPERTIES

Because of the complexity of the FIRE TF and the NCSX coils, and other similar magnets, an attractive fabrication process for the insulation is vacuum-pressure impregnation. In this process, the conductor is wrapped with dry glass fabric during or before coil winding, and liquid resin is impregnated into the coil after assembly. The resin is then cured to form the matrix of the composite insulation system. The use of this fabrication process limits the selection of resin systems. A successful VPI resin system must have a low viscosity (less than 200 cP), a long pot-life (8-hour minimum) at its processing temperature, excellent wetting and flow characteristics, and a reasonably low curing temperature. Cyanate esters were selected as the basis of these new VPI resin systems because they offered the promise of radiation resistance [7] and because they have desirable properties for use in VPI processing. Some liquid cyanate ester monomers have very low viscosities at room temperature and can be processed such that they can be blended with radiation-resistant polyimides and bismaleimides to form hybrid systems [8,9].

Three resin system formulations were used throughout this study, a single cyanate ester system, a cyanate ester/polyimide hybrid system and a cyanate ester/bismaleimide hybrid system. These three systems are designated CTD-403, CTD-406, and CTD-410, respectively.

To evaluate each system's processing properties, the initial viscosity of each system was measured at 25°C and at its respective processing temperature. Each system's viscosity was also monitored over time at its processing temperature to evaluate pot-life. For this program, pot-life was defined as the time it took for the resin's viscosity to double from its initial value. These characteristics are detailed for the three resin systems in Table 1. Processing properties for CTD-101K, a standard epoxy VPI system used extensively in the magnet industry, are also included in Table 1 for comparison.

As seen in Table 1, all three of the new resin systems meet the processing criteria outlined above, with all systems having viscosities less than 200 cP and pot-lives with a minimum of 8 hours. Of note are the relatively low processing temperatures of all three systems when compared to that of CTD-101K, particularly that of CTD-403, which is processed at room temperature.

Another important aspect of a resin system's processing properties is the ability to achieve full cure at reasonably low temperatures. This is important since curing of the insulation resin requires the entire magnet coil to be heated to the curing temperature and maintained at that temperature for several hours. This is a non-trivial issue for all but the smallest magnets. The other material property related to the cure, that has become an important issue for FIRE and NCSX coil insulation, is the cured insulation's ability to operate not only at cryogenic temperatures but also at elevated temperatures approaching 373 K (100°C). The insulation's ability to withstand elevated temperatures is related to its cure schedule since a material's glass transition temperature, T_g, is usually related to its highest cure temperature.

To address these cure-related issues, several alternate cure schedules were developed for each of the three resin systems based on previous experience with similar systems, manufacturer's recommendations, and information from the literature. Subsequently, the candidate cure schedules were evaluated through Differential Scanning Calorimetry (DSC) analysis of neat resin samples. DSC testing entails monitoring endothermic and

TABLE 1. Processing Properties of Cyanate Ester and Cyanate Ester Hybrid Resin Systems

Material Designation	CTD-403	CTD-406	CTD-410	CTD-101K
Material Type	Cyanate Ester	Cyanate Ester/ Polyimide	Cyanate Ester/ Bismaleimide	DGEBA Epoxy
Initial Viscosity at 25°C	100 cP	800 cP	400 cP	2100 cP
Initial Viscosity at Processing Temperature	100 cP	130 cP	100 cP	100 cP
Processing Temperature	25°C	40°C	40°C	60°C
Pot-Life	24 hr	13 hr	8 hr	24 hr
Glass Transition Temperature	182.1°C	230.6°C	237.6°C	112.9°C

exothermic reactions throughout the curing cycle, and can be used to determine the degree of cure. Full cure is very important for insulation resin systems in superconducting fusion magnets, since it has been noted that incompletely cured resin systems evolve more gas under irradiation than fully cured systems [10]. The exotherm profile produced during DSC analysis reveals if the resin system has achieved full cure, the temperature at which full cure is achieved, and the glass transition temperature.

With one exception, all of the materials achieved full curing with all cure schedules tested. It was found that the cyanate ester resin systems appear to require a minimum temperature of 135-140°C to reach full cure. These temperatures are very similar to the curing temperatures of current epoxy-based insulation systems.

An interesting discovery from the DSC analyses was the fact that the glass transition temperature of the cyanate ester and cyanate ester hybrid resin systems, as shown in Table 1, was consistently 25 to 30°C above the final curing temperature. This is a vast improvement over standard epoxy systems, such as CTD-101K, where the glass transition temperature is lower than the final cure temperature. This characteristic of the cyanate ester-based systems could prove very valuable for magnet insulation applications, especially when the insulation must perform at 373 K.

Another important processing characteristic of a VPI resin system is the ability to wet the fabric reinforcement within the composite insulation. Most superconducting magnets utilize a glass fabric wrap around the conductors to act as the structural reinforcement for the insulation system. The glass fabric must be fully impregnated by the resin to achieve the desired insulation structural and electrical performance. Complete impregnation is made more difficult because the fiber volume fraction is kept as high as reasonably possible, usually around 50 percent, to obtain the highest insulation strength and to minimize the organic resin content, and thus, the potential for radiation damage. Therefore, it is imperative that candidate resin systems for vacuum impregnation of superconducting magnets are able to flow and wick through tightly packed glass fabric.

To address this issue, one-dimensional flow experiments were performed for each candidate resin system to evaluate this aspect of the material's performance. For comparison, a similar test was performed using CTD-101K. The 1-D flow experiment used a closed mold that was filled to a nominal 50% volume fraction with a satin weave, S-2 glass fabric and impregnated using the VPI process. The mold was constructed of a 12.7-mm thick aluminum back plate and a 19.1-mm thick Pyrex glass cover plate in order to monitor the resin flow during the impregnation. The results of the 1-D resin flow experiments are illustrated in Figure 1.

As can be seen, the flow behavior of CTD-403 is quite similar to that of CTD-101K epoxy. Surprisingly, the cyanate ester/bismaleimide hybrid system (CTD-410) performed significantly better than either CTD-403 or CTD-101K, impregnating the full mold height in just over ten minutes. The other two systems attained that same impregnation height in approximately 25 to 40 minutes. CTD-406, the cyanate ester/polyimide hybrid material, which has a slightly higher viscosity than the other systems, achieved slightly less than half the impregnation distance of the other three systems over the same time period. However, it is significant that two of the three resin systems impregnated at least as well as CTD-101K, which is well known throughout the magnet industry for its superior impregnation performance.

MECHANICAL PERFORMANCE

Composite magnet insulation materials are typically required to maintain electrical

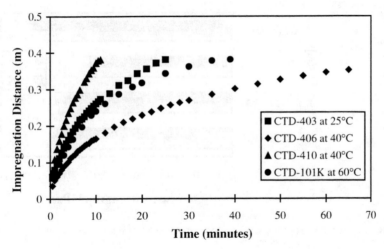

FIGURE 1. Results of 1-D resin flow experiment for candidate resin systems.

integrity when subjected to shear stresses up to 50 MPa and compressive stresses from 100 to 500 MPa while operating at cryogenic temperatures and after exposure to high doses of radiation. Often, the shear and compressive stresses act simultaneously on the insulation. Although compressive stresses are dominant in most magnet designs, it has been found that radiation exposure significantly degrades the fiber/matrix interface in most conventional magnet insulation materials [11], making interlaminar shear strength a significant parameter that must be evaluated for fusion magnet applications.

Therefore, the cyanate ester and cyanate ester hybrid resin systems were tested for interlaminar shear strength and compressive strength to evaluate their ability to withstand the stresses expected in an operating fusion magnet. Each of the candidate resin systems was mechanically tested at 76 K and 4 K. Subsequently, these systems were irradiated at the TRIGA Mark-II reactor in Vienna, Austria, made available to CTD through the Atominstitut der Österreichischen Universitäten (Atomic Institute of the Austrian Universities). The reactor provides a fast neutron flux density (E > 0.1 MeV) of about 7.6 x 10^{16} n/m^2/s, with total (neutron and gamma) absorbed doses of up to 500 MGy. The irradiation temperature for these materials was approximately 340 K.

The short-beam shear (SBS) test, in accordance with American Society for Testing and Materials (ASTM) Standard D2344 [12], was used to evaluate the apparent interlaminar shear strength of the candidate resin systems. The SBS test, described in detail elsewhere [13], employs a three-point bend test fixture and has been used extensively as a screening test for cryogenic materials. All test specimens were nominally 3.2-mm thick and were reinforced to a nominal 50 volume percent of S-2 glass fabric. A span ratio (width-to-thickness) of 5.0 was experimentally determined to provide the desired interlaminar shear failure mode.

Figure 2 illustrates the results of the SBS testing at 76 K and 4 K. Comparison data for CTD-101K were taken from the literature. [14]

As seen in Figure 2, CTD-403 and CTD-406 nearly match the strength of CTD-101K at cryogenic temperatures. CTD-410, the cyanate ester/BMI hybrid exhibits a shear strength approximately 20% less than the other systems, but is still of sufficient strength to meet or exceed most magnet requirements.

Compressive strength testing was also performed at both 76 K and 4 K on the same

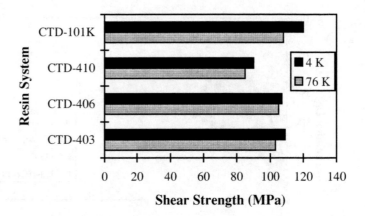

FIGURE 2. Short-Beam Shear test results for cyanate ester based resin systems at 76 and 4 K.

candidate materials tested in short-beam shear. Test specimens were made from the same 3.2-mm thick flat plate laminates that were used for the SBS test specimens. The 3.2-mm specimen thickness was selected since the actual insulation thickness in most magnet systems is in the 1 to 4-mm range. Testing was performed in the through-thickness direction, normal to the glass plies within the laminate, since this is the loading direction that the insulation will experience during most magnet operations. Testing procedures followed ASTM Standard D695 [15] as closely as possible. Specimens were nominally 6.4-mm square and 3.2-mm thick.

The results of the compression tests are shown in Table 2. As seen in the Table, the cyanate ester and cyanate ester hybrids all perform very well in compression, where all systems compare favorably with CTD-101K at 76 K and exceed it at 4 K. The compressive moduli of the cyanate ester-based resin systems are also comparable with that of the epoxy system, with the modulus of CTD-403 exceeding that of CTD-101K by a significant margin at 4 K. Another significant conclusion from the 4 K testing is the lower moduli of CTD-406 and CTD-410 when compared to CTD-403. The compressive moduli of these two systems average 13 percent lower than that of CTD-403. This may be explained by the polyimide and BMI additives, which have previously been shown to have significantly lower moduli at cryogenic temperatures than epoxy resin systems. [14] However, it should be noted that all of the cyanate ester hybrid systems still exhibit higher compressive moduli than that of the epoxy system.

TABLE 2. Compression Test Results at 76 K and 4 K

Resin System	Test Temperature (K)	Compressive Strength (GPa)	C_V	Compressive Modulus (GPa)	C_V	Strain-to-Failure
CTD-403	76	1.23	0.02	15.1	0.02	0.060
CTD-403	4	1.50	0.07	23.6	0.04	0.064
CTD-406	76	1.23	0.05	14.9	0.03	0.065
CTD-406	4	1.42	0.07	21.1	0.04	0.068
CTD-410	76	1.26	0.04	14.9	0.01	0.073
CTD-410	4	1.39	0.04	19.8	0.04	0.070
CTD-101K	76	1.30	0.04	16.7	0.05	0.078
CTD-101K	4	1.36	0.04	19.7	0.05	0.069

The SBS and compression specimens were irradiated to three different levels in the TRIGA reactor in Austria. The three levels corresponded to fast neutron fluences of 10^{21} n/m^2, 10^{22} n/m^2, and 5×10^{22} n/m^2. These levels equate to total dose levels of 4.7×10^6 Gy, 4.7×10^7 Gy, and 2.3×10^8 Gy (4.7×10^8 Rad, 4.7×10^9 Rad, and 2.3×10^{10} Rad), with the highest level surpassing the expected lifetime dose of 1.41×10^{10} Rad that the insulation is expected to encounter in the FIRE TF coils. All of the irradiated compression tests were performed at 76 K, and as such, are compared to the 76 K unirradiated data. The irradiated compression test results are presented in Figure 3 and are compared to the unirradiated results. The results of the irradiated SBS tests are presented elsewhere [16].

When examining the compressive strength results following irradiation presented in Figure 3, the effects are quite dramatic. All three cyanate ester based systems show virtually no change in strength, even out to the highest radiation level. Likewise, the effect of radiation is even more dramatic on the compressive modulus of the three cyanate ester hybrid systems. As seen in Figure 3, the modulus of all systems steadily increases as the radiation level increases, resulting in an overall increase of over 20%.

CONCLUSIONS

The material processing characterization testing in this program demonstrated that the cyanate ester and cyanate ester hybrid materials compare very favorably with current epoxy-based VPI resin systems. The viscosities are acceptable and very similar to epoxies, and in some cases are achieved at a lower processing temperature. The pot-life of CTD-403 meets or exceeds that of most epoxy-based VPI systems. The cyanate ester hybrid systems do have shorter pot-lives, but are still acceptable for most superconducting magnet insulation applications. Finally, the curing temperature of the cyanate ester and cyanate ester hybrid systems is easily achievable using standard industry practices, and is no different than other VPI systems.

Mechanically, the cyanate ester and cyanate ester hybrid resin systems were shown to provide adequate shear and compressive properties to meet the structural requirements of typical superconducting or fusion magnets, even after high doses of neutron and gamma irradiation. Measured shear strengths were generally comparable to that of CTD-101K, whose shear properties are among the best of current cryogenic insulation systems. The compressive properties of the cyanate ester-based systems were shown to be equal to or

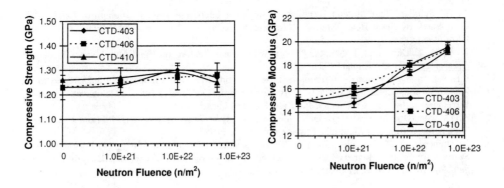

FIGURE 3. Effect of 340 K radiation on cyanate ester hybrid insulation systems at 76 K.

better than those of currently used epoxy resin systems. This was the first time that cyanate ester resin systems designed for superconducting magnets were subjected to radiation levels of this magnitude. Their performance offers very promising service to the fusion and superconducting magnet industries.

ACKNOWLEDGEMENTS

Composite Technology Development, Inc. wishes to acknowledge Prof. Dr. Harald Weber, Dr. Karl Humer, and their staff at the Atomistitut der Österreichischen Universitäten (Atomic Institute of the Austrian Universities) who performed the irradiation on these materials and provided valuable expertise in this area. Funding for this material development program was provided in part by the US Department of Energy, grant number DE-FG03-98ER82554, and internal CTD funds.

REFERENCES

1. R.P. Reed, P.E. Fabian, and J.B. Schutz, "U. S. ITER Insulation Irradiation Program, Final Report," Report to MIT Plasma Fusion Center on Contracts FCA 359063 and FCA 251317-003C, Composite Technology Development, Inc. and Cryogenic Materials, Inc., (31 August 1995).
2. T. Brown, et al, "Fusion Ignition Research Experiment (FIRE) Engineering Status Report for Fiscal Year 2000," Report No. 81-001030, (2000).
3. P. Titus, Massachusetts Institute of Technology – Plasma Science & Fusion Center, personal communication, February 2000.
4. R. Thome and P. Heitzenroeder, "Engineering Features of the Fusion Ignition Research Experiment (FIRE)," presented at the 21st Symposium of Fusion Technology, Madrid, Spain, (2000).
5. B. Nelson, Oak Ridge National Laboratory, personal communication, March 2001.
6. S. Spießberger, "Bestrahlungseffekte in Faserverstärkten Kunststoffen für Fusionsanlagen," Ph.D. dissertation, Atomic Institute of the Austrian Universities, Vienna, Austria, (1997).
7. P. Willis and D. Coulter, "Recommendation of Composite Materials for 100 K Use," Jet Propulsion Laboratory Report, Pasadena, California (1993).
8. D.A. Shimp, *Chemistry and Technology of Cyanate Ester Resins*, I. Hamerton Ed., Chapman and Hall, New York, pp. 282-327, (1994).
9. D.A. Shimp, "Cyanate-Cured Epoxy Resins: Chemistry, Properties and Applications," *Preprints of SPI Epoxy Resin Formulators Conference*, Boston, Massachusetts, (1994).
10. D. Evans and R. Reed, "Gas Evolution from Potential ITER Insulating Materials," Report to MIT Plasma Fusion Center, Cryogenic Materials, Inc., (15 November 1995).
11. T. Okada, S. Nishijima, and T. Nishiura, "Radiation Damage of Glass-Fiber-Reinforced Composite Materials at Low Temperatures," Advances in Cryogenic Engineering - Materials, Vol. 38, p. 241, (1992).
12. ASTM Standard Test Method D2344-84, "Apparent Interlaminar Shear Strength of Parallel Fiber Composites by Short-Beam Method," American Society for Testing and Materials, Philadelphia, USA, (1987).
13. R. Reed, J. Darr, and J. Schutz, "Short-Beam Shear Testing of Candidate Magnet Insulators," Cryogenics, Vol. 32, No. 9, (1992).
14. P.E. Fabian and R.P. Reed, "Candidate ITER Insulation Materials Characterization," Report to MIT Plasma Fusion Center on Contracts FCA 359063 and FCA 251317-003C. Composite Technology Development, Inc. and Cryogenic Materials, Inc., (30 January 1994).
15. ASTM Standard Test Method D695-85, "Compressive Properties of Rigid Plastics," American Society for Testing and Materials, Philadelphia, USA, (1987).
16. K. Humer, et al., "Characterization of Reactor Irradiated Organic and Inorganic Hybrid Insulation Systems for Fusion Magnets," to be presented at the International Cryogenic Materials Conference held July 16-20, 2001 in Madison, Wisconsin, USA.

AUTHOR INDEX

El-Kawni, M. I., 589, 595
Evans, D., 287

Qiu, M., 755

U

Ueno, Y., 1110
Ullmann, J. L., 1201

V

van der Laan, D. C., 639
Vargas, J. M., 429
Vedernikov, G. P., 891
Vlasenko, A. V., 398
Vlasov, Y. A., 429, 437

W

Wada, H., 703
Walsh, R. P., 53, 115, 186, 204
Wang, C. Y., 832
Wang, F. R., 787
Wang, J. R., 832
Wang, M., 654
Wang, Q., 994
Wang, W. H., 71
Wang, X. L., 795, 824
Wang, Y. S., 755
Washio, T., 444
Watanabe, K., 360, 368, 383, 564, 600
Watanabe, T., 659
Watson, D. R., 1118
Weber, H. W., 221, 261, 352
Willis, J. O., 1201
Wong, T., 1041
Wu, X. Z., 832
Wu, Y. L., 3

X

Xiao, L. Y., 755
Xin, Y., 659
Xiong, Y. F., 71

Xu, Y., 654

Y

Yamada, R., 933, 1001
Yamada, Y., 631
Yamaguchi, M., 1183
Yamaguchi, T., 211
Yamamoto, R., 761
Yamamoto, T., 581
Yamanaka, A., 1176
Yamanouchi, M., 1161
Yamazaki, S., 211
Yan, G., 832
Yanagitani, T., 34
Yang, K., 165
Yin, D. L., 787
Yoshida, K., 1126
Yoshimura, M., 581
Yoshino, D., 1176
Yoshizawa, S., 211, 761
Yu, D., 717
Yu, Y. J., 755
Yuasa, T., 631
Yuri, T., 97, 146

Z

Zeisberger, M., 817
Zeitlin, B. A., 949, 978
Zeller, A. F., 255
Zhang, G. M., 755
Zhang, J. C., 1067
Zhang, P. X., 732, 832
Zhang, Y. H., 832, 968
Zhang, Z., 229
Zhao, L. Z., 71
Zhou, G. E., 832
Zhou, L., 832
Zhu, W., 1118
Zimmer, S., 469, 864
Zlobin, A. V., 45, 933, 941